NORTHEAST AND MIDWEST UNITED STATES

Other Titles in
ABC-CLIO'S
NATURE AND HUMAN SOCIETIES SERIES

FORTHCOMING

Australia, New Zealand, and the Pacific, Donald S. Garden

The Mediterranean, J. Donald Hughes

Northern Europe, Tamara L. Whited, Jens I. Engels, Richard C. Hoffmann, Hilde Ibsen, and Wybren Verstegen

Sub-Saharan Africa, Gregory H. Maddox

NATURE AND HUMAN SOCIETIES

NORTHEAST AND MIDWEST UNITED STATES
An Environmental History

John T. Cumbler

A B C · C L I O

Santa Barbara, California · Denver, Colorado · Oxford, England

Copyright © 2005 by John T. Cumbler

All rights reserved. No part of this publication may be reproduced, stored in a retrieval system, or transmitted, in any form or by any means, electronic, mechanical, photocopying, recording, or otherwise, except for the inclusion of brief quotations in a review, without prior permission in writing from the publishers.

Library of Congress Cataloging-in-Publication Data
Cumbler, John T.
   Northeast and midwest United States : an environmental history / John T. Cumbler.
      p. cm. — (ABC-CLIO's Nature and Human Societies Series)
   Includes bibliographical references and index.
   ISBN 1-57607-909-0 (hardback : alk. paper) — ISBN 1-57607-910-4 (e-book)
1. Human ecology—Northeastern States—History. 2. Human ecology—Middle West—History. 3. Nature—Effect of human beings on—Northeastern States. 4. Nature—Effect of human beings on—Middle West. 5. Northeastern States—Environmental conditions. 6. Middle West—Environmental conditions.
I. Title. II. Series: Nature and human societies.
   GF504.N86C85 2005
   304.2'0974—dc22
                                        2005001409
07 06 05   10 9 8 7 6 5 4 3 2 1

This book is also available on the World Wide Web as an e-book.
Visit http://www.abc-clio.com for details.

ABC-CLIO, Inc.
130 Cremona Drive, P.O. Box 1911
Santa Barbara, California 93116–1911

This book is printed on acid-free paper.
Manufactured in the United States of America

CONTENTS

Series Editor's Foreword vii
Preface xi
Map xv

CHAPTER ONE: THE PEOPLE AND PLACE BEFORE COLUMBUS 1

CHAPTER TWO: FIRST ENCOUNTERS 23

CHAPTER THREE: GOODS, TRADE, MILLS, AND DAMS 55

CHAPTER FOUR: MOVING WEST: THE INTERIOR 77

CHAPTER FIVE: MILLS, TOWNS, AND CITIES 111

CHAPTER SIX: POLLUTION AND HEALTH 147

CHAPTER SEVEN: PROTECTING THE PLACE 179

Important People, Events, and Concepts 217
Chronology 243
Documents 249
Bibliography 291
Index 305
About the Author 323

SERIES FOREWORD

Long ago, only time and the elements shaped the face of the earth, the black abysses of the oceans and the winds and blue welkin of heaven. As continents floated on the earth's mantle they collided and threw up mountains, or drifted apart and made seas. Volcanoes built mountains out of fiery material from deep within the earth. Mountains and rivers of ice ground and gorged. Winds and waters sculpted and razed. Erosion buffered and salted the seas. The concert of living things created and balanced the gases of the air and moderated the earth's temperature.

The world is very different now. From the moment our ancestors emerged from the southern forests and grasslands to follow the melting glaciers or to cross the seas, all has changed. Today the universal force transforming the earth, the seas, and the air is for the first time a single form of life: we humans. We shape the world, sometimes for our purposes and often by accident. Where forests once towered, fertile fields or barren deserts or crowded cities now lie. Where the sun once warmed the heather, forests now shade the land. One creature we exterminate only to bring another from across the globe to take its place. We pull down mountains and excavate craters and caverns; drain swamps and make lakes; divert, straighten, and stop rivers. From the highest winds to the deepest currents, the world teems with chemical concoctions only we can brew. Even the very climate warms from our activity.

And as we work our will upon the land, as we grasp the things around us to fashion into instruments of our survival, our social relations, and our creativity, we find in turn our lives and even our individual and collective destinies shaped and given direction by natural forces, some controlled, some uncontrolled, and some unleashed. What is more, uniquely among the creatures, we come to love the places we live in and know. For us, the world has always abounded with unseen life and manifest meaning. Invisible beings have hidden in springs, in mountains, in groves, in the quiet sky, in the thunder of the clouds, and in the deep waters. Places of beauty from magnificent mountains to small, winding brooks captured our imaginations and our affection. We have perceived a mind like our own, but greater, designing, creating, and guiding the universe around us.

The authors of the books in this series endeavor to tell the remarkable epic of the intertwined fates of humanity and the natural world. It is a story only now coming to be fully known. Although traditional historians told the drama of men and women of the past, for more than three decades many have added the natural world as a third actor. Environmental history by that name emerged in the 1970s in the United States. Historians quickly took an interest and created a professional society, the American Society for Environmental History, and a professional journal, now called *Environmental History*. American environmental history flourished and attracted foreign scholars. By 1990 the international dimensions were clear; European scholars joined together to create the European Society for Environmental History in 2001, with its journal, *Environment and History*. A Latin American and Caribbean Society for Environmental History should not be far behind. With an abundant and growing literature of world environmental history now available, a true world environmental history can appear.

This series is organized geographically into regions determined as much as possible by environmental and ecological factors, and secondarily by historical and historiographical boundaries. Befitting the vast environmental historical literature on the United States, four volumes tell the stories of the North, the South, the Plains and Mountain West, and the Pacific Coast. Other volumes trace the environmental histories of Canada and Alaska, Latin America and the Caribbean, Northern Europe, the Mediterranean region, sub-Saharan Africa, Russia and the former Soviet Union, South Asia, Southeast Asia, East Asia, and Australia and Oceania. Authors from around the globe, experts in the various regions, have written the volumes, almost all of which are the first to convey the complete environmental history of their subjects. Each author has, as much as possible, written the twin stories of the human influence on the land and of the land's manifold influence on its human occupants. Every volume contains a narrative analysis of a region along with a body of reference material. This series constitutes the most complete environmental history of the globe ever assembled, chronicling the astonishing tragedies and triumphs of the human transformation of the earth.

Creating the series, recruiting the authors from around the world, and editing their manuscripts has been an immensely rewarding experience for me. I cannot thank the authors enough for all of their effort in realizing these volumes. I owe a great debt, too, to my editors at ABC-CLIO: Kevin Downing (now with Greenwood Publishing Group), who first approached me about the series;

and Steven Danver, who has shepherded the volumes through delays and crises to publication. Their unfaltering support for and belief in the series were essential to its successful completion.

—Mark Stoll
Department of History
Texas Tech University
Lubbock, Texas

PREFACE

Historians have always seen the natural world in history, but they have too often relegated place, or landscape, to the mere backdrop for the "real human drama" of history. For most historians humans create history by interacting with other humans. Sometimes, of course, lip service is given to the larger environment upon which that historical act plays itself out, yet far too often, having given note to the forest abundance of the New World, or the fertility of the open prairie, historians march on ahead with the story of people creating, destroying, loving, hating, killing, or saving other people, as if the landscape upon which all that action takes place was a universal given, an immutable, unvariegated blank slate.

Environmental historians see the playing out of history as more grounded. It involves people to be sure, but for environmental historians the physical place plays a more vital role in the historical drama. It is not a backdrop, but central to the plot itself. The physical landscape and how people see and understand that landscape are factors in the process of historical change. They help form and give direction to that change. The environmental historian is interested, for example, in how people saw and understood the forest abundance. For Native Americans, the forests were not abundant any more than the air was abundant. Thick forests were part of their world to be manipulated to produce nuts, berries, and provide habitat for game. Europeans coming to the western Atlantic shore from nations whose forests were to a great extent gone and whose woodlands were overcut and heavily managed, saw the thick forests of the Atlantic coast as resource extravagance as well as wildness.

Environmental historians are also interested in how the physical environment affects human interaction with place. Centuries of forest growth led to an accumulation of rich fertile loam that provided good soil for the plants of the Native Americans (beans, squash, and maize) as well as the crops the Europeans brought with them. The forests that created the rich fertile soil also hindered the planting of crops. Bringing an understanding of land use, agricultural methods, and rules of property from Europe, the new settlers were at first forced to accommodate themselves to a land of forests, while at the same time they attempted to alter the landscape to more closely approximate the one they left behind.

Historians are interested in time and change. Environmental historians share these concerns. Where environmental historians deviate from their more traditional colleagues is that environmental historians are also interested in how the physical world changes humans and how humans change the physical world. That interaction is a central part of the environmental historical story. People interact with the physical world to survive. At the most basic level, humans—like all animals—live by eating, breathing, and excreting. They share with all other animals the fact that in doing so they alter the world. Unlike their fellow creatures, humans know their physical world and know about change and changing it. Humans consciously labor to change their world to survive. They build fires to keep warm, they construct shelters for protection, and they trap, stalk, and catch game. They pick and dig plants. And for the last several thousand years, they have raised animals and planted and harvested plants. Environmental historians are interested in the interaction between people and their world and how through that interaction people alter the world. Environmental historians are also interested in how the world affects how people alter it. To plant in a forest one needs to cut trees to open the ground to sunlight. On a prairie one needs to break up the thick mass of matted grass roots. In a desert one needs to irrigate. People survive by extracting resources from the environment by labor, but who controls what resources also determines who and how one survives. Struggles over resources are also the concern of environmental historians.

Like the relationship between groups of people, the relationship between people and the environment is fluid. It is constantly made and remade by its very history. And, like the relationship between groups of people, the people-environment relationship does not always follow the direction intended by its creators. Today, as we in the Western world find ourselves living within an overconstructed environment, it is easy to forget that the construction itself originated in human interactions with nature. The constructed world has become our natural world, but no matter how we build and destroy, we are still constructing a nature from a nature. Even when our attentions turn to warring with each other, we should not forget that we war on landscape created through an interaction with nature. Often as not, despite the massive constructs we have placed between ourselves and our ancestors' world of the immediacy of the struggle with nature, we still find ourselves battling over who controls what resources in our neighborhoods, our towns, our cities, our nation, and our world. And win or lose, we still must find a way to live within the larger constraints of nature.

Even as I talk about human interaction with nature, it must be understood that nature is an idea, a human creation with its own history. Yet although the word *nature* is just a word, a creation of language and linked to a common thread of ever-changing understanding, it claims to make reference to an external reality. It is that world outside of us, the world that impinges on us as we collectively need to consume calories and shelter our bodies, that is not totally of our creation. It has a reality that transcends our understanding, but that we are forced to understand as our human condition makes us labor in that world to extract what we need to survive.

Places have their characteristics, which are what make them pleasurable (or unpleasant) to visitors. The unique physical landscape of a place fascinates those from afar. Landscapes also carry the history of a region, both its geologic and its human history. Most of the physical world we know has been remade by the human hand, often over and over again. Each remaking leaves its imprint on the landscape. Careful observations reveal traces of those past imprints. Environmental history is the history of the interactions of humans with the natural world. It looks to understand how the physical world affects human affairs and how human actions affect the physical world. A good starting point to begin that investigation is with the imprints left behind from both the forces of nature and human interactions with nature. The imprints of human interaction with the natural world often need close examination to discern: a walk through a woods, along a riverbank, or even a look out a kitchen window. The larger geologic imprints can be seen even from afar as one travels through or even over a region.

Just as place carries the impact of the past that we tend to overlook as we rush about our daily lives, the past as a creation of human activity is often underappreciated. Yet for many students of history, the past is remote; remote not only from them but also from real human beings. To be sure, historical characters occupy historical narratives. But too often there is a gulf between the present and the narrated historical past. The past and present are presented as separate, as if they existed not at different times, but in different dimensions. In practice, we all know that the past flows into the present. It is no more separate from the present than the water flowing out of the mouth of a river is separate from the headwaters. It is not so much that there is a connection between the past and the present as that the present is an extension of the past. History is made by us, and our past history was made by our biological and cultural ancestors. As we come to know history recorded in texts, convention tends to remove the teller of the tale from the story. Yet people tell history. History, as known, is created by historians. If they are professionals, the story they tell is constructed from the sorting, weighing, and evaluating of evidence. It should have a truth-value that can be judged independent of the historian who created it. Nonetheless, it was created by a historian.

In this history, I have consciously not hidden my authorship. I have also tried to bring home the reality that history is made by real people dissimilar from us only in that their world is dissimilar from ours. But they are our past and we are linked to them by living history. To emphasize the human nature of historical creation, I have brought real people into this work. By adding members of my own family to the text, I hope to remind the reader not only that I exist as an author, but that my past and the past of those who read this work are linked to real people, who lived their lives as best they could given the resources available to them. They had children and grandchildren, and the offspring of those people live their lives as best they can and live amongst us today as we live our lives.

READING THIS BOOK

This work is organized chronologically, geographically, and topically. Readers interested in a particular region or time or topic can go to the appropriate chapter or section of chapter for the information needed. The bibliography suggests easily accessible primary sources for the reader interested in exploring source material in environmental history. The bibliography sections also list important secondary works by historians addressing the various issues and topics covered by the preceding chapter. The glossary defines and explains terms, ideas, names, and unusual words used in each chapter.

Many people have helped me with this work. Most importantly was Malcolm Fleschner who read over everything carefully and guided me to a more coherent style. Mark Stoll, the editor for the series, gave the manuscript a careful and critical read. I would also like to thank the project editor, Laura Stine, and the copy editor, Lauren Arnest. Acknowledgment is also due to media editor, Giulia Rossi, for finding the art. My immediate family, Ethan, Kazia, and Judith, all encouraged me to do this project. Rusty Scudder not only does good environmental work in Vermont, but also provided help to me, as did Sam Warner. I would also like to thank my sister Libby and her husband, my fishing and snorkeling companion Bruce Bartolini, who provided me with much of this story. Thanks also goes to my chair, Tom Mackey, for his support. I would also like to thank my doctor, Jane Seale Nevitt, who caught my cancer at an early stage and enabled me to see this book in print.

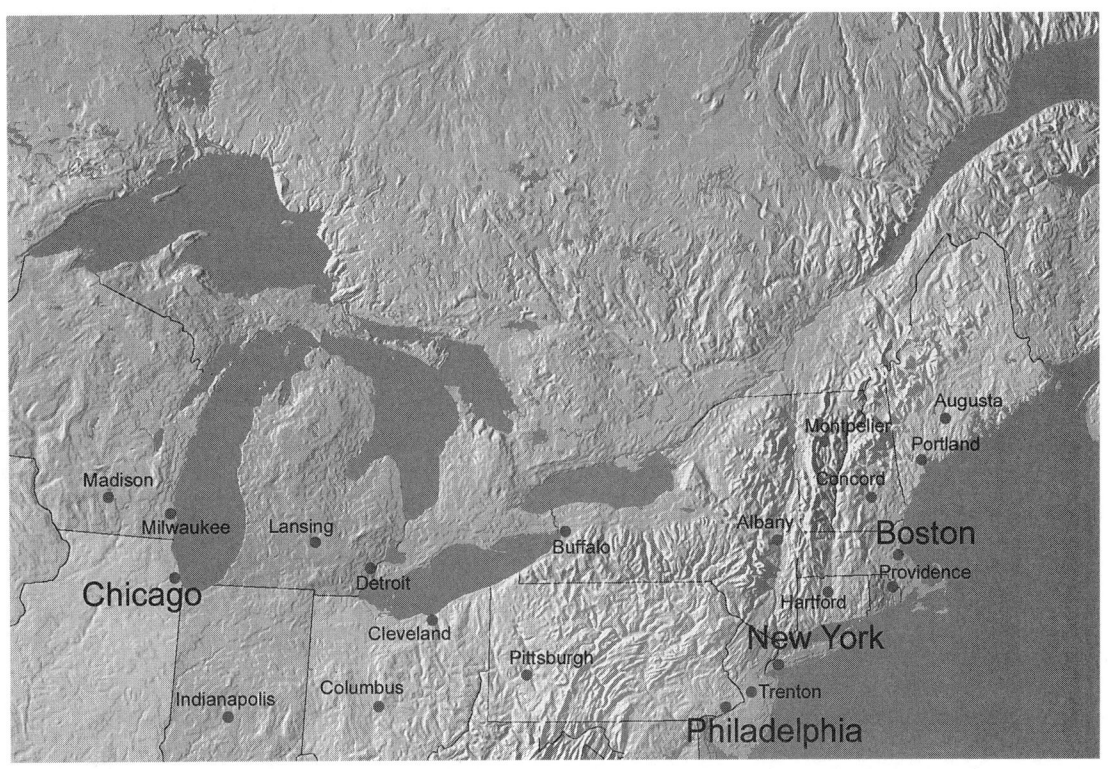

The U.S. Great Lakes and Northeast.

1

THE PEOPLE AND PLACE BEFORE COLUMBUS

I live on the south side of the Ohio River, the southern boundary of the region encompassed in this volume. To visit my sisters who live in New England, I cannot fly directly to Boston. First I have to fly to a hub city, Chicago typically, and transfer before flying on to New England. My flight oversees territory we investigate in the ensuing chapters. On my way to Chicago, I fly over the rolling lands of the central Midwest, beginning with the Ohio River valley delineated on its northern edge by small hills or knobs and then over the plateau covering much of central Indiana. Crossing the Wabash River valley, which separates Illinois from the Hoosier State, the plane then descends along the eastern-most edge of the plains to the west before landing in Chicago. Heading east out of Chicago, I first fly over the great marshes of northern Indiana, where hundreds of thousands of small birds, ducks, and geese rest and take refuge during their yearly migrations. Looking out the window to the east, I see the rolling land of forests and now farmland of northwest Indiana and Ohio before the plane crosses over Lake Erie, the shallowest of the Great Lakes. Erie sits against the western edge of the Appalachian Mountains, a range beginning just north of Atlanta, Georgia, that runs northeast to northern Maine. As the dividing point between the East and the Midwest, the Appalachians have always played a central role in this nation's history.

In 12,000 B.C.E. the land the plane today flies over to the west of these mountains was covered with the ice of the Wisconsin Glacier, one more chapter in an astounding history. The past 40 million years have seen great climatic shifts in North America. Thirty-three million years ago, even Canada's northern reaches were covered with warm tropical forests. But shifting land masses in the distant southern hemisphere (the breaking away from Antarctica of Australia and Tasmania) created the Antarctic Circumpolar Current, which cut Antarctica off from warmer northern waters. The polar ice that began to cover Antarctica reflected sunlight away from the earth. The earth cooled dramatically as tundra slowly replaced the North American rainforests. About 2.8 million years ago, North and South America joined at the Panamanian isthmus. The warm, west-flowing North Equatorial Current turned northeast, creating the Gulf Stream. This shift brought warm, moisture-laden air into contact with colder arctic winds. The result was significantly more rain and snow. Forests spread across the north while snow began to pile up into glaciers and slowly push south. Digging out huge troughs and filling in natural depressions as it melted, the Wisconsin Glacier not only transformed the land but also left in its wake hundreds of lakes, the five Great Lakes being the most impressive. The southern-most Great Lake, Michigan, was rich in fish but was bordered on the eastern side with huge sand dunes and sandy soil. In northern Indiana the succeeding glaciers, the last of which was the Wisconsin Glacier, created a vast wetlands, stretching for hundreds of miles, teeming with wildlife particularly in the early spring and late fall when migratory birds, ducks, and geese filled the marshes. West of the vast stretches of reeds and marsh grasses were the lands of high bluestem, grama, and other prairie high grasses. Although it rained here as much as in Ohio and eastern Indiana, dry summer winds blew down from Canada and fires regularly swept through, burning off tree saplings. Saplings that escaped the fires provided meals for prairie rabbits. Fires and grazing animals did not hinder prairie grasses. While trees and shrubs concentrated their energy in above-ground growth, grasses concentrated energy in their root systems. Fires swept through and burned all above ground, and animals grazed upon grass stems, but from their roots grasses quickly threw up new growth to capture the sun's energy.

If it is a clear day as I fly over the Appalachian mountain range, I will first pass over thick forests of mixed oak that give way to spruce, pine, hemlock, maple, beech, and other northern hardwoods as the plane flies north over New York State. Here the forest-covered mountains occasionally grant passage to narrow river valleys, until eventually I see the deep narrow channels of central New York's Finger Lake district. To the north lie the broad and fertile Genesee and Mohawk River valleys. These provided the path for the Erie Canal to link the East to the Central-West by joining the Hudson River at Albany with Lake Erie at Buffalo. Crossing over the wide Hudson River valley, the plane next flies over the thickly forested Green Mountains of western New England. In autumn these mountains and their hardwoods give New England its reputation for fabulous fall colors.

The Green Mountains, along with the rest of the Appalachian mountain range, formed 400 to 350 million years ago when the plates of earth's crust collided. The range changed shape again when the massive glaciers passed over what is now New England and New York, shearing off mountain tops; filling in valleys with glacial till, outwash, and water; and creating plateaus, lakes, ponds, marshes, and river channels. Fifty miles west of Boston, as the plane descends into the city, it leaves the mountains behind. Turning to make the loop south of

Boston to approach the airport from the east, the airplane passes over the South Shore, where lakes and marshes dot the narrow coastal plain. In this northern region much of the coastal plain lies submerged below the bay, providing a shallow ocean floor that combines with colliding ocean currents to create the rich fishing banks that gave Cape Cod its name.

Although there are rich deposits of alluvial soil along the river valleys, New England generally lacks deep, rich soil. In the region's forested mountain areas the podzol soils are thin and acidic, and their nutrients wash out easily. The brown podzolic soils farther south are less acidic but still limited in nitrogen and phosphorus. The sandy river valley soils drain well but also easily leech out their nutrients. In general the soils are shallow, two to four inches deep, and acidic. Despite the poverty of New England's soil, the region remains rich in geological, ecological, and historical diversity.

In New Hampshire, north of Boston, one of my sisters lives in a semirural community about two miles inland from the coast. That her house is situated on an old farmstead becomes evident from a walk about her property. Overgrown stone walls that the early farmers built crop up among tall stands of second-growth hemlock and pine, interspersed with oak and maple. By clearing fields of stones and depositing them along the fields' edges, farmers not only found a convenient place to put the obstacles they removed from their plow's path, but they also created walls that delineated fields and offered protection from wandering farm animals. Deeper into the forest surrounding my sister's home, the land slopes down into a willow and cattail marsh, likely the legacy of an industrious beaver who departed before the stone walls were built and the first forest was removed for homes, barns, and fireplaces and replaced with crops. Out my sister's kitchen window sits a rugged granite boulder the size of an old Volkswagen Beetle. This boulder was created by the slow cooling of material below the earth's surface 600 million years ago. Later, glaciers, winds, rains, and human shovels exposed this ancient piece of granite.

The boulder reminds one that this is a scarred landscape—scarred certainly by people who cut trees, plowed fields, and dug basements, but also in a larger sense it is a land scraped over. The scraping was done within the last 2.5 million years when large glaciers, one to two miles in depth, slowly bulldozed south, moving forward and then retreating several times before finally moving back to their original arctic home more than 15,000 years ago. The soil of the land has been built and rebuilt by a succession of plants after each glacier, but the memory of the scraping remains in the thin soil and the rugged terrain. This geologic sequence explains why the stone walls around my sister's home are deep within a forest, rather than bordering fields of grain. Hers is a forest that has grown up amidst fields abandoned for more fertile pastures. Today, the land grows suburbs

The Presidential Range of the White Mountains in New Hampshire. (David Muench/Corbis)

rather than crops. Even now the glaciers' scrapings, or till, factor in my sister's life. Every summer she and her family drive south to the beaches of Cape Cod. Those beaches—indeed, Cape Cod itself and its sister leavings to the south—are the glacier plow's remains. Coming to a rest along a line running northeast along Long Island, Block Island, Martha's Vineyard, and Nantucket, the glacier dumped a huge chunk of its accumulated scrapings, sand stones, rocks, and mud in moraines. It then retreated back to the present boundary of Cape Cod where it dumped another load to create a second line of moraines.

On her trip south to Cape Cod, my sister's path parallels the Merrimack River. The Merrimack begins in the big northern lake of Winnapesaukee, which is fed by mountain streams swollen with rainwater that flows out of the White Mountains. The White Mountains form the northeastern ridge of the Appalachian chain. There is no shortage of rain in the Northeast. It receives between forty and forty-eight inches of rainfall a year; almost twice Minnesota's average and three times Colorado's. The prevailing weather patterns flow east. From the south the winds pick up warm moisture from the Gulf of Mexico, while the northern winds bring down cool, dry air. Off the coast the cold ocean waters contribute moisture to the thick fogs. As these weather patterns collide over the region and especially as they ascend up the New England mountains, which hug close to the coast, they dump loads of rain, snow, and mist onto the hills and mountains. The rains and snows falling through the trees are picked up by the thick forest ground cover, settle or melt into marshes and swamps in the high Appalachian plateau, sink into the aquifer, or tumble down the hillsides into streams and brooks that feed source lakes and rivers. Like the Penobscot, Kennebec, Connecticut, Housatonic, Hudson, Delaware, and Susquehanna rivers, the Merrimack picks up this flow and describes a southern route running parallel to the Appalachian mountain chain crowding the coastal plain in the northern half of the United States. Many of the northeastern rivers follow a similar southerly path before dumping their waters into coastal bays and estuaries. Though it also flows south, the Merrimack does not make it all the way to the coast. When my sister crosses over the Merrimack, just south of the Massachusetts–New Hampshire border, the river runs east. It does so because at Pawtucket Falls at Lowell, Massachusetts, the river hits the Nashoba Thrust Belt, a ledge of hard stone. To the west the Connecticut River follows a similar trajectory traveling south from spring-fed marshy lakes along the Canadian border for some 350 miles before turning east for the last 50 miles. The Housatonic, the Hudson, the Delaware, and the Susquehanna also confront rocky outcroppings before finally reaching their destinations, but they hold truer to their north-south flow. Farther south, as the coastal plain widens away from the Appalachian Mountains, rivers run more easterly as they descend from the mountains and flow gently through the wide coastal plain.

Like the eastern rivers that find their origins in the hills of the Appalachian mountain range, rivers to the west of the mountains also find their origins in mountain streams and creeks. The prevailing weather pattern brings cool fronts in from the northwest. Traveling northeast up into the central-eastern section of the nation flows warm air that has gathered up moisture from the Gulf of Mexico. These colliding weather fronts bring rains to the eastern Midwest. As these fronts move up over the Appalachian range they drop moisture as snow in the winter and rain the rest of the year. The rich tree growth of the mountains helps to hold this moisture in the hills. The tree canopy breaks the fall of the rain and holds some of it in the leaves. Then, as the water drips and flows to the forest floor, thick beds of moss and humus act to sponge up and hold the water, slowly leaching it out into streams and creeks. Flowing out of the mountains, the streams merge into rivers. In Pennsylvania the Allegany River flows out of the

Pittsburgh skyline (showing where the Allegheny and Monongahela meet to form the Ohio River). (PhotoDisc, Inc.)

mountains from the north, and the Monongahela flows down from the southern mountains to join in Pittsburgh to form the Ohio River. The Ohio flows southwest out of the Appalachian plain, picking up waters from other mountain-fed rivers flowing down from the western front of the Appalachians. The rivers flowing down from the hills of western New York run north to Lake Ontario and then continue northeast out the St. Lawrence River to the Atlantic.

Driving south through Massachusetts, my sister passes a landscape that retains the influences of those long-ago glaciers. The glaciers rounded hills and dribbled bits of their load throughout the region, much of it being elutriated out with melting ice as the glacial outwash. Along the southern New England coast, these streams from the melting glaciers carried with them clay, sand, and gravel to form low-lying plains that evolved into bogs, ponds, and marshes. The glaciers also left behind huge chunks of unmelted ice buried beneath loads of glacial till. Later, when the surrounding soil and debris settled, the ice slowly melted, creat-

The People and Place Before Columbus 7

Doane Rock, Cape Cod. (Courtesy of Kazia Cumbler)

ing and leaving behind deep depressions, or kettle holes. Eventually filling with water, these kettle holes were the sources of the many regional ponds. Just as they leveled hills in their paths, the glaciers also created others as they piled up mounds of clay. Many of these same hills would prove important positions of defense during the Revolutionary War. To the west along the southeastern and northeastern edge of Lake Michigan and the southern edge of Lake Erie, the same patterns of low-lying bogs, ponds, marshes, and sandy dunes appeared in the wake of the glaciers' retreat.

As my sister crosses the Sagamore Bridge onto Cape Cod, the landscape changes again, this time to one of rolling sand dunes covered with scrub oak, pitch pine, and—especially on the outer reaches—locust trees. In the low areas among the kettle ponds and marshy rivers and stream wetlands, cedar, maple, willow, and birch trees thrive. Driving slowly through these low-lying areas, she might even catch sight of a farmer's orchard. Like those in the orchard, the small windblown trees on the Cape are all recent growth. By the end of the nineteenth century, the landscape had been cut and recut many times over. The locust trees represent a twentieth-century effort to reforest an area that had been nearly denuded of its original cedar, oak, maple, and hickory.

As my sister travels up the Cape's northern spine, she passes a large boulder, known locally as Doane Rock. Doane Rock is larger than the boulder outside of her kitchen window but it is also very different in appearance. Both are granite, but while my sister's New Hampshire rock is rough, rugged, and moss covered, Doane Rock is smooth and rounded like an egg. Doane Rock appears out of place, and with good reason. On its turbulent path to Cape Cod it was twisted and turned, rolled and battered, until its rough edges were ground smooth. When the glacier receded, Doane Rock was dumped along with the till to be polished and weathered by the strong winds blowing in from the cold Atlantic. Just a few miles from Doane Rock, my sister passes the sign for First Encounter Beach. This beach was so named for the initial contact between the Pilgrim settlers and the native Wampanoags.

Despite the beach's name, Europeans and the native people of New England had previous interactions. And although Europeans referred to the lands on the western Atlantic as the "New World," it was hardly a new world to the people who had lived there for some 12,000 years.

THE NATIVE WORLD

Although archeologists keep finding older and older signs of human habitation in North America, the best estimates show that between 14,000 and 15,000 years ago, Paleo-Indians following game traveled from Siberia into Alaska and northwestern Canada. Around 12,000 years ago, warming ocean waters rose, closing off the land bridge connecting Asia to the Americas. The hardy populations that survived in the cold arctic regions migrated down into the game-rich lands of mastodons and mammoths south of the glacial cap. These paleolithic wanderers (known today as Clovis hunters) were skilled hunters, but their years in cold climates inhospitable to germ microbes and their lack of contact with domesticated animals harboring disease-bearing pests meant that these wanderers and their descendants also lacked immunities to the diseases that had spread throughout much of the rest of the settled world. Between 8000 and 6000 B.C.E. these hunters moved into the midwestern and northeastern reaches of the continent. There they found an abundance of food. Huge slow animals were plentiful on the ground. Climate warming and extensive hunting ultimately took its toll on these animals, leading to their extinction some 8,000 years ago. Fortunately for these early archaic people, warming produced more diverse plant and animal populations. Oak trees began to appear where previously there had been a park tundra. By 4000 B.C.E. modern hardwood forests with nut-bearing trees covered the region east of the Mississippi; anadromous fish (fish such as salmon that

Members of the Hopewell culture gather crops of maize and squash. (North Wind Picture Archives)

spend their mature lives in salt water but travel up rivers to spawn in fresh water) traversed coastal rivers and streams; fruit-bearing bushes and shrubs spread across the landscape, particularly in low-lying marshy and open areas; and small quick-footed animals, birds, and waterfowl multiplied along with their food supply. As the large animals went extinct, and more diverse flora and fauna emerged, the peoples of the Americas were forced to adopt more intricate systems of survival. They learned to gather nuts, fruits, and berries, to catch fish, and to harvest coastal shellfish. They also developed nets, traps, and sharper and more accurate projectiles. These innovations helped the mobile bands of hunters become hunter-gatherers and to survive and grow in number.

Acorns from oaks and hickory and beech nuts were nutritionally valuable contributions to the Native American diet. Acorns were particularly rich in starch and oil, but they also contained bitter tannins. By grinding and leaching the acorns to remove the tannins, acorn flour became an important source of fall and winter food. The region north of the Ohio River and west of the Appalachians was made up of thickly forested rolling hills. The land was primarily covered with mixed mesophytic and deciduous forests of maple, beech, oak, and hickory. Roughly 2,000 years ago paleo-Indians of the Hopewell culture began to settle communities in the rich fertile land in the broad Mississippi watershed. Along the river valleys they began cultivating in a limited fashion squash, sunflowers, sumpweed, and goosefoot.

Although these crops had nutritional potential, all containing high protein and oil, they never became central to the Native American diet. Goosefoot seeds are small and require much labor to extract their nutritional value, the sunflower seed is encased in a hard shell, and sumpweed irritates the skin to handle. Hopewells also picked wild berries and nuts, hunted deer and bear, trapped birds, and caught fish. Around 300 B.C.E. knotweed, maygrass, and a native barley were added to the cultivated crops of the Americans of the Northeast. However, outside of the Hopewell culture north of the Ohio River that flourished in the period 200 B.C.E.–400 C.E., this cultivation represented only a small part of Native American diets.

While the hunters and gatherers of the Northeast learned to trap and hunt game, catch fish, and gather nuts and berries, their cousins to the south and west observed that some plants produced larger seeds than others and that these seeds themselves grew into plants with larger seeds. To benefit from this observation, these innovators needed containers to protect the seeds from pests and moisture and for storage between planting seasons. They also needed growing seasons long enough to successfully experiment with selective breeding and rich soil, sufficient moisture to guarantee germination and mature plants, and enough food in the habitat to allow the innovators to stay put during their experiments. Available wild plants that could be converted into cultivatable ones through the application of selection had to be present in the environment. With these conditions prevailing in Mesoamerica, sometime before 3000 B.C.E. the communities there began cultivating maize, beans, and squash (including a gourd squash suitable for use as a container).

This horticultural revolution spread slowly northward. Sometime after 200 C.E. Mesoamerican maize began to find its way along trade routes into the Midwest and areas of the present eastern United States. Around 900 C.E. a new variety of maize adapted to the short summers of the northern region was developed. Two hundred years later beans arrived. By the second millennium C.E. the native communities of the southern Northeast combined hunting and gathering land and sea flora and fauna with cultivating beans, maize, and squash. Except for some sunflower cultivation, they abandoned their cultivation of sumpweed, goosefoot, knotweed, maygrass, and native barley for these more prolific imports from the south. Maize had lower protein content than the older cultivated crops but it was easier to grow and harvest and, with the complement of the high protein found in beans, it contributed to a nutritional diet. Horticulture helped spur significant population growth within these communities. The introduction and adaptation of maize, beans, and squash led to the flourishing of densely populated chiefdoms along the Mississippi River and its tributaries.

The settlements of the Hopewell culture died out in the fourth century. They were replaced in the tenth century with a new civilization of mound builders. The mound builders grew beans, maize, and squash, developed in Mesoamerica, in the rich river valleys of the southern Midwest. The broad floodplain of the Mississippi Valley offered well-drained alluvial soil and sufficient rainfall to sustain prolific harvests and substantial communities. The residents of these communities built huge mounds that dominated the surrounding countryside. Mounds were used for both burials and ceremonial temples. The region was dotted with local chiefdoms, often at war with each other. The chief usually lived atop the central mound that was surrounded by thatched houses of the common people. Often the whole town would be surrounded by a defensive palisade. The dominant chief resided upon the largest mound of the region. The most significant of these mound builders were centered in Cahokia, located in southern Illinois. Cahokia was home for over 10,000 inhabitants. Cahokia had over 100 temples and burial mounds. Its defensive stockade was a wooden wall two miles in circumference. At the center of Cahokia was an earthen pyramid rising 110 feet in the air.

Sometime in the twelfth century, Cahokia began to decline. Although the alluvial river valley soil was rich, a growing population and overplanting for generations began to take its toll on the community. Game animals that had supplemented the Cahokia diet were hunted out of the region. Overcutting of forests limited the wood available for cooking fires and warmth, as well as for building material. By the middle of the thirteenth century, Cahokia was abandoned. Mound-building communities remained farther south, but the Native Americans of the central Midwest never regained the population size or power of the early mound builders.

In southern New England, Long Island, and up the coast to southeastern Maine, groups developed common cultural, social, and religious customs. They shared the Algonquian language and centered their social units in villages that moved about according to the seasons, land fertility, and the availability of fish and game. To the north and into the mountains, the shorter growing season precluded much horticulture. As a result, the communities there retained Paleolithic hunting and gathering patterns. They also differed culturally, socially, and linguistically from their southern and lowland neighbors. Estimates put the population of Native Americans in New England at the end of the sixteenth century at between 80,000 and 140,000, with most concentrated along the coast and up the region's river valleys. Mixed hunting and gathering and horticulture contributed to these settled patterns of existence and afforded significantly larger settlements. The Abenaki-speaking peoples of northern New England relied primarily on hunting and gathering. They lived thinly scattered amidst the coniferous forests of spruce and fir that dominate the mountain regions of the Northeast. This forest ends at the Berkshires at the eastern edge of the Hudson River valley. In the thin soil of the high mountains, hearty red spruce took root between the firs. In the mountain valleys along streams and lakes, the northern hardwoods of birch, aspen, and red maple grew, and in the marshes spruce and tamarack flourished. To the west in the pine forests of northern Michigan and Wisconsin, Huron, Menominee, and Chippewa communities lived primarily by hunting, fishing, and gathering, with a minimum of summer horticulture.

South of the Native American hunters and gatherers to the east were the Algonquians. During the century before the first Europeans came to New England, nine major groupings or nations of Algonquian-speaking peoples formed. In the northern and higher elevation reaches of their territory, these peoples lived among the birch, sugar maple, beech, white pine, and hemlock, while to the south and in the lower river valleys they were surrounded by oaks, hickories, and chestnuts. In southern New England (along with southern New York, New Jersey, and Pennsylvania) forests of white, red, and black oaks grew. To the south and east, chestnut and scarlet oaks predominated in the drier lands, and basswood, ash, maple, and birch proliferated in wetter soil. The sandy soils along the windy coast from Plymouth south to Delaware Bay supported pitch pine and scrub oak. From these trees and their products, the native peoples built shelters, canoes, weapons, and tools; fueled their fires; and looked for signs of good planting land and gathering places for game.

Those Native Americans who lived inland from the coast but south of 42 degrees of latitude existed amidst a broadleaf deciduous forest dominated by oak and tulip trees. Higher up in the hills and mountains, the oak and tulips gave way to forests of maple, beech, and hemlock.

The human activities of hunting, gathering, and setting fires helped to transform the landscape. But other forces affected the land as well. When the glaciers retreated, meltwater and outwash covered the land. With the weight of the glacier removed from the ground, the continent gradually rose, waters retreated, and invading grasses took over, providing food for the giant roaming Pleistocene mammals that the early Paleolithic peoples hunted so ferociously. Behind the grasses came invading seedlings. As the Pleistocene grazers declined in number, these seedlings grew into trees and then forests, first of spruce and then white pine and birch, followed by hickory, oak, chestnut, maple, and beech.

Plant succession is an ongoing process. The natural world changes constantly under a variety of influences. Hurricanes, floods, droughts, lightning-sparked fires, and running water that changes river channels can all radically transform an ecosystem from dense forests to open meadows. The meadows in turn gradually revert to brambled fields, woods, or forests. Lands cleared of forest cover increase in temperature, hold less moisture, and erode more easily. Deep forests lower temperatures. Different forest trees add diverse nutrients to the soil to encourage new forms of undergrowth.

Natural catastrophic events are not the only nonhuman forces of change in the natural world. Animals, too, play a role in creating the landscape. Henry David Thoreau noted more than 150 years ago the role squirrels played in forest succession. Birds roosting in trees dropped seeds, nitrates, and phosphorus to the soil below. Migrating passenger pigeons that every few years blackened the skies with hundreds of thousands of birds also enriched the land with fertilizer. Deer nesting for the winter cleared openings in the woods. Along with rabbits, deer also grazed on young saplings, which retarded forest growth. The world from the coast west to the prairie and beyond was altered by these interactions between fauna and flora.

But aside from humans, the landscape's most ambitious transformer was the beaver. Across the entire wooded region from the Atlantic coast to the western plains, beavers left their mark. Building dams across streams to protect their lodges, beavers flooded woodlands. The trees in these flooded areas died and fell into the accumulated backwater. As debris filled the ponds and colonies grew in size, the beavers built higher and higher dams, increasing the flooded areas, turning ponds into lakes. Dams beavers built on hillside streams spread out, slowing the flow of water. In the warmer stagnant water behind the dams new plants and insects thrived, providing food for fish and fowl alike. In the dead trees wood ducks found homes. When heavy rains sent torrents of water downstream, these ponds acted as collecting basins to hold it back. In dry periods seepage from these ponds helped maintain a constant flow of water through the watershed.

Over time, sediment flowing downstream settled on pond and lake floors, joining leaves, decaying tree stumps, and dead animals that had also collected there. The beavers found plenty of food at the water's edge, but because they transported their food by water, the beavers' varied activities of feeding their families slowly depleted the trees at the water's boundary. In order to capture more trees, the beavers were forced to build wider and higher dams, further extending the water's edge away from the lodges. Eventually the beavers would move on, allowing water to seep out of the abandoned dams. The forest that had become a lake or pond was transformed anew into a bog. Soon meadow grass grew in the rich soil, and moose, deer, and rabbits arrived to graze, keeping back the forest, while ground birds built nests in the thick grasses.

PEOPLE AND THE LAND

Native Americans depended on their knowledge of the landscape for survival. Along the coast they speared flounder and gathered lobsters, crabs, and shellfish. Boats were built from trees felled by burning them at the base of their trunks. Native workmen would set fires on the top sides of the felled logs. Using sharpened flint and stone, the canoe builders scraped out the burnt portions until they had hollowed out a section large enough to hold several passengers. For lakes and coastal areas these boats were roughly the size of the birch-bark canoes that the inland and northern communities used to navigate rivers and lakes. But for ocean travel, a single huge tree was used to create a boat large enough to hold a dozen or more men who would venture out to sea to fish for cod or to spear swordfish (which, fortunately for native hunters, had a habit of resting on the ocean surface) or whales. At the mouths of rivers and streams handcrafted weirs trapped migrating fish. Ocean and Great Lakes marshes were rich in small animals that could be trapped or brought down with a bow and arrow. Ducks, geese, turkeys, and heath hens could be snared. Meadows were thick with highbush blueberries, blackberries, and cranberries. On open hillsides throughout the Northeast wherever the soil was acidic and sandy, lowbush blueberries abounded. In the fall, chestnuts added protein to the native people's diet.

Among coastal and Great Lakes communities fishing fell off during the fall months. Increasing winds and fall storms limited the tippy wooden canoe's viability in open water. Although migrating whales occasionally beached themselves, and seals frequently swam along the shore, as winter approached gathering food from the ocean became more problematic. Storms arrived from the northeast, threatening coastal villages, while small animals became harder to snare, and berries disappeared. For Native Americans along the Great Lakes, winter storms from the northwest threatened lake area villages. Fall usually meant inland travel and resettlement. In the cold weather meat could be saved, allowing hunters to shift from small animals to deer and moose, eating some meat and storing the rest.

Rivers, lakes, and streams provided food for communities away from the coast. For native spearers and netters, anadromous fish, migrating upstream from the ocean to spawn, ran in such huge numbers that they made easy targets. Each spring salmon darkened the rivers and streams several hundred miles up from the ocean on their search for the cold, oxygen-rich pebble-bottomed headwaters where they made nests for their eggs. At each fault along the river, native fishers gathered to take what they needed to feed their families from the migrating mass. Salmon weren't alone on the journey upriver. Shad and alewives ran upstream from the ocean in numbers so vast that early commenters claimed that one could almost walk across the rivers on their backs. Each spring thousands of these species also found their way into native weirs and nets. Lakes and ponds provided freshwater trout, bass, pickerel, pike, and perch, while small mammals, fowl, and berries proliferated in the inland marshes and meadows. Lake salmon, migrating up Lake Ontario's feeder streams and rivers to spawn, provided abundant food for Native Americans living along those waters.

Despite the bounty of food the lands and coastal regions offered, the growing human population eventually began to overtax area resources. Game was limited, particularly in the winter months. As areas were hunted and fished out, and berries became scarce, communities would move into new areas. Sometime between 800 B.C.E. and 1000 C.E., when that strategy became insufficient, the peoples of the Northeast and Midcentral America added corn, beans, and squash to their diet. The shift to horticulture did not preclude other food-procuring strategies, however. Native Americans continued to hunt, fish, and gather fruits and nuts. But particularly in southern New England and other areas to the south and west, planting, harvesting, and storing crops became more central to community survival. Horticulture also contributed to a significant growth in population numbers—numbers that in turn labored to help produce this cultivated food.

Horticulture had created fairly stable settlements among the Delaware and other communities south of Massachusetts and for the Iroquois, Erie, Miami, Kickapoo, and Illinois in western New York and the central Midwest. Large fields of corn, beans, and squash surrounded settled villages. Storage units were established. Yet when pests began to plague the crops, and the fields lost their fecundity, even these comparatively stable communities were forced to move. Moving a settlement entailed much labor and offered no guarantee of preferable living conditions. The ideal location offered close proximity to cultivated fields, which made defending the plants from animals and hauling the harvest easier. Meadows created in abandoned beaver ponds provided fertile fields, but the ground was also thick with plant life and deeply embedded with roots that had to be dug up and removed before crops could be planted. The wet land was also thick with mosquitoes and other pests, making a poor place to set up a village. The deep forest had loose fertile soil and little undergrowth to clear, but the forest canopy blocked out the sun. To grow crops in that loose fertile soil, Native Americans had to clear the thick forest cover.

Native Americans cleared their fields by setting fires around the bases of trees to kill them. This was difficult because the thick bark of live trees resisted burning. Another method was to girdle the trees by cutting a ring around the trunk so that the bark was broken in a complete circle, thus depriving the tree above the ring of water and nutrients and ultimately killing it. The ground below the thick forest canopy was typically loose. After the trees were killed by this slash and burn method, and sunlight reached the forest floor, workers—mostly women and children—hoed the soil into mounds between the dead tree trunks and around the thick roots. The hoes consisted of a thick clamshell or flat stone lashed to a green stick with leather, vine, or stripped bark. Corn was planted in the center of the mounds, with the edges reserved for squash. Native women worked hard in the early planting season to weed out competing plants and protect the young shoots of corn and squash seedlings from pests. After the corn had established itself, beans were planted that could climb the full-grown corn stalks. The squash plant's broad leaves spread out over the mound, helping to keep down competing plants and providing shade to prevent the soil from drying out. As nitrogen fixers, the beans helped replenish some of the nutrients leached out of the soil by the corn. Some cultivators added cucumbers and Jerusalem artichokes to their crop inventory. In separate fields men grew tobacco.

The fields were harvested in late summer and early fall. Again, the women and children bore the brunt of the labor. This was only the beginning of a labor-intensive process. Next, the corn had to be shucked and the beans had to be husked and dried. Protecting the exposed drying corn and beans from birds and animals required constant vigilance. Once dried, they had to be stored out of the reach of both animals and moisture. Containers needed to be fashioned. Those living above the southern reaches of the glaciers constructed bowls from the thick mounds of clay left behind by the retreating ice and exposed by eroding waters. The clay containers were fired in charcoal cooking pits. Containers fashioned from dried gourds, hollowed-out wood, or animal skins also helped the cultivators store the dried kernels and beans. Besides what would be set aside for the next year's planting, this crop constituted the foundation of the community's sustenance until the late spring, when fish runs and fruiting plants brought more variety into the diet. Dried corn kernels made for an inefficient source of calories, however. To maximize caloric value, Native Americans ground the corn and made it into a palatable mush. This again was the labor of women and children. They scoured the streams and brooks for rocks eroded into natural bowl shapes by dripping water. Using the rock bowl as a mortar (clay bowls could not withstand the pressure necessary for producing meal) and a cylindrical stone as a pestle, the women would begin the backbreaking task of grinding the corn into meal that would then be boiled with the dried beans and dried squash into a mush or porridge. This combination provided the amino acids necessary for a balanced diet of vegetable protein. Meanwhile, the men remained busy fishing; or snaring, trapping, or netting game; or fashioning weapons for the winter hunt.

The Native Americans who lived along the southern New England coast and further south, as well as those along the rich river valleys south of the Ohio River, enjoyed longer growing seasons and more fertile brown podzolic soil than their counterparts to the north. They also lived at higher population densities and took a heavier toll on game and wild edible plants. Their communities were more stable and larger, as were their fields. They depended more heavily upon horticulture, which accounted for as much as 85 percent of their food intake. This greater reliance on horticulture was reflected in manifold aspects of life in southern New England and further south and west. These peoples had a greater store of hoes, rakes, and instruments to work the land. Their storage containers were also more substantial and more permanent, as were their instruments for grinding cornmeal. Stable communities required less forest clearing and less time spent lugging tools, looking for mortar bowls, and erecting shelters. But they also demanded more labor in the fields to keep out pests; hold off invading weeds, briers, and brambles; and increased effort to maintain soil fertility. Native horticulturalists in these regions achieved yields of twenty-five to sixty bushels on one or two acres of land, which compares quite favorably to the yields of modern-day farmers. Native Americans in the rich valleys of the central Midwest harvested at the high end of this scale. Periodically, the community would set controlled (and sometimes not so controlled) fires to burn away brush, clear trails, and maintain open space for game. The burnings opened ground that soon sprouted berries and small seedlings that also provided food for game and the community. Even the area of the village was periodically burned to reduce fleas and lice.

To the north and up in the hills where the growing season was shorter and the game heavier on the ground, mobility was the norm. In the early spring, just before the fish began their annual runs, groups of families came together along the coast or the banks of the rivers and lakes and set up a village consisting of bark-covered houses or wigwams. The men fashioned canoes and prepared spears, nets, and weirs. Woods were burned or girdled to open up the ground to the sun. Women cleared the fields for planting. As the village took shape, the men would go fishing, trapping, and snaring, while the women began the planting. By late spring, with the fields completed, the women set to gathering early berries and edible roots. Fish, shellfish, small game, and fowl helped fill out the village meals. By late summer the community was well fed and settled. Surplus fish were dried. Then as the season shortened, the women again concentrated on the fields, harvesting the corn, beans, and squash. Children gathered the last of the summer berries. In the late fall some of the harvest would be stored away for food in a deep hole dug into the side of a hill. The hole needed to be dry and protected from hungry animals. This store would sustain the community in the early spring before the game and wild plants became plentiful again and would also provide the seed for the next year's planting. With the stored food secure, the various families would gather their share of dried fish, corn, beans, and squash and move inland in smaller family units.

Winter months were hard. Food supplies ran low. The men scoured the forests in search of ever more elusive game. They also shifted their focus to bigger game. Deer and moose became the hunters' desired targets. One moose or deer could supply the family with protein for long periods of time. Game also furnished important materials for the community. Leather and bone were made into clothing and tools. Leather pouches and bags carried goods. Fur blankets and coverings kept out the winter cold. Even for the southern horticultural communities, the soil eventually wore out; wood for tools, spears, arrows, building material, and fuel grew scarce; pests built up; and the village would have to move to another location. Farther north the mobility was more pronounced and frequent, with villages breaking up and reconvening yearly, and families relocating in new inland areas according to the availability of game and supplies. Such movement gave rise to a fluid concept of rights to the land. Native Americans did not share the European concept of property. Native Americans did not own land; they used it. Large groups of related villages or tribes protected their right to use land and water to hunt, gather, and plant relative to competing village groupings, but this differed from property ownership. Land use rights depended upon two primary factors, present use and tribal membership. The Narragansetts, for example, believed they held rights to the land they controlled and used. The Narragansetts did not perceive that they owned the land, nor did they understand the concept of alienation, or selling, of land. They understood that, for example, if the Pequot pushed east and took control over land that the Narragansetts had previously used, the Narragansetts could lose the use of that land. So even though they traded and exchanged what they captured off of the land or grew on the land, because they did not view the land itself as a similar commodity, they did not sell or trade it as Europeans did.

Within the villages, sachems or community leaders parceled out land for fields to village families. Unlike European lords who owned the land, these leaders were appointed by consensus to oversee village affairs. They allocated community land use by what they deemed was in the community's best interest. The family could then use their allotted land and keep what they produced, but they did not own the land itself. Their use of the fields concluded, they abandoned the land and all claims to it. By this point, the fields' productivity had typically declined or pests had built up to the extent that abandoning the fields and any claim to them made more sense than expending the significant effort necessary to hold on to or continue to work them. At some future date, when the forests had reclaimed the lands and the soils had been replenished with minerals and nutrients, a new family might begin the process again. Further inland away from the river valleys and in the northern region of New England and the upper Midwest, horticulture accounted for far less of the community's diet. The peoples north of the Kennebec River in Maine relied primarily on hunting and gathering, as did the Huron, Menominee, and Potawatomi in northern Michigan and Wisconsin who lived among the conifer forests of that northern region. In the spring they migrated to the rivers' edges to capture fish running upstream to spawn. Fresh and then dried fish formed the basis of their diet until the summer game became more plentiful and the fruit matured on the berrying plants. Late spring also brought ducks and geese. Women and children gathered eggs and snared the birds. In the fall, hunters stalked moose, deer, and bear, fattened for the winter hibernation, while gatherers foraged in the forests for nuts and edible roots. In the late fall families searched out and settled in valleys protected from the fierce winter storms. There they stored the fall hunt out of reach of other predators, gathered fuel, and prepared furs and hides for the coming months. The winter was an ordeal. Food was scarce and hunger common. Snow on the ground made tracking game easier. But little snow often produced real suffering. Families survived by husbanding resources and limiting their expenditure of energy. Though winter was a trial, they knew that spring would bring new runs of fish and with them an abundance of food.

Those living in the mountains and further north where game abounded had access to meat, furs, and hides, but short growing seasons precluded much horticultural activity. To their south the communities in the broad river valleys generated horticultural produce but fewer furs and hides. Along the edges dividing the two ways of life, differing groups traded continually, with meat, furs, and hides moving south, while cornmeal, beans, and squash traveled north.

Lake Michigan from Pierce Stocking Scenic Drive. Sleeping Bear Dunes National Lakeshore, Michigan. (William Manning/Corbis)

Like their southern neighbors, Native American hunters and gatherers understood the right to land in terms of present use. The village or tribe might defend the right to hunt in an area against other tribes and the right to all the game they shot or captured there, but they did not believe they owned the land itself or the animals living there. Until killed, the animals belonged to no one. To people who saw wild game wander freely over the landscape and fly across hundreds of miles of land and water, this seemed perfectly reasonable.

Roughly 300 miles west of Boston begin the great inland waters of the Americas, the Great Lakes. These lakes cover 94,250 square miles and hold 5,440 cubic miles of water. They boast 10,200 miles of shoreline and encompass a basin of 201,460 square miles. Until the last 200 years, the land surrounding the southern segment of the Great Lakes (that part resting within the United States) was covered with a combination of deciduous forests, areas of grassland, prairie and woodland intermixed, and vast wetland marshes. The soil was deep, temperatures warm, and the growing season along the southern edge of Lakes Ontario, Erie, and Michigan lasted 160 to 200 days and 120 to 160 days further inland. Fish and game abounded. These lakes combined to form the largest contiguous body of freshwater in the world and they supported a plentitude of aquatic life. Of particular interest to native fishers were the lakes' whitefish, trout, salmon, herring, pike, and huge sturgeon. Native communities had been fishing these waters with weirs, nets, spears, and hooks for some 4,000 years before the mass arrival of Europeans disrupted their way of life. Local native communities fished cooperatively within families and villages. Women gathered nettles and hemp and made fibers into cords for netting. The men made the gill nets and did the fishing while the women preserved and stored what was brought ashore. In the southern Great Lakes region fish and domesticated plants provided the bulk of the communities' food. West of Lake Michigan wild rice and fish were the mainstays. Though they fished spring, summer, and fall, native fishers brought in the greatest catches during the spring and fall spawning runs. To reach the larger fish in the lakes and rivers of the Great Lakes water basin, native fishers relied on birch-bark canoes. These light, versatile boats were framed in white cedar and covered with white birch bark sewn together with coarse thread from the roots of tamarack, spruce, and pine. Pine also supplied the pitch that sealed the seams.

For mobile people who carried their goods on their backs or dragged them along on polls, anything that was not easily carried had limited value. Food shortages in late winter spurred hunters in their search for game, but upon encountering a herd of deer, the hunters had no reason to kill more than would meet each family's immediate needs. The dead deer had to be dressed and hauled back to camp. Killing more than was needed simply entailed extra labor in hauling back to camp more than was necessary. Native Americans knew they had to husband their strength. Later, when another need arose, the hunter would again go in search of deer. The same understanding applied to leather hides and furs. During the winter months, furs helped keep the family warm, covered cold ground, and protected children from hypothermia. But furs were also heavy, bulky, and, in abundance, burdensome. A community kept only the necessary numbers of furs and leather hides. No matter how enticing the target, few hunters would kill an animal for its fur if there was not an immediate need for the pelt. For Native Americans the distinction between "need" and "want" made in the lyrics of the Rolling Stones song, "you can't always get what you want, but if you try sometimes, you just might find, you get what you need," was an irrelevant one. For them, what they wanted was what they needed.

Scholars of the past must be careful to avoid idealizing the lives of these pre–European-contact Americans. Trade took place along borders, but so did conflict. War was not unknown to Native Americans. Late winter brought severe hunger to most communities. Lice, fleas, and mosquitoes plagued them constantly. Storms battered them and their wigwams. Yet by limiting their wants, these Americans nevertheless enjoyed a life of some leisure. Work was not a constant struggle but an intermediate activity.

Unquestionably, these early Americans altered the landscapes they inhabited. But these changes were transformative—not destructive. Their fires cleared brush and opened fields and the forest floor. Their abandoned fields created breaks in the forests that helped diversify both flora and fauna. Yet, without draft animals, they lacked both the means and the inclination for broad open fields. Their tools and mounds functioned effectively around tree stumps, rocks, and boulders. They had no reason to clear their fields of stones or delineate boundaries with stone walls. When their fields became depleted, they moved on. Without domesticated animals, they had no reason for pastures or hay fields. We may be guilty of romanticizing these early Native Americans' lives, but there is little question that they trod lightly on the land they inhabited. They never felt compelled to conquer the land and its inhabitants. The conquerors came from Europe.

2

FIRST ENCOUNTERS

As my sister drives north along Cape Cod past the sign for First Encounter Beach, she encounters a store selling beach chairs, umbrellas, towels, and T-shirts, a hot dog stand, and a real estate agency, but no Native Americans. To understand why few Native Americans remain in the area, we must first understand the conditions that prevailed on the Cape when that momentous "first encounter" took place.

Separated for millennia from the Eurasian continent, the flora and fauna of the Americas evolved in accord with prevailing local conditions. The mammals, birds, lizards, snakes, trees, bushes, and grasses that took root corresponded to similar plants and animals in Europe but nevertheless remained distinct species. It was this unique ecosystem that humans confronted when they first began spilling into the interior of the new continent some 12,000 years ago. The hardy hunters who moved south from the arctic regions arrived relatively free of disease. Living with no domesticated animals, they were not exposed to contagious diseases that migrate from pigs or fowl and spread widely among densely settled peoples linked through trade routes crossing whole continents.

Sixteenth-century Europeans actively traded throughout the Mediterranean world and had trade links to Asia and sub–Saharan Africa through the silk routes. Europeans and many of the peoples with whom they traded kept domesticated animals such as sheep, goats, cows, horses, pigs, chickens, and ducks. Limited quarters (and cold winters) encouraged close living between humans and their animals. In such environments diseases arose and evolved, especially in the warmer climates. Dense settlements facilitated the spread of diseases among people, and trade moved these infections between communities, cultures, and even continents.

Both cities and farms characterized the Europe of Columbus, Desoto, and Champlain. It was also a land of traders, traveling monks, pilgrims, and itinerant artisans. Columbus set sail from Spain but hailed from Genoa, Italy. At the time, a relatively successful Italian merchant's home would contain silks from China and cottons from India. The table of even a rural manor house in England or France would feature peppers from the Spice Islands. The ships in European harbors helped link together the people of the town and city with every far-flung community where these ships put in to harbor, as well as the individuals who loaded and unloaded cargo, sailed the ships, and repaired them with their brethren of different hues and tongues. Although diseases continued to ravage Europe, immunities and resistances built up over centuries, helping save populations from complete decimation.

NEW ENGLAND

The Native Americans who greeted the Pilgrims on Cape Cod in 1620 were openly hostile to the new arrivals. With a flurry of arrows they drove the Pilgrims back to the *Mayflower*. We cannot possibly know what motivated these Wampanoags to such a hostile response, but we can speculate. Perhaps they had an idea of what was in store for their children and grandchildren—if they lived—once European agriculture and commercial economy began radically transforming the land and culture that for centuries had remained relatively unchanged. More likely, they had learned what had happened to their cousins, the Pawtuxets, in the area the Pilgrims would later call Plymouth.

Although this confrontation between the Pilgrims and the Wampanoags was a "first encounter" for the Pilgrims, this was certainly not the first interaction between Europeans and the Native Americans of New England. In the early seventeenth century, Champlain explored the northeast coast, putting in at many bays and harbors to replenish fresh water and stores. There Champlain and his men traded with or stole from local communities. In 1605 he entered Plymouth, put down anchor, and explored the local native settlement, returning to Europe with a detailed map of the harbor and tales of a thriving community of horticulturalists. In 1614 the London Company sent John Smith to explore the New England coast for possible colonization sites. In 1619 Thomas Dermer returned from an exploratory investigation of the New England coast to announce that local opposition to colonization would be minimal because most of the natives were dead. Indeed, when the Pilgrims did finally drop anchor in Plymouth Harbor and explored the area on foot, they found cleared fields, but an empty village.

In addition to the knives, blankets, and trinkets traded to the local natives for water and stores, the European explorers furnished the native populations with a bounty of diseases. Of these, smallpox proved the most devastating, although measles, chickenpox, and whooping cough also took their toll. When locals proved uncooperative in trade, Europeans instead opted to rob native fields or kidnap people for slaves or to take back to Europe as examples of the local population. And although stealing food or healthy members of the community would mean deprivation over the long winter months, it was the European diseases that ultimately proved calamitous to the peoples of the Americas.

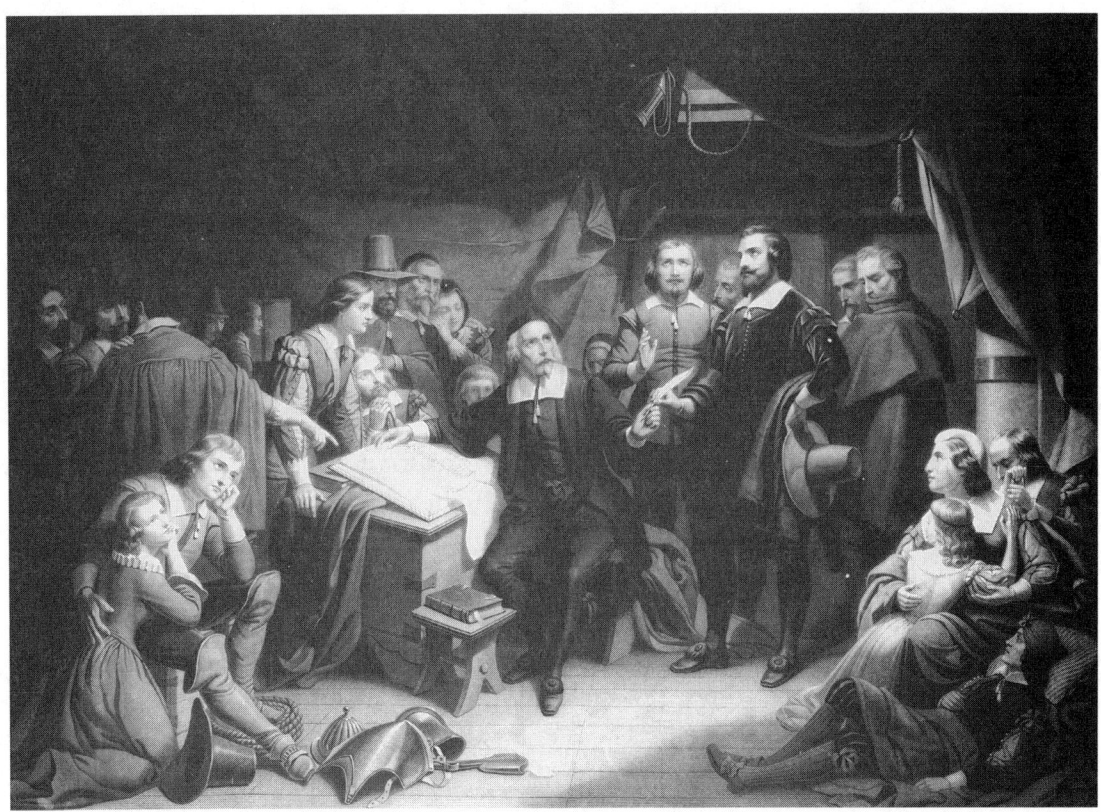

Pilgrims aboard the Mayflower *sign the Mayflower Compact in 1620. (Library of Congress)*

Smallpox is caused by a filterable virus of the category variola and is closely related to the virus that causes cowpox. A highly contagious pathogen, smallpox can be transmitted between people either in airborne droplets from a sneeze or cough or by individuals coming into contact with contaminated clothing or other items recently handled by an infected person. After an incubation period of ten to fourteen days it strikes abruptly, with chills, a high fever, headache, and muscle pains, followed by vomiting and convulsions. Days later lesions appear deep in the skin, giving rise to a spreading rash and purulent sores. Again, the victim experiences high fevers and painful sores spreading into the nose, mouth, throat, vagina, urethral passage, and rectum. Hair, eyebrows, and nails drop off. Death is usually caused by hemorrhaging of the sores and gangrene of the skin. For Native Americans it was a difficult, nasty, hard death and it wasn't quick. It took long enough for plenty of family members, caretakers, neighbors, and others to be exposed, while its contagion lingered on among things used or handled in the form of small dry scales of the lesions shed by the afflicted.

Europeans had dealt with smallpox for centuries. Exposure in childhood was common. Those who survived developed immunities. The disease could recur in persons previously exposed, but they usually suffered minimally, with few of the more extreme symptoms. Before the early and primitive vaccinations of the eighteenth century, smallpox survivors in Europe commonly bore vestiges of this disease as pockmarks on the face or body. Through their contact with cows, milkmaids often picked up the related cowpox, a strain that did not cause smallpox's severe reaction but which gave them immunity to the more virulent complications of smallpox. This is precisely how English milkmaids gained their reputation for clear skin. Smallpox, in one form or another, was so common that it was Europeans' near constant companion. The two-week incubation period followed by three weeks of active disease meant that a sailor could be exposed in England and be well out to sea before showing symptoms. By the time the ship crossed the Atlantic and reached Plymouth Harbor, the disease could have spread among any number of crew members by contact with each other or from items handled by infected crew members. At that point, it was a simple matter to pass the disease on to Native Americans on infected trade items.

With no history of exposure, Native Americans were devastated by smallpox. They had no familiarity with the disease, and native cures proved totally ineffectual. Infected persons were soon prostrate with fever and pain. Rashes spread, and lesions began to erupt deep in the skin. When lesions in the mouth and throat broke out, eating became an impossibility, while lesions in the urethra and anus made excretion an experience that brought the hope of death. Family members and caregivers were quickly struck down, and soon few remained to nurse the stricken. Gangrene and hemorrhaging finally brought death, accompanied by great agony and the smell of rotting flesh. The disease quickly spread throughout the community, so that soon few were able to stand, let alone do the work of planting, fishing, harvesting, or hunting. This loss of labor further threatened villages. Eventually, survivors fled.

Like all peoples, these Native Americans survived through communal activity. They shared tasks and depended upon one another. When an epidemic of smallpox destroyed the group, individuals without family or village sought to attach themselves to other communities. In so doing they spread the disease to other villages.

These epidemics took a staggering toll on native communities. Mortality rates of 80 and 90 percent attended the initial onslaught of the disease. A village of sixty inhabitants might leave only a handful of survivors. Sometimes only one villager stood alone among the huts where his or her dead family and friends might still lie (Tucker 2001, 11–12).

First Encounters 27

Pilgrim overlooking spring planting scene in Plymouth Colony, 1620s. (North Wind Picture Archives)

Europeans marveled at the power of disease to wipe out native peoples. Some Puritan clergy concluded that it must be God's work, removing troublesome heathens from the new Eden. The deaths certainly helped the settlers establish a foothold in their new world. Although the Europeans brought new military weaponry and technology with them, in conflict with Native Americans their weapons alone did not provide them with the ultimate advantage. Arrows were probably more accurate than the European blunderbuss and certainly easier to load and fire. Iron swords trumped wood and stone weapons, but large numbers could overwhelm that advantage. With their more cumbersome weapons, the Europeans understood the importance of fortifying positions. They were also accustomed to organized and disciplined military actions, which gave them a slight advantage over indigenous fighters. The Native American fighting style depended more on individualized acts of courage in open combat, warrior against warrior. The European's advantage notwithstanding, fortified positions could be held only so long, and against a large enough enemy even a disciplined military unit can be overrun. The Europeans came in small numbers; their boats were small and crammed tight. Just 102 Pilgrims arrived aboard the *Mayflower*. Of that group, only a minority were capable of military action, as most were farmers, women, and children. Before coming into contact with Europeans, the Wampanoags around Plymouth Harbor would have far outnumbered the 102 Pilgrims and could easily have pushed them back into the sea. But the group pushed back was not the Pilgrims but the Wampanoags.

The Pilgrims also faced hardship and death. Of the initial 102 who set out from England, only half survived the first winter. Yet one hardship they did not have to face was hundreds of angry native warriors. Native Americans, reeling from the impact of smallpox, measles, chickenpox, and the other communicable European diseases, could not mount an effective military campaign against the Europeans. Initially, Europeans faced minimal opposition as they set up colonies and trading posts up and down the Atlantic coast. For on the whole, that opposition was prostrate with disease.

The Pilgrims who settled in Plymouth were primarily farmers. They originated in Scrooby Manor, 150 miles north of London, where they tended small farms, growing wheat, rye, and oats, rotating crops, and using animals to plow their fields and haul off their harvest. They kept pigs, usually penned and fed scraps and leavings from the fields. Chickens ran freely about the yard and also ate scraps and leavings. The more prosperous farmers owned cows, oxen, and occasionally horses. Shortly after the turn of the seventeenth century, these prosperous yeomen began to feel conflicted about the Church of England and eventually resolved that it was too close to Roman Catholicism. In response, they determined that only by separating themselves from the Church of England could they retain religious purity. In 1608–1609 a group of these separatists left England for Holland where they could practice their purified religion without interference.

Holland made for an interesting choice. In addition to religious tolerance, Holland also offered the Separatists an example of some of the most intensely farmed and manipulated land in Europe. The Dutch were a numerous people with limited arable land. They were adept at capturing farmland by draining marshes and other wetlands. They built dikes to hold back inundating seawater and dug ditches to carry away unwanted water. Living in lowlands with few fast-flowing streams, the Dutch could not rely on water-powered mills. Instead, they built windmills to capture energy to mill their grain and pump water from their fields.

Although the Dutch accepted the uncompromising Separatists, the Separatists themselves were not comfortable in Holland. Under the leadership of

Illustration depicting Squanto (aka Tisquantum), from the Native American Pawtuxet tribe, who served as a guide and interpreter for pilgrim colonists at Plymouth Colony and Massasoit. He died in 1622 from contracting smallpox while guiding William Bradford's expedition around Cape Cod, Massachusetts. (Hulton Archive/Getty Images)

Eighteenth-century stone wall, Hampton, New Hampshire. (Courtesy of Kazia Cumbler)

William Bradford, the Separatists returned to England with a plan to settle their own colony in America and petitioned for a land grant from the Virginia Company of London. After stocking up on supplies—and acquiring substantial debt—in England, the group set off for Virginia aboard the *Mayflower* in 1620. Bad weather and a navigational error landed the Pilgrims not in Virginia but first on Cape Cod and then into Plymouth Harbor.

Conditions for the Pilgrims' settlement in the area were far from ideal. Some among them were not adherents to the faith but sailors and adventurers who signed on for profit rather than theology, and their supplies were short from the beginning. They did, however, have a core of true believers, a dedicated and talented leader in William Bradford, a heritage of farming and husbandry, and an accompanying supply of seeds and animals. They were also fortunate to arrive and discover, instead of hundreds of armed opponents, just an empty village and a lone English speaker, Squanto, who had earlier lived among Europeans. Landing in Plymouth in the fall of 1620, they survived the winter by living on board the ship and eating up much of their remaining supplies. Some of the men tried hunting on shore, but as plentiful as game was in the New World, these farmers were not experienced hunters.

Squanto showed the Pilgrims new ways to see the land. He directed them to plant maize or Indian corn, beans, and squash and gave them clues about hunting. Indeed, that is the story of Thanksgiving. It is a giving of thanks for the Indian foodstuffs that helped the Pilgrims survive those difficult first years, before they transformed the land away from the production of corn, beans, squash, and turkey. But the Pilgrims did not come to the New World to live like Indians. They wanted to recreate the world they left behind in England—to create settled villages and enclosed fields; to plant wheat, oats, barley, and rye, and neat vegetable gardens of carrots, beets, radishes, turnips, peas, and cabbage; to keep pigs, cows, goats, horses, sheep, chickens, and oxen. And ultimately it meant cultivating flower gardens and growing herbs and fruit trees.

Whether in Plymouth, New Amsterdam, or Boston, the Europeans' early years were marked by hardships and unfamiliar surroundings. The first boats arrived crammed with settlers with little room left for animals, tools, or supplies. Settlers soon learned of the extreme challenges they would face in this new land. Winters were cold, particularly by English standards. Mosquitoes in numbers and of ferociousness uncommon in England, black flies with bites that could drive calm men mad, poisonous snakes, and poison ivy all awaited the settlers who disembarked from the *Mayflower* in the early spring of 1621.

To survive, early settlers up and down the North Atlantic coast relied on trading with Native Americans, consuming their own stocks, and receiving new supplies from Europe. A settled people, Europeans built homes and set out fields.

They marked property boundaries. When threatened, they built castles. Faced with this new land, they settled. They built homes and forts, delineated fields, and allocated holdings. Being short of seed grain, they traded items they had—knives, blankets, axes, and metal cooking utensils—to the indigenous people in exchange for corn, bean, and squash seeds. Because space on the initial ships was limited, the settlers brought few domesticated animals and instead depended upon Indian methods for clearing and planting fields. Game and fish may have been plentiful for skillful hunters who knew the animals' habits; were attuned to seasonal shifts in habitat; and had weapons, nets, traps, and snares designed for particular prey. These Europeans qualified on none of these counts. Gathering local nuts and berries proved almost as difficult. The blueberries, raspberries, blackberries, and strawberries, the growth of which Native Americans encouraged with their periodic burning, had to be found to be collected. Europeans may have been familiar with similar edible fruiting plants from Europe, but this familiarity did not include knowing where to hunt for the New World plants.

Although the Europeans depended heavily upon trading with Native Americans, as well as copying—and sometimes stealing—their techniques and food stuffs, these recent arrivals to the New World still depended upon support from the Old World. Every returning boat carried pleas for additional supplies, particularly more seeds and animals. But as goods from family, friends, merchants, and investors flowed west across the Atlantic, those in England increasingly demanded that something flow back east in return.

The Pilgrims had little to send back in the early years. They tried to gather furs to send back, but they were poor traders. Either the local Native Americans did not have skins to trade, or the Pilgrims had nothing the Native Americans wanted in exchange. Whatever the case, when the Pilgrims' first return shipment sailed in 1621 aboard the *Fortune,* it included only two barrels of furs. What the Pilgrims did send back, however, was lumber. Indeed, they stuffed the ship with it. At first blush, lumber may seem a curious cargo for a ship supplying materials to England. In fact, late seventeenth-century England had plenty of coal, but very little lumber. Since the beginning of the sixteenth century, the English had been cutting down forests far beyond sustainable levels. Although the first couple of waves of the bubonic plague in the second half of the fourteenth century dropped England's population dramatically, by the late 1500s the numbers had more than recovered and even began a period of rapid growth, despite occasional revisits from the plague. Agricultural innovations, increased manuring in particular, along with the related growth in domesticated animals, marginally improved the diet. This in turn improved fertility and lowered infant mortality rates, particularly in the rural countryside. Growing rural populations increasingly reclaimed more land for agricultural production. In the southeast they drained wetlands, but in most of the rest of the country they cut down trees and opened fields. Population growth also increased demand for housing, fuel, wagons, ships, and tools—much of which were fashioned from downed trees. During the sixteenth century England prospered, and with prosperity came more building and more cutting of trees. England's prosperity was also linked to iron manufacturing. England defended its ships and coast with weapons made of iron. It surrounded its wheels with iron rims; it bolted and laced together its structures with iron; and until the eighteenth century, when the Darbys, a father and son team, discovered how to make iron from coked coal (coal that is heated to high temperatures in closed ovens), iron was made by smelting iron ore with charcoal from wood. As a result, by the beginning of the seventeenth century, forests in England were becoming scarce. Wood was imported from Scandinavia. Increasingly, the English gave up building homes with wood in favor of waddle and dab (mud, sticks, and straw), stones, and bricks, and took to heating them with bog bricks or coal. In seventeenth-century England trees were precious commodities. Trees were not nearly so precious on the western shores of the Atlantic. There they abounded. They cluttered the landscape. Trees even stood amidst the fields of the indigenous people. Indeed, to properly clear a field in the European manner, settlers had to remove the trees, either by burning them or cutting them down. Burnt forests produced ash that could be boiled down to potash and sent back home, but cut trees could be sent directly. Unlike other items wanted back in England, trees didn't have to be raised, tended, or traded for. They were simply there for the taking. And take the early colonists did. Because of the superabundance of forests, the early settlers blithely cut down any tree within an easy roll to a waterway that could convey them to a harbor and waiting ships. Those too far for easy transport were burned. Trees yielding the best building lumber were the first cut. Tall white pines could be used for ships' masts or cut into clapboards. Chestnut wood made good posts, foundations, timbering for houses, ax handles, and plows, for the wood was hard and resisted pests and rot. And because chestnuts sprouted from roots, cutting them down seemed to have little effect—they would simply sprout up again along the roots of the cut tree. Cedar was cut for shingles; oak for barrels and ship siding; walnut, hickory, and cherry for furniture; and anything that would burn was used to cook food and keep the settlers warm in the winter. Colonists along the coast south from Boston to southern New Jersey particularly liked pitch pine because they could turn the sap into turpentine or into rosin to caulk ships and protect ropes used at sea, while the wood burned hot and fast in their fireplaces.

American forest abundance and English lumber shortages combined to drive a thriving commerce in wood. To the Native Americans and the first settlers,

An early settler clears a homestead, 1740. (Harvard Forest Dioramas/Fisher Museum/Harvard Forest, Petersham, MA)

trees were just there. They were useful for building canoes, homes, tools, and weapons, or were just things to be removed to make room for corn, wheat, oats, barley, and vegetables. But as early settlers requested more and more goods from Europe, Europe demanded more in return from the settlers as well. And because these exchanges took place within an increasingly commercial-capitalist world, the nature of that exchange required the commodification of objects. The goods shipped to the New World had commercial value measured in commodity terms. The Pilgrims owed debts back in England to those who first financed their enterprise. The Pilgrims may have been religious purists seeking a new Jerusalem, but their trip floated on paper debt. Lumber sent back was no longer just a thing—a cut tree. Nor was it merely wood to be used. It was wood as a commodity with an exchange value. That value was certainly linked to its use and need in England, but it was also linked to a money measurement in pounds sterling. Because wood had an exchange value, it could cancel out debt.

But that was back in England, where trees were becoming scarce. In North America trees were still trees. And although the early colonies quickly began dividing up the land into estates and establishing property boundaries, the trees were so ubiquitous that few people paid attention to where the trees stood. But as the trees were transported from the land to the harbor and ultimately to Europe, the trees took on value. The person who brought the trees to a market claimed property in them and their value. Laboring over the trees, cutting them down, and hauling them to market, the Europeans claimed title to the value they added to the trees by so doing.

By the late 1630s there were several commercial lumber centers along the rivers spilling into the North Atlantic, particularly to the north in New Hampshire and Maine. Among the colonists' first acts after building their homes and clearing their fields was to build a sawmill that they could convert to a gristmill once the grain was harvested. Native American observers must have scratched their heads in curious wonder to see grown men arguing and bargaining over the value of a cut tree.

Even more curious than Europeans' habit of claiming ownership in trees was their inclination to trade things of real value like knives, cooking utensils, cloth, and blankets for common items the Native Americans considered only minimally useful, such as a beaver hide. Although beaver hides were not particularly ubiquitous in Native American villages—deer, moose, or bear hides being more common in lodges and wigwams—beavers themselves did populate the streams and rivers of North America. Beavers were clever animals that had evolved a complex system of defense. They built their lodges with exit tunnels at the base. They then built dams that flooded the area around their lodges, effectively protecting the lodge from predators. In winter when the ponds froze, the beavers were protected by the thick lodge of mud, sticks, and branches and the entrance tunnel below the ice. Living on the bark and young shoots of trees, the beavers would cut the trees, float them to the lodge, and store what they needed for winter underwater or in the lodge. Few enemies could challenge beavers. Although out of water they were slow and potentially vulnerable, with their sharp incisor front teeth and hind claws in water few predators had a chance. But although few animals had the natural advantages to successfully hunt beavers, humans did. The beaver lodge was easily identified. It was no small challenge, but in winter a hunter could cross the ice, smash the lodge with an axe, and kill the beavers at will. Beaver meat was not particularly desirable, however, and the hides, although they repelled water, were not of greater use than deer, moose, or bear hides. So although Native Americans did kill beavers, particularly in times of great hunger or when in need of waterproof hides, they did not kill great numbers of them. As a result of these factors, beavers flourished in North America. Estimates put the number of beavers, *Castor canadensis,* in the continent's streams and ponds prior to European arrival at 60 million. Then the Europeans came and the beavers' fortunes changed dramatically.

Native Americans armed with muskets and bows hunt beavers on a river in Ohio, mid-1700s. (Getty Images)

Wealthy, urban Europeans liked hats. Smooth, water repellent, and consisting of hundreds of small barbs, beaver fur suited the hatmakers' needs perfectly, and by 1600 beaver skins had become their preferred material. As a result, the European beaver, *Castor fiber*, was hunted almost to extinction, yet demand for the hides remained high. European adventurers in the New World noticed beaver hides among the skins in the communities of the indigenous peoples of the north. Correctly assuming that Native Americans knew how to locate and capture beavers, the Europeans began offering to trade for beaver hides. Wanting knives, blankets, and cooking utensils, the Native Americans quickly emptied their villages of their beaver skins. But unlike the Native American, the European trader had no immediate need or personal use for the beaver skin. Rather, the trader was interested in the skin's commodity or market value in Europe. This market value trumped the individual traders' immediate needs. It was a market that ultimately linked the Native American with some spare beaver skins to the wider European demand for those skins. Still wanting knives, blankets, and cooking utensils after the few spare beaver skins on hand had been traded away, the Native Americans were willing to go out and seek more. This link to the European market had disastrous consequences for the beaver.

For the European trader, however, it meant profit and markups of well over 1,000 percent. Profits like these led many Europeans into the trade. To the north along the St. Lawrence River, the French established trading posts. At the mouth of the Hudson and upriver and along the Connecticut coast and up the Connecticut River, the Dutch established a colony and trading posts. The English were not as successful as the Dutch and French, but they also entered the trade. Trading posts were positioned along the key waterways leading into the interior. The Europeans stationed along these posts not only traded with the Native Americans, who brought pelts downriver in birch-bark canoes (made from the bark of the paper birch tree that grew in the northern regions), they also cleared land and built homes and warehouses to store and protect their purchases. They planted crops and introduced pigs, chickens, goats, cows, and horses to the area. They cut trees to build homes and warehouses and to feed their fires, cook their food, and warm their bodies. Soon the areas surrounding trading posts became small settled communities with domesticated animals, fields, houses, and fences. But these traders were not farmers raising crops to sell or even to trade. Their commodities of trade were blankets, knives, cooking utensils, and later alcohol. Over time, as Native American families also took up residence near the trading posts, the trader might also have traded food, but growing food for trade was not his central activity. As a result, only a limited expanse of land was cleared for planting crops, while the countryside surrounding the post remained comparatively untouched, with the exception of the beaver lodges. Soon the Europeans began

Dutch trading (wampum) with the Native Americans. Undated engraving. (Bettmann/Corbis)

running out of things native hunters wanted. Native Americans were still a mobile people. To a point, cooking utensils, knives, and blankets could be easily transported when the village moved to other quarters, but too many cooking utensils became a burden. The Dutch solved this dilemma by offering wampum—shells worked into decorative ornaments—as a unit of exchange. The shells and the manufacturing of them into wampum came primarily from the Long Island Sound area. Using stone instruments, the fashioning of wampum was a laborious task. In precolonial times, wampum was the preserve of sachems, leaders of the community, or other special persons. It was highly prized and often exchanged as gifts between tribes and privileged individuals. By the 1620s, the Dutch were trading goods to the Long Island Sound tribes for wampum, which in turn they exchanged upriver for beaver pelts. Shortly thereafter, the English began using wampum as a medium of exchange.

Demand for wampum was particularly high among the peoples away from the Long Island and southern New England coasts, and killing beaver became a means of satisfying that demand. Hunters who had previously rarely sought beaver now forsook other animals for the beaver. Villagers ignored their traditional winter habitats and instead clustered around the trading posts. Hunters abandoned their families and kin to focus on the beaver hunt. Previously self-sufficient native villages became dependent upon the posts for food. The search for beaver began to change how the community related to the land and the seasons.

Beavers quickly disappeared from the coastal regions from the St. Lawrence River south. By the beginning of the eighteenth century, beavers were rare—if they existed at all—east of the Allegheny Mountains. Native Americans went deeper into the interior hunting for beavers or other Indians who would be willing to get beaver. The posts themselves moved inland as traders tried to gain a competitive advantage by setting up closer to the source of the beavers.

Fewer beavers meant fewer beaver ponds. The ecological diversity the beaver ponds created and their role in holding back floodwater was lost along with the beavers. Many contemporary observers noticed the decline of the beaver. Traders lamented the diminishing loads of pelts coming downriver each successive spring. Native Americans, who had become increasingly dependent upon the beaver trade for food and supplies, also noticed the shortage of beaver. Conflict emerged between traders and hunters and among different groups of hunters.

Europeans did not create conflict among the various alliances of Native American nations or language groupings; conflict had existed long before European traders arrived. With relatively equal weaponry and limited bounty to be won via conquest, the conflict had always remained at a low level. In the seventeenth century, that began to change. Disease wracked villages and communities, killing off men, women, and children. Before long there were too few able-bodied men left to carry on the hunt or women to plant the crops and bear the next generation. Compounding the depopulation crisis was the European avarice for pelts and the indigenous communities' desire for tools and materials, blankets and wampum, and metal weapons to war against enemies and defend against attack.

The land surrounding the Great Lakes was also rich in game, including beaver. With the pelt-bearing animals depleted in the northeast, the Five Nation Iroquois, armed with European weapons, began moving west in search of hunting grounds and captives to help repopulate the villages. They moved west armed and ferocious. Before them the Iroquois-speaking Huron-Petuns and the Algonquian-speaking Miami, Illinois, Kickapoo, and Potawatomi fled west. Among both groups were the traders, the French to the north and west with the English and American colonists just behind to the east. Although the beaver's decline was apparent to many, far fewer likely appreciated how the beaver's departure altered the ecology of the Northeast.

The loss of the beaver was not the only, and probably not even the most significant, factor in the shifting ecology of the region that accompanied the establishment of European settlements. The Pilgrims who settled in Plymouth were farmers, not traders or trappers. They tried their hand at trading, but with limited success. But farming they knew.

Their settlement lay on the shores of Cape Cod Bay. The fish that gave the bay its name was indeed prolific in the region's waters. The rich fishing grounds of Georgia's Bank, Stelwegon's Bank, Brown's Bank, Stable Island Bank, Banquereau Bank, St. Pierre Bank, and the Grand Banks had been drawing Basque, French, Dutch, and English fishermen for decades—maybe even centuries—before the Pilgrims arrived. These were the richest cod-fishing grounds in the world, and for good reason.

In the ocean, nutrients at the surface sink to lower levels. This process can leave the surface area devoid of the nutrients necessary to sustain the small marine life that is central to the food chain. But in waters into which the deep ocean currents penetrate these nutrients are conveyed back to the surface, giving rise to an abundance of life. Long summer days supply sunlight for plant photosynthesis. Deep ocean currents pushed to the surface by colliding currents and then run up against the slopes of the more shallow banks, as they do over the banks off the coast of Canada and New England, bring nutrient-rich water to the surface. In such an environment cod, *Gadus callarias,* thrive.

Cod were the fishermen's favorite because they were easy to market. Once caught, they could be gutted, dried, salted, and then shipped back to Europe in barrels. Dried cod thrown into a stew or chowder absorbed the fluids to provide a tasty, protein-rich meal. The Pilgrims weren't fishermen, however, and even if they had been, they would have faced difficulty catching cod, which prefer deeper, less accessible waters. Despite the fact that they claimed their colony would profit by fishing, the Pilgrims instead turned their backs to the sea and looked to the land for their survival and success. Not until later generations would New Englanders capture cod in numbers sufficient to enrich their communities and lead commonwealth fathers to hang the image of the cod in the state house. Although the Pilgrims were initially forced to grow native crops, their plan was to grow the familiar crops from Europe, and to grow them as they were grown back home. While Europe was a land of settled agriculture and land ownership, where the Pilgrims came ashore—far from their land grant—land abounded and had no owners, at least not in the European sense. The Pilgrims' first official act was to draw up a "compact" to create a government, which then proceeded to allocate land to the settlers. Once on land designated as theirs, families built homes, cleared fields, and put in vegetable gardens. Within a few years more animals began to arrive—cows (it took three years before the first bull and three heifers arrived), oxen, pigs, sheep, goats, chickens, and horses. With land aplenty and little time to gather food for themselves, let alone for their animals, these early settlers let their livestock roam free. The animals' ears were clipped to distinguish ownership among neighbors. The pigs flourished the most in this new environment. Left on their own, they roamed the forests, feasting on acorns, wild tubers, and other roots. Cows and horses were also left to wander and scavenge food where they could. As they were more vulnerable to predators, sheep and goats were kept closer to home. To keep the cows, horses, and pigs from tramping down or digging up plantings, fields were fenced. Pigs dug up the forest, while cows and horses trampled the ground, compacting it in the process. As herding animals from a wet climate, they gathered, usually by streambeds, eating the native grasses until nothing was left but bare ground, compacted but still subject to erosion.

As their animals began to change the land, the settlers began to alter the landscape by importing flora from home; other plant species arrived on their own. The colonists wanted flour, particularly wheat or rye flour, for their bread and for market. Wheat and rye flour sold in the markets of Europe, so that is what the early settlers assumed they would have to grow to be commercially successful in this new world. Wheat grew in cleared fields. Granted land possession by their new government, the early settlers imported oxen to pull tree stumps from their new fields. They also built wooden flats or sleds on which they loaded the stones that littered the ground. With the aid of oxen, they transported these stones to the fields' edges and used them to construct what would become the classic New England stone walls. With the fields cleared, the colonists planted wheat and rye. Closer to their homes they planted vegetables, and in the lowlands they cut down the maples and birches and planted fruit trees brought from home—first apple trees of many varieties, then pear. But as they planted the seeds they painstakingly brought or had sent from home, they also planted what they had not intended. Weeds also arrived as stowaways in the bags of seeds, in animal feed, and clinging to clothing, buckets, and barrels. Soon, around each English settlement, there were not only fields of wheat and rye, but also patches of dandelions, chickweeds, bloodworts, and nightshades. Noticing the prevalence of these new and exotic plants where the Europeans settled, the Native Americans labeled them "Englishman's Foot." These invading plants were well adapted to surviving in disturbed ecosystems. Broadleaves, like dandelions, thrived in open, plowed fields and overgrazed pastures. As European animals ate the existing native grasses like broomstraw, wild rye, or marsh grass (*spartinas*), the invading weeds moved in. Ragweed, an indigenous plant held in check by forest cover, took advantage of the opening of fields and spread rapidly. Other native grasses and plants faired poorly under the impact of European animals. And as the native flora were eaten and trampled, Old World grasses, having evolved along with domesticated animals and capable of quick rooting and seed-ing in tough, trampled ground, spread in their stead. When the Europeans first built their settlements, open ground was scarce and confined mostly to abandoned Indian fields and meadows. When domestic animals began arriving, they were turned out onto these fields and meadows. The settlers considered the indigenous grasses inferior to those back home. As the animals overgrazed these spaces, newly imported grasses, such as bluegrass and white clover, grew up. Initially these plants arose as volunteers, having come over in animal fodder. Although the appearance of the "better" haying and grazing grasses pleased the European settlers, their animals' ever-increasing demand for food could not be satisfied by these "volunteers." As a result, farmers were soon clearing more pastureland to sow fields with "English hay." But the demand for animal food increased too rapidly for the farmers to keep pace by clearing land for pasture and haying. This forced most settlements to remain close to areas with large tidal marshes or extensive meadow wetlands where fodder could be harvested, even if by necessity from "inferior" native grasses.

The Plymouth farmers and other early settlers along the North Atlantic coast planted wheat partly with an eye to a marketable commodity, but their main concern was feeding their families. This focus not only occupied their time—it governed their strategies around how to organize the land. North America offered a radically different world than the one the colonists left behind and forced them to adjust to that new setting. Heedless of what the Native Americans thought, the early settlers considered the new land free of people, or at least so sparsely populated that there was virtually no limit to the abundance of rich and bountiful land. What North America lacked, however, unlike Europe, was people to work the land. This perceived imbalance resulted from the European settlers' inclination to view issues of land and land use in European terms. Contrary to the Europeans' perception, the Native Americans were not letting perfectly good land go to waste. In fact, the Native Americans used the land extensively, practicing a form of long fallow or forest fallow agriculture. Rather than clearing areas of stumps and rocks, manuring fields, fighting back pests and invading plants, and intensively working the land, native horticulturalists instead moved to new locations, thus allowing the abandoned fields to return to forest. Left alone, the land would once again accumulate nutrients, and the pests, finding less to eat, would also move on. The dense tree canopy would eventually return, cutting off the sun from the forest floor and killing off competing brush and bramble. This itinerant approach to agriculture left large areas to lie fallow. In the interim, before these abandoned fields returned to full forest, they provided a habitat for berry bushes and plants, fresh tree shoots, and native grasses. These in turn provided a food source for such game animals as birds, moose, deer, bears, and rabbits that the Native Americans hunted. Native Americans perceived abandoned fields not as wasted or empty, but as functioning and active. A planted field furnished corn, beans, and squash, while an abandoned one provided fruits and game. Both were important for community survival. Whereas the European visitors to these shores looked at forests and saw wildernesses of untapped lumber or unclaimed farmland, by contrast, the Native Americans saw habitats for communities of animals. To the Native Americans the land was not without people; they populated the land, and given the way they used the land, they in fact populated it rather densely, at least until the European diseases took hold. Native Americans also acknowledged the plants and animals that occupied the forest, as well as the claim each laid to the space it occupied.

From their European perspective, the settlers saw the land as a promising wilderness too abundant to fully exploit with their limited numbers. But they nonetheless attempted to recreate the world of exclusive ownership and use practiced in Europe. They parceled out land and gave individual property ownership to the parcels. Once a person had a piece of land, it was his (almost never hers) to do with as he wished. This typically meant excluding others from using it. Few settlers objected to such policies. Most had their own parcels, and those who did not understood the concept of private ownership and exclusive use. It was a concept central to their worldview and deeply imbedded in their history and culture. By contrast, Native Americans did not appreciate this concept. They understood using the land—they had been doing that for centuries—but they did not understand owning it at the exclusion of other people. As a result, they often came into conflict with the settlers, when hunting on a settler's allotment or burning a section of forest to plant on—land a settler felt was reserved for his exclusive use (even if he was not using it at the time). But this conflict cannot be attributed exclusively to cultural difference. Admittedly, Europeans and Native Americans understood property differently. But Europeans could be just as obtuse about property rights when, for example, a settler's pig or cow, let loose to wander about the woods, dug up or trampled a Native American field, even when that field was clearly on land that was not the settler's "property." Culture and differences in perception factored into land conflict between Native Americans and Europeans, but so did greed, power, and racism. As the wandering cow or pig example above indicates, settlers initially adjusted their agricultural practices to a perceived abundance of land. Rather than build elaborate fences and pens and haul food to the penned-in animals, early settlers let their animals roam freely about the land to scavenge for food. It was far easier to build fences around one's crops than to fence in one's livestock. While land was plentiful and prepared fields limited, this practice continued, with farmers typically marking or branding their livestock and then gathering it up in the fall. Although practiced widely among initial settlements up and down the coast and then later up the river valleys, this proved to be an ineffective long-term solution. Despite the settlers' vigorous campaign to wipe out predators, including setting bounties on wolves, for example, livestock faced great risks in the woods. Not bred to be smart, domesticated animals, particularly the young, were attacked by predators. Before extensive hunting and trapping had removed them from the region, not only wolves, but mountain lions, wolverines, and bears especially feasted on the mostly defenseless livestock.

Conflicts also arose over animal ownership and what to do about the young born in the wild. Settlers tried to prevent this conflict by bringing in their quickened stock before birth, but they often failed. Conflicts also arose when one person's cow got into another person's field and trampled or ate crops. Battles between those whose livestock roamed free and those with enclosed fields and livestock pens continued to rage until the practice of free-roaming livestock ended. Court documents are filled with records of these conflicts as they followed the settlement line west and up the river valleys. Whether in eastern Massachusetts in the seventeenth century, central Pennsylvania in the eighteenth century, northern New England in the early nineteenth century, or in western New York and the Midwest later in the nineteenth century, communities were forever appointing wardens to round up and hold errant animals that were "breaking into a close [an enclosed field] and destroying crops," and to adjudicate these cases before the local courts.

Roaming domesticated animals also deprived farmers' fields of precious nutrients. As Europeans cleared stones and stumps from their fields and began planting, they faced the problem of declining productivity as each year more nutrients were taken up in the planted grain and then removed from the field. Some early colonists solved this problem by abandoning fields and letting them lie fallow for several years before returning the fields to crop production. While plenty of other areas remained open to production that would produce high yields, this approach made sense. It also had significant drawbacks. Clearing fields of stones and stumps was hard, backbreaking work. By hand, a farmer could dig around stones and stumps, but a draft animal pulling a plow could not. Thus, fields had to be cleared of stones and stumps. Although clearing was usually done in the summer after the crops were planted on previously cleared fields but before the harvest, when there were fewer demands on a farmer's time, no farmer looked forward to clearing a field for the first time. For those farmers without oxen, it also meant borrowing one. Yet a farmer's only alternative was to see his yields slowly decline year after year.

Practicing a bush fallow strategy of letting their fields lie fallow for six or seven years, farmers were virtually guaranteed to face a field rife with bushes, brambles, and a tangle of roots at the end of the fallow period. These fields could not be plowed or dug by hand. Animal power was necessary to pull through the entangled roots. To abandon a field and let it sit idle for six or seven years required that a farmer own a large parcel of land. But because farmers ultimately passed property on to their children, there was soon too little land to let so much of it sit fallow.

Animal manure offered a way out of the land scarcity dilemma. Loaded with nitrates, animal wastes could restore fields. But as long as the animals wandered freely, grazing wherever they found food, their wastes were scattered throughout the countryside. Because edible grasses flourished by waterways, meandering livestock often deposited their wastes alongside rivers and streams. The wastes and the nutrients contained within them would then typically be washed downstream with the next rain. And even if the wastes were not washed away, the farmer could scarcely tramp around the countryside collecting cow, horse, or oxen dung.

Containing animals in fenced pastures also contained their wastes. After harvesting the grain from fields with declining productivity, farmers would seed them with "English hay," bluegrass, or white clover. Once the grasses came in, animals could be turned out on the fields, and their manure would add nutrients to the soil. With such a plan, rather than having to wait six or seven years, farmers could replant their fields with grain crops after just two or three years. But this strategy also involved more work. Fences had to be well built to make sure the animals did not wander off. Animals confined to pasture fields ate away grass cover. They also trampled and compacted the earth, which made it more difficult for grasses to reseed or resprout. Thus, farmers increasingly needed to plow and reseed or face the problem of open dirt subject to erosion. The compacted ground was also more difficult to plow and demanded the aid of draft animals. The greater dependence on draft animals, and their concentration in fenced pastures, increased the need for winter hay. This meant the farmer either had to plant more hay fields or have access to meadow and marsh grasses. Even with the help of animals, farming was a labor-intensive activity. Early settlers worked their own land, but they regularly needed help from neighbors, friends, and others. Some English settlers were used to practicing open-field farming where farmers worked together in a large enclosed field. More common, however, were individually owned, closed fields, where farmers were exclusively responsible for their own crops. But even these farmers worked in a community environment where neighbors helped one another put down foundations, raise barns, clear fields, or lent out draft animals. Settlers also encouraged others to join them. Relatives or neighbors or even strangers from Europe, coming with few resources, lived on the goodwill and credit of those already here. In exchange, they contributed their labor to the family or community that took them in. Some of these others who came were indentured servants who owed service in exchange for passage over from the Old World. Indentured servants lived in the home of the family who paid their passage, ate with the family, and worked with the family on the farm. They were not paid for their labor. But with completion of their indentureship—usually four to seven years—they achieved their freedom and could then sell their labor. Fortunate indentured servants went on to gain land and a stake for themselves.

Indentured servants, new migrants, and to a smaller extent slaves in the North who were mostly confined to the Hudson River valley helped address the acute need for human muscle power on the farm, but they could never fully meet the needs of the growing colonies. Ultimately, farmers turned to the family for muscle. Large families meant many hands to labor. A typical farmhouse consumed several cords of firewood in the winter to keep warm. Roughly the same amount of wood kept a family of eight as warm as a family of four, but the larger family had more hands to help with cutting and splitting the cordwood. When all these children grew up they inevitably wanted farms and families of their own. This left the farmer with two options: either divide his farm or find the means to get more land for his children.

A Native American passing a typical seventeenth-century farm would find it a curious sight. The Native American would first be surprised to see men in the fields hoeing, raking, tending to crops, or planting or harvesting. Native American men did not tend crops. They considered it women's work. Men hunted, fished, and built canoes, but did not tend the ground. Fenced fields and domesticated animals would also have struck the native passerby as a curiosity. Fencing a field to lay claim to it would have seemed absurd to the Native American. The Native American would also have considered unearthing and lugging stones off the ground and uprooting stumps a foolish waste of energy, for the very ground itself seemed to grow rocks, and the trees would only regrow. Likewise, maintaining domesticated animals would seem wasteful to the Native American. Plowing fields with oxen seemed to require more work for the goods the animals produced than the goods were worth. Domesticated animals had to be cared for, penned, and—most importantly—fed. Rather than use animal power to haul stones from a field, the Native American simply had his wife construct her planting mounds around and between the rocks. Even the plow would have seemed superfluous to the Native American. It would never occur to him to plow a whole field into bare earth, when in fact one only placed seeds in small sections of the field. For the Native American, burning the entire field and then loosening the earth only where one actually placed the seed made more sense. This is precisely why Native Americans used mounds specifically built up to take seeds. The passerby would have been similarly baffled by the colonists' practice of keeping, tending, and feeding animals, only to slaughter them for meat and hides. Animals in the wild furnished Native Americans with meat and hides without the bother and labor of tending them. In the spring ducks, geese, and other nesting birds provided eggs and fowl meat. Deer, moose, bear, and rabbits satisfied the Native Americans' need for meat without having to feed them. The Native American would even consider the well-constructed home of the colonial farmer foolish because it would make moving so much more difficult in the event of a long period of drought, excessive flooding, a buildup of pests, or a decline in fish or game.

It would have been difficult for the Native American to understand the colonist's obsession with property and things, his inclination to make work for himself, or his tendency to cohabit with and tend animals. Differences notwithstanding, these colonists and their neighboring Native Americans in fact shared a range of approaches to using the land. Yes, the colonists did tend to settle, build permanent structures, and alter the land in ways radically different from the land's prior occupants, but like the Native Americans they were also dependent upon the land and its ecology even as they altered it. Like the Native Americans, the colonists also worked according to the seasons, and, though they divided labor differently, the Native Americans and the colonists still divided tasks along gender and age lines. As the colonial settlers built homes, cleared fields, and put in crops, different tasks fell to different people. Generally, tasks performed closer to the home were more likely to fall to the female members of the household. After about five or ten years, a colonial farm took on the look and pattern that, with little variation, would remain in place for the next several generations. Beside the house, the wife would plant a small vegetable garden that would ultimately expand with the family to encompass a quarter acre. Here she would raise the food for the stews and potpies that supplied the family with vitamins, taste, and added variety to meals. Typically, she would plant a perennial asparagus patch at the edge of the garden to provide food in April. By May peas began producing as well. Potatoes, cabbage, beans, onions, cauliflower, parsnips, leeks, and carrots filled out the family garden, supplying vegetables throughout the summer and fall and, in the case of parsnips, onions, and carrots, even into the winter. Melons and squash ripened in the late summer and fall, along with Indian corn, which after a few years of farming, was usually relegated to the vegetable garden. This garden was almost exclusively a female domain. Men might help in plowing the earth, but once the ground was ready, they would know the garden only from its products.

An orchard would sit beyond the vegetable garden, growing along with the family to eventually encompass an acre or more. Cherry and pear trees, originally imported from Europe, along with raspberries, blackberries, strawberries, and blueberries that the children would gather wild on the hills or from patches on the edge of the garden, provided fruit. But the orchard's mainstay was the apple trees. Throughout the year, apple trees provided fruit, whether fresh in the fall or dried and brought up from root cellars in the winter and spring. Apples were also made into cider, the primary drink of rural folk from colonial times to the end of the nineteenth century. Cider mash and rotted and bad apples were fed to cattle and pigs.

Beyond both the vegetable garden and orchard lay the fields of grain. Farmers initially planted Indian corn but soon replaced the corn with wheat, barley, and oats. These were sown broadcast style on freshly plowed fields. Men took charge of these fields but were helped by women and children at planting and harvesting time. A successful farmer planted two to four acres of grain, to be harvested in the fall. The grain was threshed on the ground with a threshing flail, then tossed in the air to separate the grain from the shaft. Once gathered, the grain would be divided, a portion set aside for the next year's planting, another for the animals to supplement their winter supply of hay, and some would be reserved for future use. Another portion would go to the country store to cancel debts, while the rest would be hauled to the closest mill to be ground into flour, with the miller keeping a portion for his efforts. Beside the grain fields lay pastureland for the farm animals and, in the lowlands and meadows, haying fields that were twice mowed—first in early August and then in late September. Farmers also set aside some land for flax that, along with wool from sheep, the female folk would spin and weave into cloth for clothing, blankets, and coverings.

A farm family relied on their labor and the materials at hand to survive. The most abundant material available to colonial farmers was wood. Woods surrounded most early farms and farmers used these woods in a variety of ways. Wood provided the building materials for homes and barns. Chestnut trees were used for fence posts, housing beams, and any other wood constructions that needed to be resistant to rot and pests. Oak and cherry became siding, barrels, and flooring. Cedar shingles protected the home from snow and rain. Handles, rakes, hoes, and even plows were made from wood, as were furniture, wagons, and carts. Most farmers built these items themselves, either alone or with the help of neighbors. The wood came from woodlots, which also supplied fuel for cooking and to keep the family warm in the winter. A typical farmer needed up to an acre of woodland a year for cordwood and building supplies, with about fifteen acres set aside, as it would take about fifteen years before a woodlot could be recut. In the late fall and early winter, the farmer and his older sons would head for the woodlots. There they would cut down and then haul the trees over the new snow to the barn to be cut up and split over the winter.

These early colonial farmers were also still greatly dependent upon the land they did not control. Wild berries picked in the spring and summer by children and nuts gathered in the fall were important complements to the garden and field produce. Domesticated animals provided the family with an important source of protein. A fortunate family could expect 250 to 300 pounds of meat from a slaughtered oxen or cow, along with 40 pounds of leather, while a large pig yielded 100 or more pounds of meat. But not all farms could spare a cow or pig. Yet a typical family could consume between 700 and 1,000 pounds of meat a year (Patten 1780/1903, 422). In the early years of settlement, when domesticated animals were in short supply, settlers obtained protein the same way Native Americans did—they hunted. And they hunted the same animals: deer, moose, ducks, geese, pigeons, and ground fowl such as turkey and grouse. They hunted with such intensity that within two generations deer and moose disappeared from settled areas. Turkey and other ground fowl soon followed. Passenger pigeons were so prolific that they persisted for another hundred years before being eliminated from the area. The passenger pigeon population may have increased in the early seventeenth century as the deaths of Native Americans that had set up their villages in river valleys opened up new areas for the pigeons to roost. The appearance of huge flocks of passenger pigeons that darkened the skies of the Northeast for days were not yearly occurrences. The massive flocks would arrive irregularly, usually once every seven or eight years, in what were known as "pigeon years." Pigeon feathers were plucked for bedding, and the meat was either cooked and eaten or given away. The plentitude of pigeons had its drawbacks, however. More than one person complained of sitting down to yet another dinner of pigeon pie. Pigeon populations may have held out longer than those of deer and moose, but declining numbers of game soon became a reality, forcing families to depend even more upon animals they raised.

Hunted game did not serve as the settlers' only source of wild protein. Like their Native American predecessors, colonial settlers also looked to the region's waters for food. Salmon, alewives, smelt, and shad ran up the rivers in prolific numbers. Before the Europeans arrived and drove them back from their fishing spots, native fishers for generations had responded to the runs by abandoning other tasks and rushing to the water's edge with spears and nets. Similarly, colonial farmers dropped their hammers and saws to scamper off to the nearest river or stream bank to catch their haul of fish. Successful fishermen brought salmon home for food, passed it on to neighbors as payment perhaps on some other item, or salted it down for later consumption. Shad followed the salmon upstream and came in such numbers that it was said that one could cross the stream on their backs or pull them up with one's bare hands. In fact, nets were used, with hundreds of shad hauled out with a single sweep. Fish were considered so important to the early settlers' survival that among their first acts the colonial governments moved to protect the fish and access to their spawning beds. One of the first property restrictions enacted by these governments allowed any freeholder to cross another person's property to gain access to water in order to "fish and fowl." While the Pilgrims were clearing fields and building homes in Plymouth, nearby other Europeans were encountering the New World. Besides the Pilgrims, other English people were growing anxious about their country's religious direction. Increasing numbers of English Puritans came to believe that the Church of England was either drifting closer to, or not drifting far enough away from, the church in Rome. Led by John Winthrop, a farmer from modest gentry in Suffolk, these Puritans formed the Massachusetts Bay Company and obtained a charter from the king to establish a colony in New England. In March 1630 a small fleet of ships left England for Massachusetts Bay. They arrived in Boston Harbor in time to clear land, build shelters, and plant crops. Unlike the Pilgrims, the Massachusetts Bay Colony had strong financial support, and its charter to the region was officially recognized by the English crown. By the end of the first year, more than 2,000 had come, and by the early 1640s their numbers surpassed 16,000 (McCusker and Menard 1991). For the most part, the settlers who came were prosperous people, mostly from the rich farming regions of East Anglia, Kent, and the West Country. They brought with them or had shipped to them, tools, animals, seed, and other necessities. They also bought goods from those already here, including the Pilgrims to their south. The Puritans also brought with them their charter and government. Chartered as a company, they had directors who met and decided policy. But because they met in Boston and their directors were the "freedmen" (free men) of the colony, they effectively governed themselves. The land granted them from the king was the company's assets. They parceled that land out as towns. People who gathered together found a minister and petitioned the General Court for a town grant. Given the grant, the town was then founded and land passed out in lots to the town's member families. Unlike the colonizers of Virginia or Maryland, the settlers who came to Massachusetts arrived as families. Families settled into these towns and began clearing land to farm, much as their cousins to the south in Plymouth had done. Within a few years the communities of the Massachusetts Bay Company and Plymouth looked very much the same, so much so that by the end of the century the Massachusetts Bay Company absorbed its southern neighbor.

The Massachusetts Bay settlers were also religious purists. As a result, doctrinal dissent soon broke out. Some settlers responded to such conflicts by moving and establishing a new colony, as in the settling of Rhode Island. Others, such as those who settled Connecticut, broke away in search of greater opportunity or because they had fallen under the influence of a particularly dynamic leader. Although there were differences in the communities that were set up according to where the settlers came from and the particular ecology of where they settled, the communities tended to follow the land use and farming patterns already seen in Plymouth and Boston.

THE MID-ATLANTIC COAST

To the south of these colonies lived the Dutch. Unlike the Pilgrims or Puritans, the Dutch settled for the purpose of trade. In 1609 Henry Hudson, an English explorer under Dutch hire, sailed up the river that now bears his name looking for a passage to the East Indies. He didn't find his Northwest Passage but he did return with reports of lucrative trading possibilities. In 1624 the Dutch set up trading posts: one they called New Amsterdam, situated on an island at the mouth of the river, and the other upriver at Fort Orange (now Albany). The Dutch intended for their colonies to make money in the fur trade, and as such these settlements drew a mix of adventurers, servants, administrators, merchants, and traders. Even though the Dutch settlers' purpose for being in the New World was to gather beaver pelts, they still needed food, shelter, and fuel for warmth and cooking. The promise of profits in the fur trade attracted more people, so that by the 1640s the Dutch settlements had grown into substantial communities. Land was cleared, fields were planted, and animals had been brought in to work the land and provide food for those engaged in trade. Though they shared many characteristics with the English settlements—certainly more so than the native villages—the Dutch settlements nonetheless reflected a variety of unique traits. Land was more likely to be worked by servants and held in large estates. To the east on Long Island, Puritan settlers coming from New England set up settlements similar to those they left behind. Because Long Island and Cape Cod had been created by the same glacial process, these roaming Puritans recognized a familiar ecology of sandy soil, scrub oak, and pitch pine with marshy tidal inlets, kettle and beaver ponds, and meadow wetland thick with cedar and swamp maple.

Up the Hudson River valley, the Dutch settled a land with fertile, brown, forest alluvial soil. Periodic floods that overran the banks deposited soil washed down from the mountains. Once the dense forests of mixed oak had been removed, the Dutch settlers were able to grow crops in the acidic, but fertile, soil. Farther up the river valley, forests of giant spruce, pine, and hemlock interspersed with maple, beech, and other hardwoods grew. In 1664 these large Dutch estates were threatened when the British sent a fleet to take New Amsterdam. In exchange for the city's capitulation, Holland received a few spice islands in the Banda Sea (part of present-day Indonesia) and recognition of the Dutch settlers' property claims.

Engraving of sailing ships and activity along the waterfront of the early settlement of New York shortly after it was seized by the English from the Dutch. The image is surmounted by a banner that reads "Novum Amsterodamum," Latin for "New Amsterdam," the city's original name before it changed in 1673. (Hulton Archive/Getty Images)

South and west of the mouth of the Hudson River, the mountains of the Appalachian chain shift westward. The land of northern and eastern New Jersey and eastern Pennsylvania combined crystalline piedmont uplands, triassic lowlands, and a rolling limestone plain, all bordered to the west by the eastern edge of the Appalachian Blue and South mountains. Here lay fertile, well-drained loamy soil, brown podzolic in the north and red and yellow podzolic in the south. In the late sixteenth and early seventeenth centuries when Europeans came to settle these areas, a rich, mixed oak forest (consisting of black and white oak, hickories, chestnuts on the drier hillsides, and soft white tulip poplar that sprouted up in abandoned Indian fields) covered the region. Although the region south of the Hudson's mouth was never covered with glacial ice, glacial outwash still affected it, especially along the river valleys. In southern New Jersey, the coastal plain begins, growing in width as one moves south. The soil here is mostly sandy coastal deposits or poorly drained red and yellow podzolic. River valleys, particularly the lower Delaware, Schuylkill, and lower Susquehanna (which flows south into the Chesapeake Bay), offer rich alluvial land. The region's gentle rolling topography, especially the limestone or coastal plains, make the soil less susceptible to erosion. Although Swedes had first tried to settle in the area of today's New Jersey, it was not until the late sixteenth century when Quakers from England and migrants from New England began moving into the region that a significant number of farms begin to appear. In search of new space, New England Puritans in the 1670s began cutting forests and setting up settlements in northern New Jersey, while Quakers moved into southern New Jersey in the 1680s. Yet neither group established much of a foothold.

The territory west of the Delaware River was given by the king of England to William Penn, the influential and dynamic Quaker. Penn began an aggressive campaign to people his grant with Europeans—mostly English Quakers but also German-speaking Mennonites, and later other Protestant Germans, Swiss, and Scots-Irish who rapidly began to take up land and cut down the forests. Smallpox and measles having wiped out most of the Native Americans before the Europeans appeared, Penn negotiated with those remaining for the land to the east of the Susquehanna River valley and attempted to keep white settlers from Indian lands. Despite his efforts, the immigrants soon disregarded these agreements and flooded into Indian lands.

The climate and soil favored the European interlopers. With summer temperatures 5 to 10 degrees (Fahrenheit) higher than southern England and 5 to 10 degrees colder in the winter, the Mid-Atlantic area experienced greater weather extremes than what the settlers from southern England were accustomed to. Yet this climate was milder than that to the north and favored the production of a variety of crops. Most livestock could easily tolerate the greater extremes in cold and heat, although summer heat hurt root crop and sheep production, while particularly cold winters took a toll on cattle. Those caveats aside, Europeans found that the 165- to 200-day growing season, rich soil, and rolling countryside offered a genial place to settle—and settle they did.

Like their New England cousins, these settlers initially adopted Indian crops, corn, beans, and squash; burnt forests to clear trees, often selling off the resulting ash to English traders as potash or pearl ash; and supplemented their diets with fish, wild fowl, and deer. But these settlers were not romantic searchers looking to return to a more "natural" interaction with nature. They were European, mostly middling farmers (of course some were also artisans, merchants, and adventurers). They were interested not in living as the Native Americans lived, but as successful European farmers lived. As such, they attempted to transform and use the land as they knew it back home. These settlers came from more diverse backgrounds than the New Englanders and settled an area with more fertile soil, a longer growing season, and less rocky and hilly land. With their more individualized and isolated farms, these colonists created a different, uniquely Mid-Atlantic colonial experience.

Despite these differences, the settlers of New York, New Jersey, and Pennsylvania shared a variety of lifestyle experiences with their New England cousins. In both regions, for example, farmers fished, hunted, sowed and harvested crops, butchered cows and pigs, and gathered and picked fruits and vegetables to feed their families. A small family of four to six members consumed between sixty to eighty bushels of wheat and corn flour a year that they would raise on four to eight acres. Orchards provided fruit and drink. The vegetable garden, as well as wild nuts and berries, diversified the diet. Meat from deer, moose, rabbits, turkeys, grouse, pigeons, cows, or pigs was baked into pies, served fresh, or salted and dried. What remained the farmer traded with neighbors or sent off to the country store. Some of these extra foodstuffs were used as payment for the processing. The miller received a percentage of the grains brought in to be ground. At the sawmill, lumber was put aside as payment for the cutting. Most of what the community produced, the community consumed. The community grew by producing more. As a result, more forest was cut, and more land went under the plow or was turned into pasture. The countryside began to open up. It also became hotter and drier. Streams grew increasingly muddier and flowed less evenly. Within a few generations, the settlers realized their dreams as the countryside began to look more like England and less like the forested wilderness they first encountered.

3

GOODS, TRADE, MILLS, AND DAMS

Driving across the Sagamore Bridge onto Cape Cod, my sister immediately encounters a gigantic windmill. Unlike the windmills common to the Cape during the colonial era, this particular windmill is of recent construction and houses a discount store. The Pilgrims who migrated to Cape Cod from Plymouth did not build windmills out of nostalgia for Holland, but because, like the Dutch, they needed to harness the wind's power. They needed the power to turn their millstones to grind their grain. Lacking enough streams with sufficient flow to turn waterwheels, Cape Codders naturally looked to the wind as an abundant power source.

Like Holland, the Cape is relatively flat but windy. Jutting out into the North Atlantic, it catches winds from the west during the summer, but in the fall the winds switch to the east and grow more intense, strong enough to easily turn heavy millstones or wood saws.

The Cape's old windmills make great tourist attractions, but they were creations of necessity when first built. Cape Codders would have preferred to use waterpower. In fact, wherever possible, settlers dammed streams, brooks, and creeks, backing up water and turning fields into millponds. Indeed, further examination reveals that the marsh behind my sister's house, mentioned in Chapter One, was created not by an old beaver dam, but rather by an abandoned milldam. This was revealed in an old town map that names an old, long-unused road that runs by the marsh, Milldam Road. For those willing to venture into a territory dominated by mosquitoes and poison ivy, the ruins of the old dam are still discernable. Small affairs, these dams were usually only two or three feet high and built of stones, timber, and rocks—all of which would often wash out and have to be rebuilt. Once the water reached a certain level, it would run through a sluice, over which a waterwheel was built. The water rushing underneath turned the waterwheel. The turning waterwheel, by way of shafts and gears, then turned the grindstone, which crushed the grain, producing flour.

The settlers needed mills. Without mills, the grain they grew was of little use. Additionally, the mills needed to be close by. Overland travel, usually along old Indian trails, was difficult, and with a shortage of horses and oxen, colonists often had to carry grain to mills on their backs or pull it in wagons. Even with

The oldest windmill on Cape Cod stands in Eastham. The windmill was built in Plymouth in 1680, and later moved to Eastham. (ChromoSohm Inc./Corbis)

the aid of horses or oxen, the trip was difficult. Recognizing the importance of mills, colonial legislatures passed mill acts to encourage mill construction. These acts allowed mill owners to build dams even when the dams flooded others' land. Under the acts, the mill owners were only required to pay the landowners a fee for flooding the land; the landowners could not take action against the mill. Although the mill acts seemed to run against prevailing conceptions of property rights, because the mills provided a clear benefit to the whole community, they were widely supported. Hence, Massachusetts Judge Sedgwick made the observation in 1805 when Sylvanus Lowell attempted legal action against Seth Spring's mill for flooding Lowell's field that, "Sacred rights of private property are never to be invaded but for obvious and important purposes of public utility. Such are all things necessary to the upholding of mills" (*Seth Springs vs Sylvanus Lowell* 1805). In the case of Seth Spring's sawmill dam flooding Sylvanus Lowell's field, the court found that Spring had the right to flood Lowell's field, but that a jury should decide how long the dam could be maintained at its full height. With the fall harvest in, the farmers packed their grain into sacks and onto wagons or their own backs and headed for the mills. Anticipating the coming grain, the miller would sharpen his millstone by grinding out the grooves and putting a headboard across the dam face to raise the water to its maximum height. Once ground, the flour would be divvied up, the miller taking some and leaving the remainder for the farmer. With the fall harvest ground, the miller could remove the headboard from the dam, allowing the water level to drop. As the water receded, the meadows at the pond's edge dried, giving the farmers the opportunity to harvest the meadow hay. In the spring, the miller also lifted the headboard to create an opening for fish to migrate upstream to their spawning grounds.

Gristmills were not the only mills the community needed. Lumber for building needed to be cut. Early colonists cut logs by hand. In the winter, loggers ventured into the woods and cut down the largest trees with axes and two-person saws. The logs were dragged by horses or oxen across the frozen snow to pits. Each log was spread across the cutting pit and—with one person standing atop the log and another in the pit—the log would be cut lengthwise into pieces. This hard and labor-intensive work was soon taken over by sawmills. The mills' arrival altered the fallen trees' journey as well. Instead of being hauled to cutting pits, the logs were transported from the woods to the nearest stream, where in the spring loggers floated the logs (either on rafts or in the water depending upon whether the log floated) downstream to the nearest sawmill. These mills were similar to gristmills (many, in fact, were converted gristmills), but instead of turning a millstone, the shafts moved reciprocal saws. The sawmills operated in the late spring and early summer, after the spring fish runs but before the fall harvest. Others besides the miller laid claim to a portion of the farmer's extra flour. Early communities existed within a dense network of home production, exchange, and trade. That system needed both diversity and flexibility. For example, Matthew Patten, who lived in the Merrimack River valley about forty miles west of where my sister's house now stands in New Hampshire, and Sylvester Judd Sr., who lived in the Connecticut River valley, were farmers as well as justices of the peace. They made and fashioned tools and traded goods. They grew grain and traded for grain. They raised and slaughtered animals and traded meat and sold animals. They traded and exchanged among other farmers, as well as with tradesmen, artisans, and shopkeepers. Goods in their communities were in constant circulation.

To a surprising degree, people acquired goods through trades and exchanges that, at other times, they themselves were trading away. Patten might at one point have traded fish for flax or grain, then turned around and at another point traded grain for meat, even though he raised flax, grain, and meat on his own farm. Within this system of exchange, goods were processed and transformed. Ashes were soaked, boiled down, and dried into potash, or processed again into white pearl ash and then into lye. Lumber was cut into clapboards or split into barrel staves. Flax was cured, dressed, bleached, and spun. Wool had to be fulled, another act often performed with the aid of waterpower, at a fulling mill. Wool also had to be carded, spun, and woven. Grain needed to be ground and sifted. Cattle and hogs were butchered, and then the meat was dressed, salted, and dried, and the leather was tanned. To prevent spoiling, fish needed to be pickled. Some of this work occurred on the farm; some took place at the sawmill, gristmill, fulling mill, or tannery. Local millers and shopkeepers handled much of the processing. Through their central place in this system of exchange and circulation, shopkeepers in particular collected much of the community's surplus goods. The country store did not handle all the goods that were traded, however. A majority of the goods moving through the community likely passed among friends, neighbors, and relatives without ever reaching the country store, but whatever surplus the region produced typically did end up there. As much as the gristmill or sawmill, the country store was critical to the community's survival.

Remembering their childhoods spent growing up in these early communities, storytellers often recalled raising their own produce, making their own clothes, and being entirely self-sufficient. These stories contained an element of truth, but memory often distorts reality. Often forgotten were the critical tools, materials, and supplies that could not be produced locally, but without which the colonists could not survive. How these items arrived on the farm was often lost in the haze of remembrance. When Sarah Wright, the only daughter of Preserved Wright, married Asahel Clapp in the middle of the eighteenth century in

Goods, Trade, Mills, and Dams 59

Cod fishermen in a small boat off the New England coast. (North Wind Picture Archives)

what was then the newly settled region of the Connecticut River valley, her parents fully outfitted her to set up house. They sent John Lyman, a local carrier, to Boston to bring back seven yards of calico, a brass kettle, a dish kettle, a frying pan, a warming pan, a few pots, a skillet, six plates, a chamber pot, a tankard, and a looking glass. In addition to items that could be manufactured on their farm, Sarah's parents also gave her a large and a small spinning wheel, some tubs, kitchen tools, and a Bible. Presumably Asahel Clapp's parents also provided him with the tools and supplies he would need to work a farm. Although the Clapps and Wrights likely fashioned some of these items at home, many of the items they needed had to be paid for, either in goods exchanged at the country store or with cash. The items purchased in Boston cost the Wrights about fifty pounds—not a small figure in the eighteenth century.

To raise that money, the Wrights had to sell goods to the local country store. Country stores purchased a wide variety of the local farmers' surplus products, including grains, flour, peas, dried and salted beef and pork, pickled fish, honey, flax, and wool. In turn, the country store provided farm families with necessities they could not raise or make at home: tea, coffee, spices, iron tools, pots, pans, gunpowder, and rum. The country store procured these items by trading out the surrounding farmers' surpluses to merchants and tradesmen in the larger towns and cities. The merchants in these larger centers then moved the goods into the local and larger world market.

The links between that larger market and the farmers existed from the very beginning of colonial times. Even the early Pilgrims sent back goods in the supply boats that came from England. In addition to collecting pelts and lumber, Puritan settlers around Boston Harbor also sent grain back to Europe. Within a few generations, Boston fishermen were working the seas as hard as farmers and loggers worked the land and forest for goods to send back in the hulls of the ships that brought more settlers, tools, and supplies from Europe. Cod fishermen spilled out of the harbors of Boston, Gloucester, Marblehead, Salem, Newport, Portsmouth, Chatham, Provincetown, and Providence to hand line for cod in small cod boats. A generation later, whalers set out from the protected harbors of New Bedford, Nantucket, and Martha's Vineyard in search of the leviathans of the deep: whales.

Dried cod and whale oil from America found a ready market in Europe. The fishermen and whalers who took their boats to the fishing banks needed food and ships. By the eighteenth century, the ships were increasingly constructed alongside of American creeks and streams. Wood floated down to ship works from upstream forests was first used to build small coastal vessels designed to carry fishermen out to the nearest banks or to transport goods to and from country stores upriver. Soon American shipwrights were building bigger ships—ships large enough to pursue whales or travel across oceans. The men who built these ships and manned them at sea were fed with food raised on American farms. Sailor, fisher, boatbuilder, trader, and artisan families were too busy or lacked the land necessary to produce enough food to live on. Farmers satisfied this need by furnishing country stores and coastal towns with surplus flour, dried peas, cider, apples, and pickled fish, as well as fresh, salted, and dried beef and pork. While these expanding town markets absorbed a portion of the farms' surplus, by the middle of the eighteenth century, farms were producing more surplus than even the towns needed.

Several factors accounted for the growing surpluses that flowed into urban markets. First, American farm families were large. A predominantly rural population drinking unpolluted well water, most Americans were not exposed to the diseases of crowded urban centers spread by contaminated water. Typhoid, cholera, and dysentery took a smaller toll on farm families, particularly the young. The heartier diet also gave these folk the stronger constitutions necessary to fight off the diseases they contracted. As a result, Americans tended to be healthier than their European counterparts and more likely to survive into adulthood. Inexpensive and plentiful land also encouraged children to marry and start families earlier. More children meant many hands to help out on the farm and contribute to the farm's productivity. These children then grew up, married, and moved on to their own farms. Soon most of the arable land near the coast was taken and under production. In search of new lands to farm, the next generation of children moved up and down the coast, to the north into New Hampshire and Maine and to the south—as we have already seen—into Connecticut, New York, and New Jersey. But farming north of Massachusetts proved problematic. The short growing season and poor soil made farming there difficult at best. Despite the difficulties, the need for land was so great that people were willing to chance it. The settlers found some fertile soil in the rolling hills between the lower Kennebec and Penobscot rivers. Fertile soil along the upper banks of the large rivers—the Seco, Merrimack, Connecticut, Hudson, and Delaware—proved farmable. The river valleys also moderated the colder weather, extending the growing season, while the rivers provided a means to move surpluses to markets.

Moving up these river valleys, newly married couples laid claim to the best available lands they could find. The lower reaches of the river valleys were initially wrested from Native Americans by traders and land speculators. In New England by the second half of the eighteenth century, many of these older estates had been broken up and sold off. Large estates worked by tenant farmers or indentured servants remained more common along the Hudson, while in Pennsylvania, as in New England, small farmers predominated. After the French and Indian War, families moved farther and farther up the larger rivers. The rich alluvial soil was highly productive. Soon wheat, oats, rye, peas, and dried beef and pork accumulated in the back rooms of country stores, waiting for rafts or smaller river-sailing ships to take them to the larger market towns with deepwater harbors. There the goods would be loaded on sloops or schooners for transport to Portland, Boston, New York, and Philadelphia.

Farm families produced goods and children. As farm children grew into adults, they needed farms of their own. They also needed the tools and materials to work and make those farms succeed. Some farms were cut from older farms. A typical early-eighteenth-century farm consisted of 100 to 200 acres. The farm family would work only a small portion of that. Even a successful and fully productive farm used somewhere between thirty and sixty acres, including the woodlot. As sons and daughters grew and married, the farm would be divided up. Daughters received land as part of their dowries, while sons were granted land as inheritance. More smaller farms put more land into production, and while much of the produce was consumed on the farm, any surplus entered the general flow of goods heading for a larger market.

Once all the available farmland was taken up, children moved away to areas with cheap available land. Initially, they moved upriver. The land was good and

Forest clearing, Petersham, Massachusetts, 1830. (Harvard Forest Dioramas/Fisher Museum, Harvard Forest/Petersham, MA)

the river provided a link to family and markets. But soon all the land along the rivers was spoken for, and new farm families moved into the hills away from the river valleys. Initially, these hill farms offered many advantages. The land was cheaper and not subject to spring floods and freshets, while the air was cooler and healthier (meaning fewer mosquitoes) than in the river valleys. The agricultural technology available to farmers just starting out—simple wooden plows, rakes, hoes, and sickles—and handwork suited the problems of the hills. Farmers could cut or burn trees, loosen the soil (not yet packed down by animals pasturing) with hoes and shovels, and then plant grain among the stones, tree stumps, and irregular terrain of the hill farms. Able to defend themselves against wolves and bears, pigs ran free through the lush hill forests, eating acorns, chestnuts, walnuts, and young shoots and roots. Because a sow dropped a litter of twelve to sixteen piglets two or three times a year, pork was always available for the dinner table. Wild game in the woods and high meadows offered alternatives to pork. But even land for hill farms was limited. Hill farmers began to work the land more intensely, leaving less of it to lie fallow. They cut forests and sold lumber to pay off debts and burned wood to make potash, which was easy to transport and in high de-

Settlers building a home in New England. (Library of Congress)

mand in the lower market towns. Overhunting and the conversion of habitat into cultivated land reduced the numbers of wild game. As the forests thinned and the land was more intensely worked, the soil lost its fertility, more land eroded, and farming in the hill country became increasingly difficult.

Whether a farm family inherited undeveloped land from a parent's farm or moved inland or upriver, they and their families needed money to set them up in farming. As in Sarah Wright's case, setting up a household was expensive. Money to buy land and materials had to be earned, borrowed, or both. Farmers intent on earning money put more land under production in order to realize greater surpluses to sell or trade to the country store. The country store had tools and supplies. The country storekeeper also had the connections to move the surplus goods into the larger flow of goods that would eventually find a market. The market that individual goods reached depended on the changing needs of a rapidly developing intercontinental commercial trade.

Whale oil, potash, and even cod went to Europe, as did tobacco, which, although identified with the tidewater colonies, was also grown in the Connecticut River valley. But the European market for wheat, rye, oats, peas, and dried beef and pork was limited by English policy and the financial resources of those hungry for food. But Europe was not the only place American surplus goods could find a market. Sugar became a major component of the European diet in the seventeenth century. Sugar proved to be a very profitable commodity; once people acquired a taste for sugar they found it difficult to give it up. Europeans' rapidly growing taste for sugar delivered an ever-expanding market and high profits to sugar producers. Sugar grew well in the humid tropics of the Brazilian coast and in the West Indies, but sugar production was dangerous, labor-intensive work. The Portuguese began producing sugar on Brazilian plantations in the 1520s and by the end of the century they were importing thousands of slaves to work on the sugar estates of the Brazilian northeastern coast. In the 1650s Europeans brought sugarcane to the Caribbean islands. Before that date the Caribbean islands produced mostly tobacco grown by small farmers. Sugar transformed labor on the islands. After 1650 the small farms gave way to large sugar plantations worked by slave labor imported from Africa. Although sugar production required more hands per acre to work the land than other forms of agriculture, the immense profits far outweighed the expense for plantation owners. Great fortunes were built in just ten years in the sugar business. With such high returns, plantation owners put every arable acre into sugar. By the end of the eighteenth century, the English market alone was consuming 150,000 tons of sugar (Taylor 2001, 310; Muir 2001, 38). Soon a sugar monoculture emerged and, as production of other crops dropped off, the islands became unable to feed their growing populations. The sugar planter could import slaves from Africa, but if he couldn't feed them the plantation would fail. The peoples of the West Indies, both white and black, needed food, but they also needed lumber, used to boil down molasses, and wooden staves to make barrels to ship the molasses. To satisfy all these needs they looked to the farmers and fishers of North America.

If planters in the West Indies needed North American farmers for food and wood, the farmers similarly depended on country storekeepers to buy their surpluses, allowing the farmers to pay off debts or accumulate cash to buy land and materials for their children. Whether wheat the farmer raised for flour that went into his wife's pies and breads, cows and pigs he slaughtered for meat, peas picked for meals, or flax cut to be made into cloth, there were frequently extra goods to be hauled off to the country store. Grist millers and lumber mill owners also brought in the surpluses they accumulated as payment for their services. By shipping out this surplus, the storekeeper paid for the goods he imported into the region and then sold to the farmers and accumulated hard currency. For most of the eighteenth century, and later in more isolated areas, storekeepers took whatever the farmers offered. Levi Shepard in Hadley, Massachusetts, accepted flax, pigeon, suet, mutton, wheat, turnips, rye, honey, beef, butter, tow cloth, flour,

View of Boston during the colonial era. (North Wind Picture Archives)

beeswax, corn, oats, wood, cider, wool, hay, salt pork, salmon, potash, or boards in exchange for the imported goods he carried. Local folks bought from storeowners on credit and then used the "country goods" they later brought in to cancel the debt. The value of those goods was typically determined by both the larger market values and local custom. Shepard would pay cash only for items in high demand in Boston and Hartford such as wheat, rye, wood boards, potash, peas, flaxseed, beef hides, and beeswax.

In addition to the goods brought in by local farmers, country store owners like Levi Shepard also furnished their customers with goods from more distant lands. These included cloth, china, and silverware manufactured in England; coffee and spices from the Caribbean; and tea, silks, and fine cottons from South Asia; as well as pots, pans, iron nails, and implements manufactured in east coast towns and books printed in Boston, Newport, New York, and Philadelphia that were brought upriver or inland and offered for sale. To acquire these goods, Shepard depended upon credit with merchants in the larger coastal cities. The merchants in return assumed that Shepard would ship down items they could sell in their markets. And just as Shepard had credit and debts with merchants in the larger trading centers, so, too, the merchants in the larger trading centers were in a credit/debt relationship with merchants abroad. The goods from England or South or Southeast Asia that flowed into Levi Shepard's store and ultimately into the homes of local farmers were brought in on debt. That debt had to be paid off with goods shipped abroad. The West Indies sugar plantations' demand for food helped cancel those transatlantic debts. The appetites, and need for building materials, of the growing populations of the coastal cities and towns helped satisfy the internal debt. New England farmers' compatriots in the mid-Atlantic region were just as involved in the international trade. The rich well-drained soils, long growing season, and gentle rolling countryside made the area west of Philadelphia perfectly suited for wheat production. By the early eighteenth century, the mills along the brooks and streams of southeastern Pennsylvania and central New Jersey were working late into the night producing flour for ships waiting to sail to the West Indies, while German, Swiss, and Scots-Irish farmers to the west cut forest and plowed ever more land for wheat production. Through such an indirect path, farmers in Pennsylvania and New Jersey, tenant farmers in the central Hudson River valley, and small freeholders in New England found prosperity in Europe's sweet tooth.

By the beginning of the eighteenth century, Americans had settled into and transformed the eastern edge of the continent into an ordered, controlled, and partially Europeanized environment. The more they refashioned the land on the European model, the more European goods Americans demanded. Benjamin Franklin noted in his autobiography that his family had eaten breakfast for years in locally produced earthenware bowls using pewter spoons. Later as the family grew more prosperous "a china bowl, with a spoon of silver" appeared one morning. As farm families grew and as their tastes for European manufactured goods increased, farmers needed more cash and more credit. To procure cash and credit they needed to grow more crops that the country storekeepers could sell outside the region. But just as farmers were being pushed to focus more on marketable crops, nature and politics conspired to add to their woes. This was particularly true for New Englanders.

New England experienced a long, cold, two-century-old miniature ice age that ended roughly around the middle of the eighteenth century. As a result, colonial settlers faced hard winters and were forced to burn a surfeit of wood. But firewood was plentiful, and despite being cold, the weather was nonetheless predictable. The warming trend that began at midcentury and continued for the next seventy-five years was wild and unpredictable. Storms were more frequent and violent. Droughts followed by torrential rains began to plague the Northeast. Unexpected flooding, exacerbated by the cutting of forests, washed out fields and drowned animals. Unusually wet springs or late snows slowed planting, while early fall storms or unexpected cold snaps cut the growing season, ruining crops in the fields before they could mature. Wheat, although in high demand, was a fickle crop. In ground too rich the wheat went to straw before fully fruiting, while land too sterile produced thin fields. To compensate, wheat farmers rotated crops accordingly. Those working the fertile loam of southeastern Pennsylvania first put Indian corn or flax into a field before moving to wheat. After several years of wheat production, the field would have to be switched to clover or grass, as wheat production typically declined from thirty-five bushels an acre to ten or fewer.

To add to New Englanders' woes, in 1787 the Hessian fly (named not because of its origins but after the German troops sent by England to put down the Revolution) descended, attacking English bald wheat, the crop that produced the highest-quality flour. The Hessian fly was actually a mite that attacked the wheat stalk. Although other wheat strains such as Jeremiah Wadsworth's variety were more resistant to the Hessian fly, the insect continued to plague farmers, especially after a few plantings of wheat in a field. Though helpful against the Hessian fly, these new varieties could do nothing to combat the scourge of wheat blast, which was known as wheat "rust." In wheat attacked by blast (a fungus that would haunt American wheat production for the next hundred years) the kernel fails to develop, and the resulting wheat crops yield insufficient fruit to produce flour. Though also attacked by blast, rye proved hardier against its ravages than wheat. By the end of the eighteenth century, blast had traveled up the river valleys into the more northern reaches of New England and New York. Faced with this plague, many farmers abandoned wheat entirely and switched to rye or oats, hoping to return to wheat some later year. Wheat production fell in New England, and by the early nineteenth century New Englanders were importing wheat from Pennsylvania.

When the Seven Years War between the English and the French broke out in 1756, farmers of the Northeast benefited tremendously. After William Pitt decided in 1757 to concentrate the English military effort against the French on the western side of the Atlantic, thousands of English soldiers and sailors began flooding into the colonies, all needing food and goods, and American farmers increased production to meet the need. Accompanying the soldiers and sailors, a fresh influx of money also flowed into the American colonies.

In America the war was known as the French and Indian War—and with good reason. More important to the colonies than yet another in a long line of conflicts between England and France, this war was being waged against Native Americans. The French defeat in 1763 deprived the Algonquian-speaking peoples of their French allies in resisting white encroachment upon Indian lands. Settlers had an easier hand in their struggle for the land against its older inhabitants. The English also emerged from the war with naval superiority. The English navy, fed on hardtack made from American flour, opened the sea-lanes for English shipping, and Americans fit uncomfortably under that umbrella. Freed from the fear of pirates and attacks at sea, ships reduced their carrying weight of armaments and filled the space with goods. Perceiving a growing American market for their goods, English merchants extended credit, and Americans took advantage. Within a few years American debt had grown out of control. Looking for cash and markets for their own goods, American merchants increased their shipments to the West Indies and brought back molasses that they converted into rum for the European and, to a lesser extent, African markets.

Americans also bought more internally manufactured products. By 1770 America had more forges and furnaces than England and Wales combined and produced more pig and bar iron. Indeed, it was to keep American iron out of British hands that led George Washington to set up camp at the aptly named Valley Forge. American axes were so preferred that English axe manufacturers began falsely marketing their own as American made. Under pressure from English iron manufacturers, Parliament attempted to outlaw American iron mills. But these parliamentary acts failed. Iron was abundant in the colonies, and ubiquitous American forests supplied the charcoal used for smelting and iron. In addition to axes, many other American-made items stocked country store shelves. Native scythes, plow tips, hammers, nails, wedges, pottery items, and clocks also tempted farmers. But the debt accumulated acquiring these goods, whether they were American or British made, eventually had to be paid. That meant putting more ground into producing goods to satisfy demand in far-off markets.

In addition to the well-known struggles over taxes and freedom, conflict over debt also sent Americans to war with their British cousins. The Revolutionary War's conclusion increased the colonists' economic hardships as well as their freedom. Deprived of the protection of the English navy and faced with the hostility of the English government, Americans found that the lucrative West Indies trade had become problematic. The debts accumulated as soldiers were off fighting the war were not easily paid off. New state governments attempted to meet their increased financial obligations by raising taxes and demanding hard currency in payment. Traders in the larger commercial towns pressed country store owners to clear back debts and would accept in payment only goods that could easily be sold for cash. Farmers found that country stores would no longer accept "country goods." In 1786 the combined pressures of taxes, debt, and uneven harvests drove the farmers of western Massachusetts to rise up in Shay's Rebellion. Shay's rebels, facing these pressures in the wake of the Revolutionary War, demanded an end to farm foreclosures and lower taxes. They subsequently seized courthouses and burned records when their demands were not met.

Although the rebels were put down in 1787, farmers continued to struggle to retain their traditional use of the land. The return of greater prosperity in the 1790s, the easing of trade restrictions with the West Indies that arose as a result of Jay's Treaty with Great Britain, and the increased demand for food among the growing population centers of the coast once again gave country store owners the flexibility to take the farmers' "country produce" in exchange for goods. Merchants Tappen and Fowle ran an advertisement in a western Massachusetts paper, the *Hampshire Gazette*, offering for sale a variety of manufactured goods, including clothes, silks, ribbons, tea, sugar, chocolate, spices, etc., for which "country produce of most kinds will be received in payment." Similarly, in a 1790 advertisement Seth Wright announced that he had a variety of goods from New York, Boston, England, and the West Indies for sale for cash, but he also noted that he would "take many articles of country produce for goods as usual." But this arrangement, which may have been business as usual in 1790, by 1810 had become distinctly unusual, particularly east of the Alleghenies. Levi Shepard, who in the 1780s was accepting as payment "most kinds of country produce," by the 1790s was announcing that "goods could be paid for by either cash or 3 months credit, or on contracts for flax" (Cumbler 2001, 17–18).

Farmers looking to set up their children on farms or to buy goods and supplies needed cash. To acquire cash they had to furnish country stores with goods specified by the store owners. Increasingly, to get credit, farmers contracted to produce goods wanted by remote merchants. Decisions about what to plant and how to use the land grew to depend on the vagaries of distant markets. Farmers took to following the prices of wheat, rye, flax, hemp, beef, and pork in the markets of Boston, New York, and Philadelphia and adjusted their planting accordingly. Farms closer to the urban centers were the first to shift production into this new market-oriented farming. Those farmers who worked the fertile alluvial soil in the lower and central river valleys or the rich, rolling farm lands of southeastern Pennsylvania and central New Jersey plowed more land and increased productivity. This way they could realize the advantage of their efforts as their goods moved easily into the expanding markets. Because declining production meant declining income, however, farmers worked their land carefully, rotating crops to maintain the land's fertility. They chose crops for each rotation according to the prices those crops were liable to fetch come harvest time. They supervised their animals more closely as well. In the fall of 1789, justice of the peace, storekeeper, and farmer Sylvester Judd Sr. took his cattle to market, thereby reducing the number of animals he had to tend and feed over the winter, a long-standing practice among area farmers. By contrast, in 1818, his son, Sylvester Judd Jr., took his cattle to market in the spring. The younger Judd's cattle came to market fattened considerably by a winter of minimal activity spent dining on rich hay feed. Judd knew that at this time of the year and in this condition his cattle would command the highest price. Additionally, a season's worth of manure accumulated in the barn could then be spread on fields to improve the soil's productivity.

Urban markets looked to farmers for more than simple foodstuffs. Demand for leather goods—whether shoes, boots, bridles, harnesses, or belts—grew along with the population. Tanners produced leather for these commodities using tannic acid from local oak or sumac trees. The hides came from local farmers (and, by the 1830s, as far away as Spanish California), who increasingly factored in the price of the hides when deciding how many cattle to send to market. Rural broom manufacturers needed broomcorn and paid farmers to plant fields with it. In 1790 the *Hampshire Gazette* noted that "the Boston Sail Cloth factory had occasion for 200,000 weight annually [of flax] and the factory in Salem together with others in the country will probably have call for as much more. The price it now bears will afford ample encouragement to those who are disposed to raise it." The *Gazette* was calling on farmers to make planting decisions based not on the needs of their families but on the price of flax. Increased hemp prices also convinced farmers to shift fields into hemp production. In northern New England where the soil was particularly thin and the growing season problematic, early-nineteenth-century farmers began to take notice of William Jarvis, or more precisely, Jarvis's flock of merino sheep. Unlike traditional farm sheep, which were raised for both wool and meat, merinos were wool sheep raised purely for their wool. As demand for wool increased, farmers began buying merino sheep, not for home use, but to supply materials to the woolen mills springing up along streams from Vermont to Delaware.

For farms farther upriver or away from the river valleys, the shift to a production based on market demand rather than need took longer or never occurred at all. These farmers were restricted by the burden of getting their goods to market and the less fertile and more erosion-prone soil that limited their farms' productivity. Even poorer farmers were increasingly concerned about the broader market. As farmers became more focused on markets, however, they also became more aware of the need to facilitate the movement of goods to and from the larger market centers.

The transportation arteries of colonial and early America were its rivers: the Penobscot, Kennebec, Merrimack, Charles, Blackstone, Connecticut, Houssatonic, Hudson, Delaware, and Susquehanna. Oceangoing vessels could sail up these rivers until the series of fault lines the rivers crossed as they descended from the hill country. Goods from above these fault lines flowed downriver, while goods from below a market center were either canoed, poled, or sailed on boats with short flat sails upriver. At each fault line, boats carrying goods had to either run the rapids downstream or be unloaded and portaged around to another boat or canoe on the other side. Each fault presented yet another obstacle to the movement of goods. Running the rapids required waiting until high water and also entailed the risk of losing goods overboard. Loading and unloading and portaging goods around the falls or rapids was grueling labor-intensive work.

Merchants, traders, and farmers interested in cutting the costs of the goods brought inland and speeding the shipment of materials out were eager to find a way to avoid the long and expensive delays involved in conveying goods past fault lines. Those upstream of the obstructions believed that building a system of dams, locks, and canals around the falls and rapids would increase the speed and lower the cost of transporting goods, bringing greater prosperity to the region. The dams that would feed these locks and canals needed to be significantly bigger and more complex than the simple earthen, stone, boulder, and log dams used to power grindstones and saws. Visions of extensive trade flowing in and out of the region and profits from operating these new locks and canals danced before the eyes of regional elites and brought them together to plan these ambitious new projects. Compared to what was needed to erect a simple gristmill and dam, these projects required much more substantial capital and engineering. The necessary capital, engineering, and support outstripped the means of local elites. Pushed by project boosters, early state legislatures incorporated canal-building corporations with the power to raise capital, buy property, and flood fields and meadows far surpassing the terms of the various mill acts. In some cases, the legislatures even gave the new companies the authority to raise funds through state-sanctioned lotteries. Although legislatures were generally sympathetic to these projects, they were also concerned about the power of corporations and demanded that the corporations serve the larger public good. These new corporations also faced restrictions and limits to their activity such as the requirement of renewing their contracts.

In the 1790s, with the prodding, organizing, and financing of leading merchants and traders, dams began to be constructed for canals around the fault lines of the major water arteries. In the case of the Connecticut River, schooners or sloops transported goods as far as Warehouse Point north of Hartford, where they were rebuffed by the Enfield rapids. Above the rapids, goods could travel by small, river-going sailboats to the falls at South Hadley north of Springfield, Massachusetts. Four other major falls obstructed the movement of goods—one at Turners Falls in northern Massachusetts, another at Bellows Falls in southern Vermont, and two farther north at Sumner Falls and Olcott's Falls. Construction for the first of the major dam-canal works on the Connecticut River began in Vermont at Bellows Falls. Vermont chartered the "Company for Rendering the Connecticut River Navigable by Bellows Falls" in 1791, and work began in 1792.

Work on the canal and lock system took ten years before the first boat passed in 1802. In 1792 the "Proprietors of the Locks and Canals on the Connecticut River" received a charter from the Massachusetts legislature to build canals around South Hadley Falls and Turners Falls. By December 1792 the proprietors were seeking 70,000 feet in "length of timber of yellow, or pitch pine, hemlock, chestnut, or oak in pieces of any length not less that 25 feet to be straight and of any size not less than one foot in diameter in the small end" (Cumbler 2001, 28). The company proposed to build a dam to raise the river 7 feet. The canal and lock system completed in 1795 featured an 11-foot-high dam and an inclined plane on which a carriage moved up and down, drawn by a water wheel. Six years later the inclined plane was replaced with a 2.5-mile-long canal using 8 locks. The Turners Falls system that was completed in 1795 was even more extensive—with 2 canals, a dam 17 feet high, and 10 locks.

At the same time that Massachusetts and Vermont merchants were building dams, and locks were built around obstructions on the Connecticut, Newburyport merchants at the mouth of the Merrimack invested in a dam and lock system around Pawtucket Falls, at present-day Lowell, Massachusetts, in hopes of capturing the Merrimack River valley trade. Unfortunately for them, soon thereafter Boston merchants constructed the Middlesex Canal from the point where the Concord River flowed into the Merrimack twenty-seven miles south to Boston harbor. As the merchants had hoped, the dams and canals facilitated the movement of goods both in and out of the river valleys. The structures also strengthened the link between the interior and the coastal market centers. Farmers found more goods on the shelves of country stores, but they also felt increased pressure to plant crops the country store would accept. The increased river traffic driven by canals and dams in one area encouraged their construction in another. Canal workers built dams, canals, and locks to facilitate the flow of goods on rivers not only in New England but also across the Northeast (and South). In 1792 the Western Inland Lock Navigation Company was incorporated to open navigation to Lake Ontario by digging a mile-long canal and a series of five locks to bypass a rapids on the Mohawk River. That same year Pennsylvanians began work on a canal around the series of rapids on the Delaware River beginning at Morristown (across the river from Trenton, New Jersey) and another one on the Schuylkill. In 1793 workers began a four-year project to dig a one-mile channel around the falls of the lower Susquehanna at Conewago, Pennsylvania.

These dams were bigger and far more extensive than the earlier milldams and they did more to transform the environment. They were substantial edifices, built of lumber and stone. As was the case with the South Hadley Falls dam, their construction demanded a great deal of lumber, which was cut from the hills and mountains upriver. Because they were significantly higher than grist-

Goods, Trade, Mills, and Dams 73

Mill Dam, Holyoke, Massachusetts, c.1850. (Library of Congress)

mill dams, these dams flooded substantially more land and kept it flooded year-round. The dam at South Hadley Falls raised four feet of water behind it for ten miles, flooding extensive meadows and replacing flowing brooks and streams with stagnant pools of water. The people of Northampton complained that their community, which had previously been "one of the healthiest . . . after the erection of the dam [it was] . . . extremely afflicted with the fever and ague," (Cumbler 2001, 29) probably the result of a rise in mosquito populations.

The dams may have offered boats a means around river obstructions, but many, by cutting across the whole of those rivers, obstructed the paths for anadromous fish heading upstream. Salmon were the most likely fish to be affected. Whereas falls and rapids impeded the upriver run of shad, salmon easily surmounted these obstacles on the way to their spawning beds. To make their nests and lay eggs, salmon needed rapidly flowing water with high levels of dissolved oxygen and pebbly beds. These conditions prevailed at the feeder brooks and streams and river headwaters. High dams stretching across the river barred the salmon from these spawning areas. Immediately after the dams went up, fishers coming to the riverbanks began noticing fewer and fewer salmon making the upriver run. On the Connecticut, the dam at Turners Falls effectively blocked the salmon from their northern destinations. With no new salmon returning to the ocean to grow to maturity, by the second decade of the nineteenth century, the fish were gone from the Connecticut.

Store owners shipped the rural surplus goods culled from the countryside downriver to the major ports along the Atlantic coast. Goods coming in on small river schooners were gathered up in Salem, Boston, Newport, Providence, Hartford, New York, and Philadelphia, where they were stored before being loaded onto the larger transatlantic sailing ships. City merchants with contacts in far-off ports handled this part of the trade. Into these coastal cities came the goods not produced locally: tea and coffee, spices, sugar and molasses, printing presses and books, fine cloth, china, and glassware. Out from these cities poured salted beef, pork, and fish; wheat and flour; dried peas; cheese; and lumber products. A forest of masts dominated these port cities' harbors.

The boats that brought the goods to these coastal ports were built in nearby boatyards. The outfittings for these boats and for the larger ocean-worthy ships were also locally produced. Rope works and, increasingly, sail works sprang up along the edge of the harbors. Soon, spread out around the busy harbor were counting houses to maintain records; warehouses to store goods; cooperages to make barrels to hold produce; printing shops to make up bills of lading, account books, and broadside advertisements; newspaper offices to carry the shipping news; insurance firms; and law offices. Taverns and boarding houses crowded in along the wharf, while homes and craft shops moved uphill, away from the busy docks. To these growing port towns and cities came sailors and captains to sail the ships; teamsters and day laborers to load, unload, and store goods; and shopkeepers, lawyers, artisans, and craftsmen to maintain the needs of the growing urban community. By the end of the American Revolution, many of these port communities had grown into major cities. Philadelphia with 35,000 inhabitants vied with Liverpool as the second-largest city in the English-speaking world. New York had almost 25,000 residents, Boston boasted 16,000, and Newport 10,000. Increased density also facilitated the spread of disease. In 1793 Philadelphia was struck with a yellow fever epidemic carried by mosquitoes that bred in stagnant water, fed upon infected individuals, and then spread the disease to others. Four thousand residents of the City of Brotherly Love—close to a twelfth of the population—died in that epidemic.

Following the War of 1812, trade increased dramatically between the new nation and Europe, as well as with the Caribbean region. American ships soon outnumbered European ships in the transatlantic trade. Goods from the South were brought to northern ports before being shipped to Europe. Captains from Salem to Philadelphia carried goods from Asia to American ports and spread American goods around the world. Reflecting the ongoing search for new goods to trade, New Englanders even harvested ice from their winter ponds, packed it in straw, and then loaded it onto ships to sell to customers in the warmer climates to the south. Although Europeans had been harvesting the New England seas—particularly the Georgia and Grand Banks—for centuries, increasingly in the nineteenth century American fishers dominated these rich fishing grounds and supplied cod to the world. While cod fishers worked hand lines over the side of dinghies on the banks, American whalers built a lucrative industry capturing the leviathans of the sea. Wealth from fishing and whaling expressed itself in huge captains' homes, many of which still stand in New Bedford, Nantucket, Martha's Vineyard, and Cape Cod. Riches culled from the sea's bounty also circulated through port cities like Boston, where fishers and whalers ordered ships to be built and goods to fill their new homes.

If you could travel back in time to 1740 and hiked up Bowman's Hill south of New Hope, Pennsylvania, and on that hill you found one of the giant oaks that grew over a hundred feet into the air, and you climbed to the top of that oak, you would see spread out around you a patchwork of farms, vegetable gardens, orchards, wheat fields, and woods. You would not see open land, but rather a forest spotted with openings of farmland. Certainly by 1740 the openings would have grown to encompass as much as twenty to thirty acres. But they would still dot a landscape otherwise dominated by forest. Looking east from Bowman's Hill, you would see the Delaware River and the point where George Washington later gathered his troops on Christmas Day in 1776 to cross the Delaware and attack the Hessian troops stationed at Trenton Falls. Looking farther southeast, past where Washington marched his troops that cold December night, you would again espy farmhouses surrounded by wheat, rye, and oat fields. Wheat and rye grew plentifully in the soil of mixed loam and sand.

Halfway between the point where Washington landed on the New Jersey side and Trenton Falls, a small creek, known as Jacob's Creek, flows into the Delaware. On the creek stood a mill, Jacob's Mill, where in the nineteenth century my great-grandmother, Sara Burroughs, grew up. Her family ground the wheat that farmers from the surrounding countryside brought to the mill and kept part of the resulting flour as payment. Her ancestors came to New England as Puritans and settled outside of Salem, Massachusetts. The family left Massachusetts at the end of the seventeenth century and settled on Long Island as farmers. Soon the Burroughs outgrew Long Island and one son was set up as a miller at Jacob's Creek in New Jersey in the late 1700s. From then on the Delaware Valley wing of the Burroughs family was out of farming.

If you went back to Bowman's Hill in 1800 and again found a tree tall enough to provide a view across the countryside, the land would appear much more wide open. Only on the highest hills would you see forests. Fields of grain, pastures, and orchards, with occasional patches of woods, would dominate the landscape. Fewer trees brought higher temperatures and dried out the land. Bucks County summers in the early 1800s were warmer and drier than they were a century earlier. From a vantage point high enough you just might make out the Cooper farm on the western horizon. During my childhood in the 1950s this farm belonged to my uncle, William Cooper. His family had owned the land for generations. Family legend held that the house, barn, outbuildings, and fences were all built with materials and wood from the farm at a total cost of seven dollars. In the late eighteenth and early nineteenth centuries, the Coopers grew wheat and raised beef cattle to sell in Philadelphia. When I was a child my family would drive to the Cooper farm from our house on the Delaware River. In those days the land between the river and the farm was planted in wheat stretching back to the woods that hugged the hills. Today that farmland, some of the richest on the East Coast, grows houses and subdivisions. By the time I was visiting my cousins, the Cooper farm no longer produced wheat and beef cattle. The farm had been turned over to turkeys and chickens. My uncle also leased some of his land to a neighboring dairy farmer, who pastured his cows in the field where hay once grew and planted corn for fodder where wheat once rippled with the breezes. Another family leased the meadows to keep horses for fun. I suspect that today the farm is surrounded mostly by houses with neat yards and flower gardens. In Berks and Lancaster Counties southwest of the area once encompassed by Cooper's farm, Amish and Mennonite farmers persist as wheat farmers, but even they are being squeezed out by suburban sprawl. The mill at Jacob's Creek hasn't milled grain in over a hundred years and for the past fifty has been an art studio.

4

MOVING WEST: THE INTERIOR

Seventeenth- and eighteenth-century Europeans wanted two things from America: furs and land. Native Americans had access to furs and they lived on the land. To acquire the furs, Europeans offered Native Americans axes, knives, pots, pans, blankets, wampum, and guns. In exchange for these goods, Native Americans brought the Europeans pelts. Soon the beavers, bear, lynx, and other animals with pelts of value disappeared from the land east of the Appalachian Mountains. In search of more pelts, the eastern Native Americans traveled farther inland, with European traders following. Trailing them were the children of the east coast farmers and new European immigrants. Instead of pelts, these groups came in search of land they would hold for their own exclusive, permanent use. But the lands to the west of the mountains were not uninhabited. Animals with fur lived there, but so did people who hunted those animals.

In the years following the American Revolution, considerable differences of opinion arose among the Europeans over land use in the new nation. Some with titles from the crown, a colonial company, or the new United States government had visions of setting themselves up as gentry. The English gentry owned enough land that they neither had to work the land themselves nor work at all, for that matter. They either had tenants, who rented and worked the land, or farmhands, who were paid to work land directly controlled by the lord. The image of reproducing the English gentry system twinkled in the eyes of many Americans of means. Two obstacles stood in their way: the nagging seeds of egalitarianism that exploded into full bloom during the revolutionary period and the vast quantity of land occupied by native peoples (although their hold on that land seemed tenuous at best in the eyes of the Euro-Americans). Aspiring members of the American gentry rationalized their desire for elevated status by either favorably comparing the gentry (or landowners) to the aristocracy (titled nobility) or, south of the Mason-Dixon line, by using race to argue for a commonality of whiteness that obfuscated the contradiction between revolutionary ideology and such clear class distinctions. Other Americans had different visions of both the American ideal and how the land should be used. Many New Englanders looked to self-sufficient communities of freeholders—working farmers who held land—as the model for the new nation. Those fleeing a Europe of landed estates also saw in America the

Native Americans bringing beaver pelts to white traders. (North Wind Picture Archives)

possibility of becoming independent yeomen farmers. Despite the differences of opinion, on one point all the Euro-Americans agreed: the indigenous peoples and their traditional land use had no place in this new America.

Following the Revolution, the aspiring gentry began a vigorous land grab. New York's Hudson River valley was already occupied by large, landed estates worked by tenant farmers. To the north, aspiring landlords attempted to spread this pattern of land use into Vermont. In territory that is now part of Maine, General Henry Knox acquired more than half a million acres of land he hoped to settle with tenants. Dutch land speculators bought thousands of acres in northeastern New York. The state of New York sold off over 5 million acres of land in the broad valley south of Lake Huron to a handful of buyers. Farther west, in the Northwest Territory, the federal government was selling huge parcels of land. Those who were accumulating the land in the hope of setting themselves up as local gentry were soon disabused of the notion, however. The farms of New England did not turn out much produce; the soil was too poor and the growing season too short. The families on those farms did turn out children, however. As we have seen, these children grew up and wanted farms of their own. By the end of the eighteenth century, they were ranging farther afield. Those along the northeast coast and up the Merrimack River valley moved into Maine. Where possible, they bought land. When they couldn't buy, they squatted. Many squatted on Henry Knox's land. When Knox attempted to collect rents, he encountered active resistance. In the end, he was forced into an accommodation allowing ownership of the land to move into the hands of the settlers who had improved it.

Children from western Massachusetts moved into Vermont and fought New York landowners for control of the land. But it was the land farther west that called even more insistently to these children of New England. The lower and central Hudson River valley was already well settled by the end of the eighteenth century, although renters and owners continued to contend over the land. Stephen van Rensselaer's estate, for example, had as many uncollected rents as obedient tenants. West of Albany stretched the Mohawk River valley with its rich alluvial soil. Following the Mohawk, 100 miles from the Hudson the land spreads out into a coastal lowlands stretching 300 miles west along the southern shore of Lake Huron to Lake Erie. Large landowners tried to hold on to their land and find tenants, but willing tenants were scarce. Settlers wanted land of their own, not tenancy, in the west. Unable to rent out their 300,000 acres of land on Lake Champlain, the Beekman family was forced to divide it up into parcels and sell to settlers.

More typical of the settlers who moved west was Hugh White of Middletown, Connecticut. In 1784 White sold the family farm in Connecticut and bought 1,500 acres of land outside of Utica, New York, in the Mohawk River valley. There he cut trees, let his pigs run wild, pastured his horse and cows in the meadowlands, and began clearing land to farm. He planted an orchard and put in wheat, rye, and oats. He divided his land into farms for his sons, who also began clearing land and growing grain. White's land was forested with spruce, hemlock, maple, beech, and other northern hardwoods that grew in the loam-rich, gray-brown podzolic soil. The forests on his land and the Mohawk River as a link to the Hudson and eastern markets encouraged White to build sawmills and ship out lumber. As the forests were cleared, wheat was planted, allowing flour to follow the lumber to eastern markets. As more of White's and his sons' lands were cleared and opened to production, White began making decisions about how to use the land according to market prices. He started rotating crops in his fields and fertilized them according to the latest scientific theories.

New England farms kept producing children, and New Englanders continued to move west through the Mohawk River valley into the Genesee River valley and then southwest into Ohio. Along the way, the children of farmers from New York, New Jersey, and Pennsylvania joined the movement. The new government encouraged this settlement process, if only to help secure the region

Presettlement mixed forest, 1700. (Harvard Forest Dioramas/ Fisher Museum/Harvard Forest, Petersham, MA)

against the British, French, and Spanish. In 1785 Congress passed the Northwest Land Ordinance that authorized the surveying of the territory north of the Ohio River and west from the middle of Lake Erie to the Mississippi River. The Northwest Land Ordinance captured much of the spirit of the new nation. Authored primarily by Thomas Jefferson, its regulated form reflected Jefferson's belief that America should throw off antiquated feudal patterns and embrace the New World of science, order, and liberty. Jefferson believed that the liberty of the nation would be secure only if the majority of the nation's population lived on land they owned. Jefferson felt that independent farmers made for citizens who loved and defended liberty. To that end, he did not want the lands west of the Appalachian highlands and north of the Ohio (an expanse the revolutionary states had originally claimed and finally agreed to turn over to the new national government) to fall into the hands of a few large speculators or become the estates of a new American landed aristocracy. Jefferson's concerns meshed with a popular desire for land. He envisioned surveying this new public domain and then dividing it into a grid of squares with parallel lines running east and west crossed with parallel lines running north and south. This created a rectilinear cadastral survey, known as the Jefferson grid. The survey used the latitude and longitude lines of navigators and divided them up into square units. Each square was a township, six miles by six miles. The townships were subdivided into lots, 640 acres each. Each lot was then divided up into half sections of 360 acres, then quartered into 160-acre "quarter sections." The idea assumed that each quarter section would provide enough land for a farm and would be sold off to farmers. Although Jefferson understood that the surveys would incorporate topographical information about waterways, hills, and valleys, he was primarily concerned about an ordered grid that would facilitate the sale of land.

Of course, Americans had been buying and selling land long before Jefferson thought up his grid. One of the first issues the colonial governments confronted was how to mark and secure land ownership. Without a guaranteed title, landowners could not comfortably improve the land and be sure that they could then sell it without inviting disputes regarding ownership. Land title battles raged through the early colonial legislatures. Kentucky during its early history was typical. Settlers like Todd and Nancy Lincoln bought land and worked it, only to have their title contested by a claimant who had bought the same land from a different seller. Losing that fight eventually sent the Lincolns into the Northwest Territory, where land titles were more secure. Hoping to avoid a repeat of the Kentucky mess, Jefferson's surveyors divided the land into the townships of thirty-six-square-mile lots. The smallest size offered for sale was one lot (640 acres); a dollar an acre was the lowest price for which lots were offered. To facilitate sales, land offices were set up in each state. Teams of surveyors (many with questionable ability or training) spread out across the Northwest Territory's vast and varied land, complaining the whole while of mosquitoes, swamps, and flies. Beginning in southeastern Ohio and spreading west and north, they imposed a huge grid on the land. The survey grid maps—with each lot appropriately numbered—facilitated land purchasing by speculators who could buy land without even seeing it, based on the map in the land office. At a dollar an acre, the land was cheap, but nonetheless expensive for poor immigrants or plain folk like the Lincolns.

Not all of the lands of the Northwest Territory were so easily integrated into this neat and ordered system. The land of the Western Reserve, on the northeastern edge of Ohio, was claimed by Connecticut to distribute to the state's war veterans. When the British burned Connecticut towns during the Revolution, the state granted more land to fire victims. Virginia claimed land in southwestern Ohio for its veterans. Other chunks of land went to land companies and big speculators. Yet the land did move into the hands of farmers. Speculators did not farm. They made money, but they only made money when they sold land. Farmers only wanted to work land they owned. Ultimately, this reality kept land prices within reach for enough Americans to create in the Northwest Territory a land where Jefferson's ideal could be realized.

In 1787 the Northwest Ordinance was passed, creating the process by which the new lands could become states and enter the Union. Clearly, leaders of the new nation viewed land distribution as a compelling purpose. How the actual distribution process would occur remained unclear, however, as did what would happen to the squatters and Native Americans already on the ground.

As whites from New England and New York spilled into the territory south of Lake Erie, and children of poor southern farmers moved northwest from Virginia, Maryland, and North Carolina into the area north of the Ohio River, conflict flared between whites and Algonquian villagers. Energized by the egalitarian rhetoric of the Revolution and the promise of land ownership, Scots-Irish immigrants and second- and third-generation farmers from the back country of Virginia, Maryland, and North Carolina moved west into Kentucky where they hoped to find cheap land. Good, arable land was already in short supply in Kentucky, however, and much of what was available was tied up in litigation with multiple claimants to land titles. Losers in these land title conflicts, like the Lincolns, and latecomers to the region began crossing the Ohio into the rich, rolling lands to the north, where they squatted whenever possible. Along the way, these settlers frequently attacked native villages, particularly when the native inhabitants resisted the incursions.

By the late 1780s, intermittent military skirmishes at the village level had become common throughout the lower Midwest. In 1790 Native Americans defeated General Josiah Harmar and in 1791 they defeated Governor Arthur St. Clair's 1,400-strong army. Three years later, however, the Native Americans were defeated at the battle of Fallen Timbers and were forced to cede much of the territory of Ohio. Over seventeen years, beginning with the Battle of Tippecanoe and ending in 1815, Native Americans and white American troops and militia again clashed over land and land use, this time in Indiana. And, as in 1794, the whites' military victory effectively drove the Native Americans from the land they had occupied for generations.

The settlers and squatters who moved in believed that by clearing the woods and planting crops and orchards they had transformed the land from wilderness to valuable property. Although few read John Locke, they nonetheless subscribed to his theory of property. Locke argued that labor gave value to land and thus transformed it into property. Locke was interested in discovering in natural law the basis for property and government. In his *Two Treatises on Government* published in 1689 and 1690, Locke proposed that land was initially abundant, wild, and free. Locke claimed that although in the wild no one owned land, in nature humans did own their own labor. As owners of the labor they used to improve

Portrait of seventeenth-century political philosopher John Locke. (Bettmann/Corbis)

the land, humans could reasonably claim property rights over the improvements they made to the land and in turn the land itself. Once they had the right of property ownership, they could pass that property on or sell it to others. Locke considered this the legitimate origin of property. He went on to argue that the right to property was thus a natural, inalienable right, and that governments were created by humans to protect basic rights and had their legitimacy in protecting rights, including the right of property. Locke even cited the American interior (as of his writing in 1690) as an example of land without property value. Because no one had improved it (as a European understood the concept), it was wild and free, and as there was so much of it, the land had no value. By the end of the eighteenth century, Locke's ideas had become part of the new nation's intellectual currency. Jefferson borrowed freely from them in writing the Declaration of Independence. From the European perspective, Native Americans had not worked the land to improve it. Thus, they had little right to a property claim over it. Squatters and settlers who cleared fields, fenced in pastures, planted crops, and built homes put their labor into the land. This, they felt, legitimized their claims to the land.

In 1805 the state courts recognized this principle in the famous *Pierson v. Post* case (New York, 3 Caines reports, 175). In that year Lodowick Post and his dog were tracking a fox on a wild "unpossessed and waste land." Just as Post moved in for the kill, Jesse Pierson shot the fox and claimed it for his own. Post sued, arguing that the fox was rightfully his. The court ruled in Pierson's favor because, as the judge noted, although Post had chased the fox, by killing the fox Pierson had had transformed it and put it under his total control. Now codified by law, property ownership rested in the process of taking control over that which was wild. This Lockean idea was later formalized in the National Homestead Act of 1862.

The land of western New York was fertile and productive, and the soil of Ohio even more so. Usually, every year 30 to 40 inches of rain fell throughout the whole region south of the Great Lakes, whereas the area of Michigan, Wisconsin, and farther west to Minnesota experienced 24 to 30 inches of annual rainfall. Most of the region averaged a long growing season of between 160 and 200 frost-free days, with Michigan, Wisconsin, and parts of Indiana and Ohio averaging 120 to 160 such days. This land took to the plow easily. Once the hardwood trees had been removed from a field, a farmer with horses and a metal-tipped plow could easily plow and seed several acres of land in a planting season. As the farmer moved through the fields, stones and pebbles in the soil scraped dirt from the plow. By the 1820s, farmers in western New York were growing so much wheat that the mills of Rochester and Syracuse were running at full capacity, driving millers to search desperately for more streams to dam. Soon the need for land was replaced by the need for markets.

Whether they grew food primarily for their families and only sent their surplus to the market or they based their decisions on the dictates of the market, farmers needed to get their goods to an available transportation route. Cattle could be driven to market. Indeed, the Sylvester Judds, both senior and junior, did just that. They drove their cattle across Massachusetts to the butchers of Brighton, just outside Boston. But grain or flax could not be driven to market, nor for that matter could chickens or even pigs for much of a distance. A farmer who had to travel overland more than thirty miles to reach a navigable waterway or market center found his profits evaporating along the way. Thirty miles overland by wagon on rough roads—little more than Indian trails—made for two to three days of rough travel. Ruts, mud, and downed trees made the trip a test of endurance for even the hardiest traveler. Grain was frequently lost or ruined, while wagons and draft animals were strained to their limits.

Even those farmers who lived near waterways faced difficulties. Moving downriver was relatively easy, provided one did not confront a rapids or falls. Rafts made of pine logs strapped together, with primitive steering devices, held goods traveling downriver. Small, sailing schooners could sail upstream in areas where the river was wide enough, but the going remained difficult.

Settlers in southern Ohio enjoyed the advantage of the wide Ohio River. Goods put on a raft or keelboat at the mouth of the Scioto or Miami River could be floated downstream all the way to New Orleans with only one interruption—the fault line that runs across the Ohio River at Louisville, Kentucky. Once portaged around the falls of the Ohio (or, for the risk takers, shot over the rapids during high water) the goods' trip downriver to the wharfs of the Crescent City was slow but sure, particularly after Jefferson bought the lands west of the Mississippi from France in 1803. Moving goods upriver was far more problematic, however. In the early years of the nineteenth century, most farmers selling goods downriver simply sold their rafts or keelboats along with their grain, dried beef, and pork. They would then buy a horse or mule, load it up with supplies, and begin the upriver trek home.

Moving goods out of the eastern interior was equally difficult. But then, in 1807, Robert Fulton sent his steamboat, the *Clermont*, 150 miles up the Hudson River to Albany. Fulton's steamboat dramatically cut the time needed to ship goods up and downriver. Soon steamboats were moving up and down rivers all along the East Coast. In 1809 a steamboat traveled from Pittsburgh, Pennsylvania, to Louisville, Kentucky, and back. Before the steamboat, it took over twenty days to float that distance; the new boats could make the trip in seventy hours or less. While traveling upstream had heretofore been effectively prohibitive, steamboats made it scarcely more difficult than downstream movement. By 1815 steamboats were traveling downriver from Pittsburgh and New Orleans and by 1817 they were moving regularly between New Orleans and Pittsburgh. By that same year, steamboats were transporting people and goods around the Great Lakes.

Steamboats revolutionized river and lake travel. By the 1820s, steamboats had reduced the cost and travel time of upriver shipments by 90 percent. Steam-

The Clermont, *a steamboat designed by engineer Robert Fulton. (Library of Congress)*

boats transformed the Mississippi and Ohio Rivers into central arteries transporting goods out of Ohio, Indiana, and Illinois to waiting ships in New Orleans.

The steamboat's arrival corresponded to the white settlement of Indiana and Illinois. Whereas Ohio joined the Union in 1803, Indiana in 1816 and Illinois in 1818 entered carrying a wave of migrants who flooded into the Indiana Territory after the war of 1812. Not only were farmers of the Northeast running out of land, the land they had was running out on them as well. Although valley farmers along the Merrimack, Connecticut, Hudson, and Delaware rivers prospered as the urban centers and markets downriver grew in size and appetite, hill farmers did not fare so well. Their soil was thinner, rockier, and more prone to erosion. Then in 1816 and 1817 came the cold years. In 1816 it snowed all summer in New England, and in the hills there was frost twelve months of the year. The next year was not much better. Hill farmers lost their crops and had to slaughter their animals. Farmers whose children had worked with them on worn-out land found their offspring abandoning the farms. Few children wanted to take over a marginal, rock-strewn farm when good land was available in the west. And west they went, there joining immigrants from Germany and northern Ireland who

Moving West: The Interior 87

The opening of the Erie Canal in 1825 after eight years of construction. (Library of Congress)

were also looking for good, cheap land. Between 1810 and 1840, the population of Ohio, Indiana, and Illinois rose tenfold. Ohio's population concentrated in the river valleys flowing into the Ohio River and along the southern shore of Lake Erie. Indiana's population was almost all concentrated in the southeast and along the Ohio. Initial settlers in Illinois preferred the lands in the southern and western part of the state bordering the Mississippi.

The flow of goods downriver to New Orleans explains many of these early settlement patterns. But for the farmers of northern Ohio and western New York, the route downriver to New Orleans was not an option. Besides, Europe and the coastal cities lay to the east, not south. These northern farmers wanted an easy means of moving their goods east. East Coast merchants and traders shared the sentiment. New Yorkers concerned about losing business to New Orleans or their other East Coast competitors had a champion in Governor Dewitt Clinton. Clinton pressured the New York legislature into funding a 364-mile-long canal project stretching from Albany to Buffalo on the eastern tip of Lake Erie. The Erie Canal was a massive undertaking. Because the nation had no engineering schools, engineers had to be trained on the spot. Irish immigrants were hired to labor in gangs for less than ten dollars a month, some of which was paid in whiskey as part of the daily ration. These gangs did backbreaking work, lifting boulders and digging by hand, excavating hundreds of tons of dirt and rocks, building locks, and clearing and grading paths for the mules to tow the barges. Conditions were harsh. The laborers lived in crude camps. Over a thousand fell victim to fever in the marshes southeast of Lake Ontario, many fatally.

When finally completed in 1825, the canal represented a signal victory. The cost of shipping goods from New York to Buffalo fell by 90 percent. Within two years, New York was shipping out more flour than Baltimore or Philadelphia, despite the rich wheat fields of southeastern Pennsylvania, western Maryland, and north-central Virginia. The completion of the Erie Canal also opened up the northern Midwest for market agriculture. Population and agricultural production rose dramatically in western New York and northern Ohio. By 1830 there were between forty-five and ninety people per square mile in western New York, a population density comparable to that of the East Coast, excepting the urbanized areas around Boston, New York, Philadelphia, and Baltimore. Northern Ohio achieved population densities of eighteen to forty-five persons per square mile (U.S. Bureau of the Census, Historical Statistics of the United States 1975).

Buffalo was overwhelmed by the changes wrought by the canal's completion. Wharfs were built jutting into the lake to handle the boats bringing in wheat that then went into newly constructed grain elevators and mills. Buffalo's success with the Erie Canal spurred a boom in canal building. Philadelphia merchants launched a canal west to Pittsburgh, followed by Baltimore canal boosters who began building a canal north and west into Pennsylvania. Ohio began building canals linking Lake Erie to the Ohio River. Illinois started work on the Michigan-Illinois canal to link Lake Michigan at Chicago with the Mississippi by way of the Illinois River. The canals of the Midwest did much to integrate the region to the larger transportation arteries, but the canal projects designed to compete with the Erie Canal failed to capture the western market. The Erie Canal ran through the Mohawk River valley, the only natural opening in the Appalachian Mountains wide enough to send a canal through. Smaller eastern canals such as the Raritan, the Middlesex, and the Blackstone did succeed in moving heavy bulk goods over shorter distances until the railroads reduced their importance.

Others besides midwestern farmers benefited from the Erie Canal. Riches in addition to land lay west of the Allegheny mountain range. The Great Lakes teemed with fish, with Lake Ontario home to the king of fishes: the freshwater Atlantic salmon (*Salmo salar*). Every year hundreds of thousands of these salmon swarmed into the rivers and streams feeding into Lake Ontario. Big fish weighing between eight and thirty pounds, these salmon were prized for their bounty of red meat that was both delicious fresh and easily salted and dried. The fish runs were so prolific that contemporary commentaries described people

Currier and Ives American farm scene. (Library of Congress)

pulling fish out by hand, or using shovels and spears. By the time the Erie Canal was completed, commercial fishing operations were running along the major American salmon rivers—the Salmon, Oswego, and Genesee—and on countless smaller spawning streams. Initially, white settlers followed the example set by Native Americans and caught fish for their own use, eating some fresh and salting down the rest for later use. Salmon was an important part of the early settlers' diet and kept many from hunger. By the early nineteenth century, Rochester had already become an important market for barreled, salted, and dried salmon.

The Erie Canal dramatically increased the profits for Lake Ontario salmon fishers. Piled up on the docks of Rochester, the abundant salmon catches vied with dried pork and beef for space on barges heading east. As profits from fishing grew, commercial fishers began to assume control over the better fishing spots, edging out the farmer/fisher who occasionally fished to supplement the family diet and add a little extra to the surplus going to market. Commercial fishers and the farmer/fisher attacked the salmon with ever-greater zeal as hundreds of thousands of salmon moved from the lake to dinner tables. During the boom years between 1853 and 1857 those who controlled the mouths of rivers and streams reaped huge cash windfalls by catching the ascending salmon in trap nets. By 1860, however, they had wiped out the southern Ontario salmon populations. But much as fishers prospered in the early years of the nineteenth century, the richness of the water was not the primary factor attracting thousands of migrants who streamed westward in ever greater numbers. It was land they wanted.

The farms of western New York, Ohio, and southern Indiana fronted forested land covering a rolling countryside. Arriving settlers were used to dealing with trees. They cut and burned out forests to clear land for planting. At first they planted corn around the tree stumps, built log cabin homes, and put down split-rail fences. Though they required three times as much fencing wood as traditional fences, split-rail fences could be put in three times as fast, and these farms abounded with extra wood. Pigs, which could hold their own against predators, were set free to forage acorns, nuts, and roots in the forests, while cattle, sheep, and horses were kept closer to home. With ten years of hard labor, a farm family could put eighty acres of land into full production. Cash crops soon dominated the best fields, with wheat, barley, oats, and rye replacing corn. Pastures and haying fields were set off for use by draft animals and beef cattle destined for distant dinner tables. Farmers brought in their pigs to fenced sties and fed them surplus corn and meal scraps.

Land in western New York, Ohio, and southern Indiana was flatter, richer, and less rocky than that in New England. It produced more per acre. Once the steamboats and canals offered a means to transport their goods to market, farmers began looking for ways to increase production. A farmer concerned with feeding a family and selling his surplus was also concerned about holding down his costs. As much as possible he made his own tools, frequently from the wood of the surrounding forests. Costs could also be kept down through household industry. Women made blankets, clothes, candles, and soap, while men built rakes, shovels, plows, and harnesses. But all these tasks took time—time farmers preferred to spend plowing fields.

Plows themselves represented a serious concern. A handmade wooden mould-board plow, even with an iron blade or rake, was clearly not as useful as a metal one. Farmers whose land produced twenty to forty bushels of wheat an acre did not want to spend time making, breaking, and fixing plows. In 1819 Jethro Wood patented a cast-iron plow, which, when pulled by a mule or powerful draft horse, could absorb more pressure and plow through fields cleared of tree stumps significantly faster than the old wooden, iron-tipped plows. With this new plow and other innovations, farmers turned increasingly to iron manufactured machinery to increase their productivity in growing market crops. By the 1830s, nearly half the nation's iron production was being fashioned into agricultural implements, and the old Northwest had the greatest concentration of those implements. By midcentury, while a typical American farm had fifty dollars worth of equipment per worker on the farm, farmers in the old Northwest had 20 percent more invested in equipment (North 1968).

Ohio farmers' success put additional pressure on the farmers who hung on in the East. In 1836 New York's grain production was surpassed by Ohio's grain crop, 1.2 million bushels of which were unloaded in Buffalo. The Erie Canal brought midwestern grain to eastern markets at prices that even the thriftiest, hardest-working Yankee hill farmer could not match. Except for those who worked the rich alluvial river valleys or who found convenient markets in nearby cities, eastern farmers could not achieve the efficiency or per-acre productivity that their midwestern counterparts enjoyed. As their debts rose and the land eroded, eastern farmers lost first their children, who moved west or to the growing eastern cities, and then, increasingly, their farms as well. Some survived by looking to new products and new markets, particularly milk, cheese, butter, eggs, and wool, while others abandoned their farms and headed west.

PRAIRIE WEST

By the 1830s, almost all of Ohio's arable land was already being farmed, so migrants continued to trek west. To the southwest along the White and Wabash rivers were rich, well-drained forestlands. Into Illinois, the Mississippi River valley offered rich soil covered in oak and hickory. But to the north, the settlers found a very different terrain. Here was land with no trees to clear, but the soil was too tough to plow. In addition to concerns about the soil's intransigence, many migrants from the tree-rich East worried that soil that didn't grow trees might prove too weak to grow good crops.

But, in fact, under the tall prairie grasses lay rich soil, indeed. The deep and tangled mess of grass roots held the soil firm during the spring snowmelts and heavy summer storms. Periodic fires broke down vegetable matter and added nutrients to the soil. Decaying grasses provided a bed for new growth and nurtured colonies of worms. Microbes ate food, digested it, and died, all the while adding more nutrients to the soil, which were frequently trapped in the thick bed of humus and grass roots at the soil's surface. The white Americans who first walked onto this prairie stood atop eight feet of some of the world's richest topsoil.

Deciding that the prairie soil would grow crops as easily as six-foot-tall prairie grass, the farmers had to figure out how to break up the thick mat of roots that locked the soil in place. Prairie soil had few rocks and lacked decaying tree roots. The plow-cleaning process in the East, where stones and tree roots cleared the older mortarboard plow and the more modern cast-iron plow of accumulated dirt as the plow moved through the ground, did not occur in the prairie soil. Cleared forestland lacked the thick mat of roots of the prairie soil and as a result was easily cut and turned by a mule or draft horse pulling a cast-iron plow. On the prairie, however, the old wooden plow with an iron blade could not handle the pressure needed to break the ground. Between the plow and the sod, the plow usually broke first. By contrast, the cast-iron plow was strong enough to take the pressure, but it couldn't hold a sharp edge well enough to effectively cut through the prairie roots. When it did, dirt clung to its rough surface, adding weight to the already heavy plow. Even teams of horses and mules had difficulty pulling cast-iron plows through the thick, root-entangled sod. Even without trees to cut down, to clear a field of prairie sod proved to be quite literally a tough row to hoe.

Eastern plows failed to break the thick mat of prairie sod. Local farmers and blacksmiths attempted to jerry-rig reinforced iron "breaking" plows to cut through the soil. These unwieldy plows with wheels on the front and a gigantic shear on the back attached to a heavy central beam were pulled by teams of five or six yokes of oxen or four strong horses. They often needed two strong men working the plow and horses or oxen. The plows were big and expensive, as were keeping teams of oxen and horses capable of breaking prairie sod. Many farmers looked to specialized custom breakers who had the machinery, knowledge, and livestock for the job.

In 1839 John Deere developed the steel plow. Lighter and stronger than the cast-iron plow, the steel plow could hold a sharper edge for cutting prairie sod. Deere's plow was called the self-cleaning plow because its smooth surface was also less likely to accumulate heavy dirt buildup that would weigh it down. Through a process of trial and error, prairie farmers learned that if they allowed their fields to be heavily grazed during the season and then plowed their fields in the early spring, they could, with a team of horses or mules, pull the steel plow through the tough prairie sod and get a crop off the land. John Deere's plow removed the final obstacle from farming the flat, treeless prairie. Instead of working ten years clearing fields to put a farm into production, farmers could now plant twenty or thirty acres of wheat within two years. And contrary to earlier concerns, crops grew prolifically in the newly plowed prairie soil. Corn and wheat were domesticated from grasses that thrived in open fields. In cool dry weather they grew quickly and stored carbohydrates in their seeds. Once the native grasses of the open prairie had been plowed under, these crops took over.

Though a vast open space, the prairie was not empty. The region around southern Lake Michigan was occupied by the Potawatomi, who hunted and trapped the prairies for birds and animals, fished the lake, and planted corn. French and Anglo traders and Indian-white families inhabited a small settlement where the Chicago River flowed into Lake Michigan. Natives called the place

Chicago for the smelly garlic that grew there. To the southwest lay the lands originally occupied by the Sac, Fox, and Kickapoo Indians. These lands had been "treatied" away to the U.S. government in 1804. In 1832 Black Hawk, a leader of the Indians then residing on the western side of the Mississippi, renounced the 1804 treaty, claiming that "land cannot be sold." He then led a group of his followers across the Mississippi into Illinois to reclaim the land that they believed was theirs, not as a possession, but for their perpetual use. The U.S. government had different notions of land use and possession and sent government troops and Illinois militiamen against Black Hawk, soundly defeating him in August 1832. Although they were not involved in Black Hawk's war, the Potawatomis to the north were also doomed by his defeat. The national government decided to take the lands around southwestern Lake Michigan. In the fall of 1833, government agents called the Potawatomis to a conference in Chicago. At that conference the U.S. government gave the Potawatomis a deal they could not refuse: sell the land or be forcibly removed. The Potawatomis left the prairie of southern Wisconsin and northern Illinois. From the Anglo perspective this was a propitious moment—within six years John Deere's plow would appear on the western prairie, and wheat and corn would begin to replace tall grass, bluestem, gram, and side oats as the dominant flora of the land.

As farmers grew more crops, so did markets grow for American grain. The east coast cities were growing at prodigious rates. New York grew from 135,000 people in 1820 to 349,000 in 1840, and had over 1,072,000 souls by 1860. Philadelphia grew from 135,000 in 1820 to 258,000 in 1840, and had almost 600,000 residents by 1860. Baltimore jumped from 63,000 in 1820 to over 102,000 in 1840, and counted over 212,000 by the eve of the Civil War (U.S. Bureau of the Census). The residents of all these cities ate bread made from American wheat. With the development of short-fiber cotton, the American deep South shifted to a cotton monoculture. Cotton prices drove alternative land uses to the margin, and as a result the South began importing food from the Midwest. Farmers from Ohio, Indiana, and southern Illinois sent corn and pork staples down the Ohio and Mississippi Rivers to the new markets of the cotton regions of Mississippi, Louisiana, and Arkansas. Corn production in the southern part of the state made Illinois the number one corn producer in the Union, and feeding local corn to pigs helped Illinois and Indiana lead the nation in hogs. The West Indies continued to import American foodstuffs as well, and although Britain's blockade tried to keep out American grain, with the conclusion of the Napoleonic Wars, American wheat was also finding its way into European markets.

While the open prairie eliminated the arduous task of clearing fields before plowing, it also deprived farmers of trees as a ready source of wood. Historically, Americans had been blessed with trees. Trees were among the first sights Euro-

White pine forest, 1910. (Harvard Forest Dioramas/Fisher Museum/Harvard Forest, Petersham, MA)

peans settlers noticed and over which they rejoiced. Trees on new farmsteads provided settlers with materials to build their farms and pay off debt. Ashes from burnt groves of trees could be boiled down to potash to be exchanged at the local country store. Cordwood could be sold at the riverbanks to steamships for fuel or rafted to downriver customers short of firewood. Chestnut, oak, and hickory were split for fencing or building farmhouses, barns, and sheds. Wooded lots provided much of the building material for fashioning farm tools. The barrels, buckets, and baskets that contained goods and held the water that washed bodies and clothes were all made from wood. Americans made furniture of walnut, maple, poplar, and wild cherry trees. Houses were shingled in cedar or oak, or enclosed with cherry or maple clapwood. Tools, waterwheels, clocks, gears, yokes, and stocks were wooden. Along the eastern seaboard and west to the prairies, sawmills worked continuously through the late winter and early spring cutting over 1,000 feet of lumber a day. Americans used wood profligately. Early saws were wide with spread teeth, leaving almost a third of the wood discarded as sawdust and scrap. Even with the new English band saw with its thin blade, American sawmills daily cast off over 1.5 feet of sawdust for every 100 feet cut. Americans turned this abundance of wood to their geopolitical advantage. At the time of the Revolution, one-third of British ships were built of American wood, and seventy-five years later American wooden ships dominated the seas. The famous clipper ships were floating testaments to the vitality of American wood for shipbuilding. The broad, strong whaling boats, like those that Melville's famous fictional character Ahab sailed out of New Bedford to confront the leviathans of the sea and then hold massive boiling cauldrons of whale blubber and oil, were framed of hefty oak and chestnut beams. Wood heated homes, cooked meals, powered steamboats, and fueled furnaces and foundries. America was a woodworking as well as a wood-consuming society.

When these wood-working, wood-consuming Americans spilled out onto the western prairie, they discovered no trees there. Homes, fences, sheds, and barns needed wood to be built. To find the lumber they needed to make homes and build their futures on the prairie, the settlers who had marched west had to turn back east. The wood they needed came from the forests of the Great Lakes states, particularly the mixed hardwood and white pine forests of northern Michigan and Wisconsin. But to get that wood they had to pay for it. Prairie farmers had no surplus of trees to convert into cash or credit. Their deficit of trees meant they had to buy more than their eastern ancestors. And that usually meant borrowing more.

My parents moved to northern Wisconsin in the late 1950s when I was growing up. As an adult I lived in southern Michigan and would periodically return to Wisconsin to visit. Had I made the trip in the mid-nineteenth century, I would have first crossed settled farmland in southeastern Michigan. From there west, the soil becomes sandy and dry, and the land would have been less populated. Reaching the shore of Lake Michigan, I would have been confronted with miles and miles of dunes. To the north of this route the land was sandy, the summers drier and shorter, and the winters colder. Rounding Lake Michigan and heading north, I would have passed prairie land converted to farms.

Farther north, prairie farms gave way to farms interspersed among forests of elm, basswood, sugar maple, oak, and hickory. During the centuries following the glaciers' retreat, the prairie and the forests shifted. Prairie grasses store their energy in their roots. Fires and droughts that kill off the aboveground growth seldom kill these grasses' roots. New growth quickly sprouts up again following fires or with new rains. The grasses' vigorous root systems, years of dry weather, spring grass fires, and abundant rabbit populations favored the prairie over the forests, such that the prairie pushed into the northern reaches of Wisconsin and Michigan, while wet weather and periodic rabbit die-offs in turn encouraged the forests to push south. Even in the prairie, groves of burr oaks, whose thick bark enabled them to survive the prairie grass fires, stood as holdouts against the reign of grass. In the creek beds and marshes grew tamaracks and swamp maples.

As I traveled on into northern Wisconsin, oak, hickory, elm, and basswood would have given way to forests of maple, hemlock, and yellow birch. On the sandy, well-drained soils, groves of white pine grew mixed with the hardwoods. These trees towered above their hardwood neighbors. In some sections of northern Wisconsin and Michigan white pines dominated the landscape for miles. A century later, my parents lived on the southern edge of this expanse in an area of struggling dairy farms where the soil was not as rich as in the south and the growing season was short. Their house sat on a piece of property that in the middle of the nineteenth century bordered on lumber country, a place where lumberjacks and Paul Bunyan legends flourished.

The early prairie farmers needed lumber, and white pine was light but strong wood. It was easy to work with and, cut in uniform lengths, was a versatile building material. With the coming of winter, lumbermen gathered up their axes and winter gear and headed for the forests of the north to look for white pine. Mature pine trees reached between 50 and 100 feet in height with bases 4 to 6 feet in diameter. Sawmills cut through the soft wood, and because it floated, transportation posed no great logistical challenge. Compared with the other Great Lakes, Michigan is long and narrow. Streams and rivers—notably the Muskegon and Manistee in Michigan and the Menominee, Peshigo, and Oconto in Wisconsin—flow out of the north into the lake or its bays. In the period before the Civil War, winter loggers formed working companies of ten to fifteen men. During the rest of the year, many of the loggers were farmers. But in the winter they went out into the woods because the cut trees could be chained and hauled over the frozen snow to waiting sleds. Stackers loaded the wood onto the sleds that were then pulled by oxen over icy logging roads, down to the nearest frozen river or stream. Men then piled the logs in stacks beside the river and waited for spring. With the melting of snow, river and stream water rose, allowing the lumbermen to push the logs into the rushing water to float them down to Lake Michigan. In the lumber district around the lake, hundreds of small sawmills cut wood throughout the spring, summer, and fall. Cut wood was then loaded onto steamboats bound for Milwaukee or Chicago, where it would be sold to waiting farmers from the prairie.

Demand for white pine drove an avaricious lumber industry. By the 1870s, the easily accessible pine forests of northern Michigan were already overcut, and the lumber barons began building logging railroads, as well as moving men and equipment, west into northern Wisconsin and Minnesota. Lumbermen preferred the biggest and most marketable trees. They left behind stumps, branches, and broken trees. Poor farmers willing to take up cheap, marginal land with a limited growing season bought the land from the lumber companies or from the state or county if the land had been forfeited for failure to pay taxes. In attempting to

Loggers with a train load of white pine logs. Oceana County, Michigan, c.1900. (Corbis)

burn off the waste brush, they set fires that quickly rushed out of control. The Black Year, 1871, gained its name for these horrific fires that swept through the north country. These fires that followed massive logging continued throughout the rest of the century.

Americans had always consumed a lot of trees. In the 1830s, Americans used over 1.6 billion board feet of lumber each year. By the end of the 1860s, Americans were using up over 12.8 billion board feet of lumber each year. Eastern farmers consumed wood from their own land. Houses and barns were framed in heavy beams that were cut and notched to fit together. Hoisting the beams into place was difficult, but getting the beams to the building site was not, because the trees that supplied the beams grew on site. The first white settlers in western New York, Ohio, and Indiana built their original homes with logs cut from their farms. Once established, these early cabins may have been covered over with cut clapboards, but many farmhouses began with log construction.

Prairie farmers were confronted with having to import their building lumber from afar. In 1833 Chicago builders developed the so-called balloon-frame innovation in housing and barn construction. This new means of construction saved time and lumber. Instead of using heavy timbers, which would have been prohibitively expensive to ship to the prairie farms, the balloon-frame house was built with standard-cut framing lumber. Using 2 × 4-, 2 × 6-, 2 × 8-, 2 × 12-, and 1 × 6-inch cut wood, builders nailed the sides into self-contained units, with studs spaced every 18 inches apart. These sides were then nailed together. The house gained its strength not from heavy timbers, but from being linked together on its sides. Balloon-frame construction not only reduced the labor and time needed to build a house, it also meant that wood could be cut in specified units and shipped out to waiting farmers. Farmers could order so many 2 × 4s, 1 × 6s, or 2 × 8s and have them shipped by barge to the nearest port. The farmer could then pick up the lumber in a wagon and haul it to his house or barn site. Lake Michigan linked the farmers of the prairie to the forests of the north country. Along the way, the forests were transformed into logs and then lumber before they became homes, sheds, and barns. Balloon-frame construction and standardized cuts of wood contributed to the link between the forests of the northern Midwest and the prairie as much as did sawmills and lumberjacks. Money, of course, was also made in the process. Lumberjacks risked their lives for wages. Wages were paid by companies that made money by using the lumberjacks' labor to satisfy the prairie farmers' need for lumber. Sawmills cut lumber and steamship companies shipped it because prairie farmers were willing to pay to bring wood to the treeless prairie. They paid for that wood not only because they needed it, but also because they believed the prairie soil would turn up gold dollars once the harvest was in. But bringing in the harvest was no sure thing. While eastern farmers had cash *in* lumber, prairie farmers needed cash *for* lumber. They borrowed to get that cash, and that forced them to focus even more intently on a market for their crops. Unlike earlier settlers in the East who grew for themselves and took their surplus to the market, these prairie farmers from the beginning grew more for the market than for themselves.

Homes and barns were not the only lumber concerns for the prairie farmer. For eastern farmers, fencing, like housing, was constructed from materials at hand: stones or wooden split rails. Prairie farmers had less to work with. An Illinois farmer was lucky enough to find a stand of timber only ten miles from his farm, but it took him months of hauling the cut and split rails from the timber stand to his field before he could build a seven-rail fence around a forty-acre field. But he considered himself lucky to be within ten miles of a stand of trees. Other prairie farmers were not so lucky and had to buy lumber for fencing. The problem of fencing continued to plague prairie farmers throughout the early years of settlement. The pages of the *Prairie Farmer* newspaper contained as many complaints and suggestions about fencing as they did about breaking prairie sod.

Farmers planted corn because it was easy to plant and to harvest. They could feed corn to their livestock, make it into whiskey, or grind it into cornmeal for

Moving West: The Interior 99

The McCormick binding reaper in 1878. (Hulton Archive/Getty Images)

biscuits and bread. Into the late 1840s, midwestern farmers raised over 500 million bushels of corn per year, compared to just 100 million bushels of wheat (Bogue 1963, 216–233). Prairie farmers also raised pigs and cattle and drove them to markets. But the greatest demand was for wheat. Americans, and others who would buy American grain, wanted to eat bread made from wheat flour. A farmer's wheat crops paid for the farm and the wood for the farm buildings. Though it commanded a good price in the market, wheat was a finicky crop. Harvesting wheat was particularly tricky. Corn, which stayed on its heavy and hearty stalks until late fall and protected its kernels from damage in husks, could be left in the fields to dry and be harvested over time. Wheat was far more delicate. If left too long in the fields, mature wheat plants had a tendency to fall to the ground, especially if battered by heavy winds or rain, or they dropped their seeds before the fields could be harvested. If harvested too early or wet, wheat tended to rot. Once the wheat crop was ready, a farmer needed to get his crop in fast.

In the 1840s, Cyrus McCormick developed the horse-drawn mechanical harvester with a self-rake reaper. A farmer using a McCormick harvester equipped with a self-rake reaper pulled by a team of horses could harvest twelve acres a day. Using the traditional scythe with a cradle frame to catch the wheat, a farmer needed more than six days to harvest the same twelve acres. Owning a

McCormick harvester allowed farmers to plant more wheat and worry less that it would drop in the fields before the family could reap it. Buying a McCormick reaper was expensive. Farmers either had to borrow more money, join together to share a reaper, or hire someone with a reaper. All these options tied the farmer more closely to the market place.

Balloon-frame construction and standardized lumber cuts, John Deere's plow and Cyrus McCormick's harvester, the forests of northern Michigan and Wisconsin, the lumberjacks and sawmillers, rivers and lakes, and the world's appetite for flour and the canals that moved it all contributed to the explosion of settlement on the prairie. If cotton was king in the South, wheat ruled the prairie. But in the 1840s, midwestern farmers were just beginning to plant wheat on the prairie.

On our imagined mid-nineteenth-century journey from Michigan to northwestern Wisconsin, the prairie would likely appear as an endless sea of grass. Upon closer examination, we might notice islands of trees meandering southward through this sea. These trees grew along the riverbanks breaking up the prairie. The prairie itself was not an endless expanse of grass. Woods and groves of burr oak frequently arose amidst the grassland prairies, particularly at the edges between forest and grass. Most farm families sought land in the river and stream valleys where there was access to the oak, maple, and basswood needed to build their fences, homes, and farm structures, as well as a means of floating their goods to market. In the 1840s, new arrivals first settled and farmed land with both open prairie and forests, while the previously forested lands of Ohio and south-central Indiana were pumping out wheat. Where there were no natural waterways, farmers in the established western states lobbied their legislatures to build them. The Miami and Erie Canal and the Wabash and Erie Canal linked the grain-producing areas of Ohio and Indiana to Lake Erie and the Erie Canal. Prairie lands to the north and west of Chicago were still waving in shoulder-length grasses, while the fields of Ohio, Indiana, southern Illinois, and Missouri were already dominated by amber waves of grain. In the 1830s, the rich farmland of southeastern Michigan began to feel the bite of Wood's cast-iron plow. Michigan wheat farmers could easily move their grain down the Huron or Raisen rivers to Lake Erie. As Michigan entered the Union in 1837, its farmers continued to spread west across the southern third of the state, cutting down trees and turning up soil for corn and wheat.

By 1840, 4 million bushels of midwestern wheat a year were unloaded in Buffalo. Wheat soon overwhelmed the Buffalo stevedores. On a good day, Buffalo dockworkers could unload 2,000 bushels of grain, or roughly one boatload. On one day in 1842, forty boats carrying 100,000 bushels of wheat attempted to unload in Buffalo. Responding to the crisis, the city's leaders fell in behind Joseph

Dart's plan to build a steam-powered bucket-and-pulley system that would move wheat from the lake boats to waiting elevators. Dart's elevator was soon moving 2,000 bushels of wheat an hour (Geldman 1993).

In 1846, twenty-three years after David Ricardo's death, his vision of free trade was finally realized when England eliminated the high tariffs protecting British grain producers by keeping out American grain. With the so-called corn laws abolished, American grain flooded into England, providing bread to the country's growing working class. During the Crimean War, when supplies of wheat from eastern Europe, the Ukraine, and Russia sharply fell, American wheat picked up the slack, flowing copiously into English and western European ports. The amount of wheat moving through Buffalo increased from 5.5 million bushels in 1842 to 22 million in 1852. The northern lake system and Erie Canal carried the heaviest burden of wheat travel, but similar increases were also experienced through the Chicago–St. Louis–New Orleans network.

In 1849 Daniel Muir left Scotland with three of his children, two boys and a girl (leaving behind his wife and four other children to join him later) and moved to the wilds of central Wisconsin. On their way they joined other migrants to the west—immigrants from Germany, Norway, and Sweden fleeing agricultural depression and land shortages; immigrants from northern Ireland, England, and their neighbors from Scotland; and migrants from the worn-out New England farms. Daniel Muir picked Wisconsin because a grain merchant in Buffalo told him that most of the wheat he handled came from Wisconsin. Based on that information, the Muir family took steamboats west to Milwaukee. There they hitched a ride from a farmer who had just delivered wheat to the Milwaukee market and traveled northwest over the Wisconsin prairie to Kingston on the edge of a great glacier meadow. When the glaciers were moving back across the American heartland, a huge chunk of ice lodged in the outlet gorge of the Wisconsin River. Glacial melt flowed into the river. Blocked at the Baraboo Range, the water backed up into a huge lake, then cut a new exit east of the range. When the water drained out of the lake, it left behind a giant wetland. Bogs of sphagnum moss absorbed water and became floating islands. Sedge, leatherleaf, tamarack, and spruce moved in. Marshes and meadows formed, while trees—protected from fires by the wetness around them—grew along the edges of the marshes. It was to this land that Daniel Muir took his family, including his young son John, who would become by the end of the century one of America's most famous naturalist writers and the founder of the Sierra Club.

Daniel Muir did not leave Scotland because he was poor or because he hungered to own land. The Muirs were middling town folk. Daniel Muir left Scotland to escape religious orthodoxy. He wanted to settle where he could practice his stern religion freely and without official meddling or paying taxes to support a church in which he did not believe. On his way to the New World, he purchased goods he believed his family would need: a traditional beam scale with a complete set of cast-iron counterweights, iron wedges, carpenter's tools, and other equipment any sensible townsman would expect to need in the wilderness. The family took the train from Dunbar Castle to Glasgow, where they booked passage to America. In Buffalo, Daniel bought more goods for his new life: a cast-iron stove, pots and pans, and a scythe and cradle for cutting wheat. Once in Wisconsin, he had to pay thirty dollars to a farmer to load these goods onto his wagon for the trip across the prairie. Far from being poor refugees, the Muirs were, in fact, willing to spend a substantial amount of money for the opportunity to make their way by performing backbreaking labor in this new land.

Daniel Muir traveled westward to Kingston, Wisconsin, where he went to the land agent and arranged to look at land. He eventually found a quarter section of a lot (a measurement from Jefferson's original land ordinance), or 160 acres, with woods and meadows beside a lake and put down his money. Like many new arrivals, Muir wanted land that was both open and wooded. Coming back to Milwaukee, Daniel and his two sons (with the paid help of the farmer and his wagon) loaded their goods and traveled northwest to the great glacial marsh of central Wisconsin where the prairie was broken by a mixed flora of prairie grasses and trees. It was springtime, so Daniel put his two sons, John and David, to work clearing land. They burned brush and cleared fields. Muir purchased a pair of oxen, two cows, a pony, a dog, and a pregnant sow he let forage in the woods. The father and two sons worked hard to plant corn, pumpkins, and later, wheat. The corn and pumpkins fed the family and were used to supplement the meadow hay fed to the animals. Once the family had become established, Muir bought cut lumber and built a frame house, turning the original shanty into a shed for the pony. He then brought over the rest of his family from Scotland.

Within a few years, the Muir homestead had brought order to a wilderness. The framed, two-story farmhouse stood on a hill overlooking fenced pastures and ordered rows of corn. A sturdy barn provided protection the animals needed through the hard Wisconsin winters—except for the unfortunate pony the Muir boys foolishly tried to wash in the middle of winter, leaving the poor animal soaking wet and hanging in icicles. A crib protected the corn and an orchard provided food and drink for the family and food for the livestock.

If Daniel Muir was not typical of a mid-nineteenth-century Wisconsin farmer, he was typical, if not particularly successful, in his farming. The Muir farm grew wheat to sell and corn and potatoes to eat, feed to animals, and sell. It raised cattle to work and sell, cows to milk, and chickens to eat, and grew vegetables for family meals.

Before the arrival of pioneer farmers, hungry rabbits and grass fires had sustained the prairie and eliminated the farmer's need to cut trees before planting fields. But the fires also threatened farm buildings. (Ironically it was a grass fire that took the life of one of America's other great naturalists, Aldo Leopold, in the very same county where John Muir grew up.) Plowed fields reduced the fire threat, as did cut clearings and grazed-over pastures. Prairie fires, like prairie grasses, retreated before the western advance of farms. Fire's departure allowed trees to grow on former prairie lands. Desiring wood for fuel, farmers encouraged tree growth. Within a generation land that had been open prairie was dotted not only with farms but with woodlots as well.

Farmers in Ohio, Michigan, and western New York needed markets for their farms to succeed, and so did Wisconsin farmers like Daniel Muir. In 1848 a canal was completed linking Chicago and the Illinois River, which then flowed into the Mississippi. This waterway not only connected Chicago to the wharves of St. Louis and New Orleans; it also linked Chicago to the prairie that lay to the southwest of the city. Prairie farmers looking to sell wheat, corn, or hogs could now move their goods northeast to Chicago or southwest to the market towns on the Mississippi. Chicago offered the advantage of cheap, lake steamer connections to Buffalo and through the Erie Canal to New York City. Chicago also had the largest emporium in the region. Steamboats emptied of wheat or corn in Buffalo were filled with goods from around the world and returned to Chicago loaded with merchandise to stock the shelves of Chicago's stores. The completion of the Illinois and Michigan Canal encouraged still more investment in transportation. By the 1850s, railroads spread out from Chicago like a giant spider web, funneling hundreds of tons of grain a year from the prairies to the city's waiting elevators. Out from the city on those same lines flowed pots, pans, stoves, plows, axes, shovels, and McCormick harvesters, as well as cut lumber ready to be hammered into homes, barns, sheds, and fences. Throughout the 1850s, Indiana and Illinois together added more miles of track than the five mid-Atlantic states.

Wheat flowing into Chicago was gathered into elevators by steam-powered lifters developed in 1842. Previously, farmers had loaded their grain in sacks to be stacked in the holds of steamboats with each sack bearing the identity of the farmer. The railroads and elevators encouraged farmers to forgo the sack and send wagons of grain to the railroad terminals. Grains from different farms would be sorted by grade and type—spring, white winter, and red winter—and mixed together, with the farmer receiving credit for the volume and grade he sent into the flow of grain moving into elevators. In 1848 Chicago merchants established the Chicago Board of Trade to rationalize the grading, sorting, storing, and selling of wheat. Increasingly, wheat was less a product of an individual

The skyline of Chicago behind a railyard, 1909. (Corbis)

farmer and more an abstract commodity to be bought and sold. By the 1860s, a futures market in grain developed in Chicago. Traders began buying and selling wheat at a certain price even before the wheat was grown.

Meanwhile, pioneer farmers soon discovered that even fresh prairie soil wore out. Within four or five years after they opened up new fields, the farmers' wheat harvests began to fall from twenty to twenty-five bushels an acre per year to less than ten. Initially, farmers switched these fields to corn, but corn also generated low productivity on overworked fields. Farmers then turned to green fertilizing fields, a practice that had arisen in England and among contemporary eastern farmers. Farmers planted English clover in exhausted fields and then plowed under the clover. Both corn and wheat then prospered on these fields. Wheat again went to market, while the corn was fed to cattle and hogs, fattening them up for the growing Chicago market.

Wheat was the farmers' cash crop, but its cultivation was difficult. Weather played particular havoc with wheat. In the late 1840s and early 1850s, a succession of cold winters with little snow, as well as attacks of rust and insects, dramatically reduced the harvests. In response, farmers switched to spring wheat and corn. Corn was a more reliable crop but it was a tougher sell in the market. Corn prices were lower than wheat prices. Corn's low price encouraged farmers to feed it to their animals to sustain and fatten them for the market. Fattened livestock did command a healthy price in city markets. Although pork and beef had always been part of the farmer's market crop, the growing populations of east coast cities encouraged a far greater focus on meat production. Initially, hogs comprised the majority of market-oriented meat production. Pigs had always been the farmer's favorite animal. They foraged for themselves in the woods. A sow reached bearing age at one year. She would drop several piglets and could become pregnant again before a new year was out. Pigs also put on weight fast, and the meat could be salted, smoked, and stored. Once established, most pioneer

Hog-slaughtering and pork-packing in Cincinnati. (Corbis)

farmers acquired hogs. After initial settlement, farmers began penning the pigs in fenced sties and feeding them surplus corn and scraps. In the fall of the year, the pigs would be hauled to the nearest market for sale. Until the Civil War, that market for western farmers was Cincinnati. Each fall thousands of pigs were floated down river on barges to the city's docks or were driven in from the surrounding countryside. In Cincinnati they were slaughtered by the city's butchers and packed and shipped downriver to customers in the south or to boats waiting at New Orleans wharves. Cincinnati proudly proclaimed itself "Porkopolis," as more and more of its citizens found employment "disassembling" hogs in an assembly-line process developed long before Henry Ford was even born. Corn may not have commanded much respect in the marketplace, but pigs in the form of bacon, pickled pork, ham, and sausage did.

Just as Cincinnati became a hub for hog farmers of the southern Midwest, Chicago soon followed suit for farmers farther north. Like most of their neighbors, the Muirs initially hauled wheat by wagon over the rugged prairie roads to Milwaukee, from there to be shipped to eastern markets. The railroads' arrival helped move wheat more easily to market, but hard weather and pests plagued the wheat crop. In addition to growing wheat, these farmers also grew corn and raised hogs. Rich prairie soil grew corn as readily as it grew wheat. Faced with the challenges of growing wheat and the low prices corn fetched at market, the Muirs and their neighbors increasingly turned to feeding corn to their hogs and shipping the hogs to Chicago. The railroads brought hogs and cattle to Chicago as well as wheat. Soon Chicagoans found themselves "disassembling" hogs with the same efficiency as their competitors along the Ohio River. With an expanding rail network reaching west, south, north, and east, Chicago became the destination for more and more hogs.

The Civil War was a boon to the expanding pork industry. The southern market's evaporation initially hit Cincinnati hard. At the same time, the mustering of some 2 million Union soldiers created a huge demand for meat, particularly salted or smoked pork. Before Cincinnati's butchers could redirect their attention to this ballooning market, Chicago meat packers were already filling the breach. Hogs by the thousands poured into Chicago en route to the stomachs of Union soldiers. Soon Chicago surpassed Cincinnati as the world's largest meat packer as the city's stockyards became overwhelmed with squealing pigs. In 1864, confronted with the confusion of hogs as well as cattle, horses, and sheep, the Chicago Pork Packers' Association coordinated with influential citizens to develop a plan to build a gigantic common stockyard southwest of the city. Within two years, the Union Stock Yards were built and the torrent of animals moving to Chicago grew apace. For hogs, the yards were the first step toward "disassembly." For beef—which Americans preferred fresh—it was the beginning of the journey along the rails to east coast butchers. The opening of Chicago's Union Stock Yards encouraged midwestern farmers to look to their livestock as well as their wheat fields as a source of cash. These farmers had always kept livestock. Animals helped work the farm; oxen—and later horses—helped clear fields, pulled plows, cultivators (shallow plows that ran between the rows of corn and turned the soil to keep down the weeds—a welcome replacement for hand hoeing), and McCormick reapers. The fall slaughter of pigs and cattle fed the family. By the 1850s, prairie farmers were fencing in feedlots to concentrate the animals. Feedlot cattle gained weight faster than those left to wander through prairie pastures. Savvy farmers dumped bushels of corn into feedlot troughs for cattle and let hogs scavenge the waste. Corn production for animal feed rose dramatically. By the 1870s, over a million hogs a year were dying in what the poet Carl Sandburg called the "Hog Butcher for the world." Cattle also entered the market flow and the stockyards themselves became a market for farmer's corn and hay.

Chicago also became a major market center for the dwindling natural fauna of the region. In 1873, Chicago dealers bought 600,000 prairie chickens at $3.25 a dozen (Leopold 1949, 15). Throughout the 1870s and early 1880s, midwestern

Refrigerator car of Armour and Co., Chicago. (Library of Congress)

prairie chickens joined passenger pigeons and ducks on dinner tables from Chicago to New York, but each passing year saw fewer roaming the prairie to be hunted. In 1872 the last Wisconsin turkey was killed, followed in 1899 by Wisconsin's last passenger pigeon. The huge flocks of passenger pigeons collapsed within a few decades. In 1878 there was a huge nesting of pigeons in Michigan. But this was the last such massing of these birds that had been such a part of the American diet for two centuries. Passenger pigeons needed places to nest and large numbers to survive, because each female bird laid only one egg. Habitat destruction and overkilling drove down the size of the flocks to a point where the birds simply could not survive. But killing passenger pigeons, ducks, and prairie chickens was a game for boys, not the activity that built agricultural prosperity. Brought to the Chicago market, they may have made up an important source of income for the individual hunter, but these birds were never substantial enough to become a significant factor in the developing regional economy. Buffalo also lived on the prairie at the time, but were not a significant factor east of the Mississippi River.

After Gustavus Swift perfected the refrigerated railroad car in the late 1870s, the beef-butchering industry in Chicago expanded, as the meat could now be shipped to eastern markets already "dressed." For decades live cattle had been moving into Chicago's stockyards on their way east to the urban butchers waiting in New York, Philadelphia, or Boston. The development of the refrigerated railroad car meant that cattle could be slaughtered in Chicago and the slabs of beef shipped east. As Chicago's meat market expanded, prairie farmers joined the plains states' ranchers in raising cattle to be butchered and dressed by poorly paid workers in the Windy City's slaughterhouses.

To keep this meat fresh, Swift and later Armour, hired gangs of ice cutters to cut huge blocks of winter ice off northern lakes. The ice was stored in heavily straw-insulated warehouses and then loaded onto insulated railcars carrying the meat to eastern markets. Swift and Armour maintained a series of ice storage units along the rail lines between Chicago and the final market destinations. Lakes cut out by retreating glaciers over 12,000 years prior now provided cold for America's most innovative industry. Chicago's workers, whether poorly paid meat packers or comparatively well off railroaders, had growing families.

By 1890, Chicago had surpassed Philadelphia as the nation's second largest city, with over a million people within its metropolitan reach. For the farmers of the surrounding area, that meant a greater market of mouths to feed. And feeding those mouths became more important for the financial survival of the surrounding farmers. By the end of the 1870s, farmers of the central Midwest had given up trying to raise wheat. Chinch bugs, grubs, and that ever-problematic rust persuaded farmers to finally abandon their wheat fields. Cheaper wheat from the virgin prairies farther west ended wheat's half-century reign as the region's principal cash crop. Feedlot hogs, cattle, and in Wisconsin, Ohio, Indiana, and southern Michigan vegetables, orchard fruits, milk, and dairy products became the new focus of farm activity. In 1890 University of Wisconsin Professor Stephen Babcock invented the Babcock Milk Tester for determining the percent of butterfat in milk. The test permitted the rapid and accurate grading of milk at markets and discouraged adulteration and thinning. The Babcock Milk Tester gave large-scale buyers confidence in the quality of milk delivered from specialized dairy farms. The test also allowed farmers to test individual cows' milk and helped them choose which cows to keep in production and to breed. By the end of the century, Wisconsin farmers were selecting and breeding specialized milking cows and shipping their dairy products not only to Chicago's growing market but, increasingly, with the advantage of refrigerated railroad cars, to urban markets even farther away.

Even city horses represented an important market for prairie farmers. Before the era of automobiles and bicycles, people and goods moved through city streets by horse power. Urban horses could not be pastured, and hay fields had long ago been appropriated for factories, rail yards, houses, and stockyards. Urban horse owners depended on oats for animal feed. By the 1860s and 1870s, oats for horse feed had come to represent a significant part of the flow of agricultural goods from the region's farms to its cities. The "land of stinking garlic," where Native Americans had for generations gathered wild rice, fished the lakes from canoes, planted small rotating fields of corn, and migrated according to the seasons, game availability, and the soil's productivity, became a land of calculating farmers. The wild and natural world that seventeenth-century Jesuit missionaries described was not as wild and natural as they had supposed. The peoples they came to convert had already left an imprint on the landscape. Clearings for fields, Indian trails, burnt woods, and extensive fishing camps all attested to human activity. The Jesuits' mistake is easily understood. The human imprint these native peoples left on the land was feathery compared to the later settlers' leaden footprints. But the Native Americans were not settlers. By moving from place to place, they allowed the land to recover much of its original form. Frequent movement also limited what they chose to carry. Carrying less meant having less; having less meant taking less, minimizing their impact on the land's flora and fauna. The settlers settled in. They went about transforming the land and leaving their heavily indented imprint. They adopted Indian grains and crops, particularly corn, squash, and pumpkins; hunted Indian game—prairie chickens, deer, and ducks; and fished Indian fish—salmon, trout, whitefish, and sturgeon. But they planted, hunted, and fished in a traditionally European manner. Their cornfields were vast open spaces, planted only in corn. They replanted fields trying to produce ever more grain from the same land. They hunted prairie chickens, deer, and ducks until there were no more to hunt. They fished the lakes until there were almost no fish left. The settlers did not look back with regret at the wasting of the natural world, however. Fewer deer, passenger pigeons, or prairie chickens meant losing less corn or wheat to marauding birds or wild hoofed animals.

Although some fishers lamented the elimination of salmon from Lake Ontario, others moved on to overfish the other Great Lakes species. Depleted fields were rejuvenated with green manuring, clover planting, and plowing under. Areas emptied of deer or buffalo were restocked with cows, pigs, and sheep. The wandering livestock that farmers brought into fenced pastures and corralled enclosures nibbled grasses down to their roots, cutting off the fuel for prairie fires and eventually eliminating the prairie grasses themselves. In their stead, farmers grew hearty English hay, shedding no tears for grasses that could not tolerate heavy grazing. Concentrating animals into pens concentrated wastes. Wastes in turn were returned to the fields as manure to stimulate future harvests. As people packed into the towns and cities, the demand for food grew, which encouraged more plowing, planting, pasturing, and penning. Fields of grain replaced fields of grasses. Trees began to grow around farmhouses and in adjoining woodlots. Streams' compositions changed as the manure runoff from fields increased nitrate concentrations, while woods cut from the streambeds exposed the streams to more sunlight, thereby increasing the streams' temperature. In transforming the land, the settlers settled and many prospered. Because they understood their prosperity to be a product of this remade landscape, the settlers celebrated as they continued to enlarge their range of settlement.

Immigrants from the British Isles, Germany, Holland, and Scandinavia joined Yankee families who, as John Muir reflected about his community in Wisconsin, were "all alike striking root and gripping the glacial drift soil as naturally as oak and hickory trees; happy and hopeful, establishing homes and making wider and wider fields in the hospitable wilderness. The axe and plough were kept very busy; cattle, horses, sheep, and pigs multiplied; barns and corn-cribs were filled up, and man and beast were well fed; a schoolhouse was built, which was used also for a church; and in a very short time the new country began to look like an old one." (Muir 1916, 169).

5

MILLS, TOWNS, AND CITIES

Increasingly by the second half of the nineteenth century, for farmers like the Coopers it was not the Caribbean or Europe that absorbed their goods, but the neighboring cities that needed milk, eggs, poultry, fruits, and vegetables. By 1870 the North American coast was dotted with substantial urban communities. New York's population exceeded 1,478,000; Philadelphia had over 670,000; and Boston had more than a quarter of a million residents (U.S. Bureau of the Census, 14). The growth of these urban giants created an integrated commercial interdependence, as goods moved from the interior to distant markets and manufactured goods were carried to interior markets. In the early national period, the goods moving through these ports came exclusively from the immediate or regional interior, while the manufactured goods traveling inland originated abroad. The goods that arrived from the interior (or, in the case of New England, from the sea) for the most part reflected the regional ecology. New England sent out ice, lumber, and fish, as well as dried beef, wheat, rye, and peas. The Hudson River valley with its rich fertile soils sent meat, wheat, and orchard products to New York, while the hinterlands surrounding Philadelphia produced apples, peas, milk, and eggs, as well as wheat, rye, and oats. By the second half of the nineteenth century, goods from the vast interior covering multiple ecosystems moved to the port centers and the manufactured goods flowing back in return were increasingly American made.

New York, like Boston and Philadelphia, had a deep, protected harbor that could accommodate oceangoing ships. In 1817, New York merchants created the Black Ball Line, establishing a regular schedule of packet ships across the Atlantic. The packet service gave New York a competitive edge in the transatlantic trade. That same year the New York legislature voted funds for the Erie Canal. The canal was an engineering marvel: forty feet wide, four feet deep, and containing eighty-four locks. New York City was finally connected to the Great Lakes from Albany on the Hudson to Buffalo on Lake Erie. The port of New York now had access to the resources not just of the Hudson River valley, but also of the rich fertile lands of upstate New York's Genesee valley and the rolling forestlands of western Pennsylvania, northern Ohio, Michigan, and Indiana. The cost of shipping goods east fell dramatically to one-twelfth the previ-

The Port of New York by Currier and Ives. Bird's eye view from the Battery, looking south. (Library of Congress)

ous rate. The average rate for moving a ton-mile of goods from Buffalo to New York City fell from nineteen cents in 1817 to between one and two cents by midcentury (Bass Warner 1972, 618). Farmers relocated to the cheap western land by the hundreds of thousands. Ohio, Indiana, and Illinois grew tenfold between 1810 and 1840. Improved axes, scythes, and Jethro Wood's cast-iron plow, developed in 1819, reduced plowing and harvest time and allowed a farmer to put far more land into cash crops, particularly wheat. Farmers in these newly opened regions sent their produce east to market, where they found ready buyers among overseas traders as well as hungry urban Americans. In 1846, England repealed the corn laws, thereby opening the British Isles to American grain. With this new infusion of cash, farmers bought eastern products; nearly one-half of the nation's iron production went into agricultural implements, most of which were destined for western farmers. Goods from Europe and eastern manufacturing establishments could now move cheaply and efficiently into the deep interior.

Canals were not the only means of moving goods cheaply across regions. By 1820 steamboats were navigating the Great Lakes system and the nation's major rivers. Steam-powered land transportation, already developed in England, became a reality in America when construction began on the Baltimore and Ohio Railroad in the late 1820s. By 1840 there were 3,000 miles of railroad and canals, mostly in the Northeast and Midwest. These canals and railroads turned the North into an integrated, commercially interdependent region.

New York City came to dominate the interregional trade and its port became the national port of entry. Goods from the newly accessible west flowed into the expanding entrepôt. Europeans looking to sell their goods in American markets shipped to New York. Young men, eager to make a name for themselves in business and commerce, flocked to the city, joining others just looking for work. And work was plentiful in the new metropolis. Warehouses, docks, homes, and offices needed to be built; goods needed to be loaded and moved; legal matters needed attention; loans needed to be secured and insurance issued.

A city of over a million people, New York could no longer feed itself. Local farms had long since given way to homes and factories. Beef cattle were brought in from afar to be slaughtered in Brooklyn abattoirs; fishers no longer sold directly to fishmongers, but sold their catch off their boats to wholesalers who carted the fish to the Fulton Fish Market. Central markets for vegetables emerged. Tons of food moved through elaborate systems of producers, jobbers, wholesalers, retailers, and hawkers before appearing in kitchens for final preparation.

These cities were also places of interdependences. Peter Cooper, a New York City grocer and manufacturer of sheep shears, realized that Henry Astor's slaughterhouse was throwing away as waste hooves that could be used to make glue. Locating his factory next to Astor's slaughterhouse, Cooper bought cheaply Astor's wastes and developed an inexpensive glue of consistent quality. The city's growing number of bookbinders, shoemakers, and furniture manufacturers were nearby customers for Cooper's glue.

Busy ports ran on human power, both directly and in support roles. In the increasingly interdependent and complex urban system, more and more work within the home was taken on by specialists—bakers, canners, tailors, and others. Large numbers of the city's denizens lived on their own, away from their families, and purchased goods that the household would have provided in an earlier era.

As local forests were cut for fields, building, and firewood, fuel for heating and cooking became a concern for urban communities like New York, Boston, and Philadelphia. Boston first began to harvest cordwood from the islands off its shore, and then from forests up and down the coast. In the winter, woodcutters went into the forest to cut trees and then haul them to the nearest river or brook, where the logs were cut into cordwood and stored to dry. In the fall, boats picked up the cordwood and transported it to Boston for sale. Because New York and Philadelphia were situated on major waterways, their cordwood could be harvested from the hills and mountains upriver and then floated down on barges and wood rafts. Despite America's abundant forests, as the nineteenth century wore on, cordwood became increasingly scarce and expensive. Benjamin Franklin's invention of a more efficient wood-burning stove reflected cordwood's increasing cost and limited supply. Awareness that forest cover protected upstream freshwater supplies needed by the growing cities and helped protect against flooding contributed to a growing debate over the need for cordwood and concern about forest cutting. As in Europe, growing shortages of cordwood encouraged homeowners to shift to coal for cooking and heating.

Anticipating this shift, anthracite coal mines northwest of Philadelphia were developed in the early nineteenth century. Anthracite coal was low-sulfur, hot-burning coal. Originally used to heat metal in wire works and to distill liquor, anthracite coal was initially avoided by blacksmiths, who preferred charcoal, and homeowners, who were concerned about the coal's ability to stay lit. In 1808, Jesse Fell, living in Wilkes-Barre, Pennsylvania, demonstrated that with a simple grate the coal could be easily burned without a bellows in any home. Abijah Smith, who opened the first coal mine in the Wyoming Valley of the Susquehanna River, loaded two boats with anthracite coal and a bundle of grates and poled them south to Lancaster County west of Philadelphia. Demonstrating along the way anthracite's ease of lighting and maintenance, he sold the load on both boats. Demand for coal grew, and by the 1840s, homeowners throughout the mid-Atlantic area were using hard coal for fuel. Demand grew so fast that canals were dug from the anthracite coalfields of northeastern Pennsylvania to Philadelphia, as well as New York City. To this day, those canals still cross the New Jersey and Pennsylvania countryside, functioning now as scenic walking and biking paths. Anthracite coal consumption jumped from 350 tons in 1820 to 425,000 tons in 1834, as mining towns sprung up like summer flowers along the coal seams of northeastern Pennsylvania (Brubaker 2002, 57).

Investors from Baltimore and Philadelphia bought land and began large-scale mining. Initially, shallow pit-mines were dug into the sides of hills, and miners went underground, chipping out coal and sending it out to be sized and cleaned. Buyers wanted coal in lumps the size of oranges. They were the easiest to light and burn. Coal smaller than a small lemon was discarded. Pieces larger than a grapefruit were broken up. Coal that came out of the mine was dumped on a large perforated cast-iron plate. Using sledgehammers, workers broke up the larger pieces on the plate, allowing the smaller ones to fall through the holes and be discarded as spoil or culm in growing mounds. In the 1840s, coal-breaking machines were developed to mechanize this process. The new machines dramatically increased the amount of coal that could be reduced into salable pieces. They also increased coal waste or spoil. While 6 percent of the coal mined was discarded as spoil with hand breaking, the mechanical breakers wasted over 15 percent of the mined coal. Mechanical breaking also involved more water to wash the coal. Water for coal washing was taken from local streams and rivers, particularly the Susquehanna and Lackawanna rivers. Demand for coal increased each year, and with increased demand came increased mining and growing mounds of spoil and mine waste. Culm dumps eroded into streams and rivers, joining the already polluted wash water flowing into the waterways of north-central Pennsylvania. By the second half of the nineteenth century, the banks and edges of the rivers and streams of the mining districts were layered in coal waste and the rivers ran black from waste and orange from mine sulfurs, but they did not run with fish.

New York, Boston, and Philadelphia grew exponentially during the first two-thirds of the nineteenth century. By midcentury they had become the Northeast's dominant ports, drawing in people from the whole transatlantic world. Although their points of origin and specific reasons for coming to America differed, all the new arrivals who built these cities came seeking the kinds of opportunities that did not exist back home. But the people who came to the cities from afar came with different histories.

Even before the Revolution, the port cities—especially New York and Philadelphia—drew immigrants from a variety of European countries. French Huguenots fleeing religious persecution in France found refuge in the more open society of the New World. German Lutheran Pietists and Scottish Presbyterians also found a welcoming atmosphere. By the late eighteenth century, Irish Protestants abandoning worn-out soil and Irish Catholics fleeing religious persecution and economic hardship had already defined so-called Irish quarters in these cities. Benjamin Franklin noted that he heard so many languages in Philadelphia that they would soon have to print the street signs in German. If Franklin was concerned about the number of immigrants in Philadelphia in 1770, he would have been astounded in 1860. From the eighteenth century through the first few decades of the nineteenth, it was hope in opportunity and freedom that brought immigrants to American shores. Although hope and freedom continued to play roles in the decisions of migrants to come to America, the catastrophic ecological crisis that hit Europe at midcentury changed the nature of American immigration.

The potato, developed by Andean horticulturalists, was brought to Europe in a period of radically fluctuating temperatures due to the Little Ice Age, 1350–1850. During an unusual warm spell between 1500 and 1560, the potato spread through Europe as the food for the lower classes. Land previously thought unusable was now found to be suitable for potatoes. As new land was put into cultivation, Europe's population grew. Beginning in 1660, the climate improved again, and the numbers again began to grow—particularly in England and Ireland. Between 1780 and 1840, Ireland's population jumped from 4 million to 8 million. This growth did not occur smoothly. A limited potato failure in 1821 and the hard winters of 1825–1826, 1826–1827, and 1829–1830 (one of the coldest ever recorded) increased migration out of Ireland. But it was the catastrophic potato blight of the 1840s that compelled the Irish to flee Ireland by the millions. While England continued to import wheat from Ireland, the Irish, dependent upon the potato for sustenance, went hungry. The English responded to the Great Hunger by encouraging emigration. Packing themselves into the hulls of lumber ships as ballast or buying passage on freighters crossing the Atlantic, some 2 million Irish came to America between 1820 and 1860. Meanwhile, on the Continent, hard winters, potato blight, and land consolidation by large landholders also drove 1.5 million Germans to emigrate to the New World (U.S. Bureau of the Census, Historical Statistics of the United States 1966, 56–59).

When they stepped onto the American shore, many of the migrants (particularly the Irish) arrived with no resources other than a willingness to work. For the cities of the Northeast, the migrants represented an economic godsend, as well as a political and social minefield. Without money and viewed by many as hostile Catholic aliens, the immigrants crowded into areas where the most wretched conditions prevailed. New arrivals were often forced to a city's edge, where they erected flimsy shanties and drew water from local streams, which were also used for washing and as a repository for refuse and wastes. Others merely crowded into basements and back alleys. By the 1850s, New York, Boston, and Philadelphia all had poor neighborhoods predominated by boarding houses, overcrowding, insufficient sanitation, and high mortality rates. Though they contained poor sections, these cities did not, however, feature clearly defined business districts, wealthy outlying areas, and poor cores. While wealthy neighborhoods such as Boston's Beacon Hill, Philadelphia's Chestnut Street, and New York's Washington Square existed, as did slum streets and quarters, these cities were compact. Most urban residents lived amidst a jumble of businesses, factories, shops, homes, tenements, and boardinghouses.

New systems of urban transportation—particularly the streetcar—along with the expansion of steam-powered factories, the creation of large industrial enterprises serving a national economy, and new waves of immigrants from southern and eastern Europe would again transform these cities in the second half of the nineteenth century. Factory districts moved away from residential areas, commercial districts emerged, and whole sections of the city became home to particular immigrant groups. At the same time, in poor, crowded immigrant slums, population densities reached the highest in the world, the middle class retreated into its own communities, and the wealthy locked themselves away from the unpleasant realities of urban life. By the end of the century, the segmented and segregated city came to typify the American urban experience.

But also, as the century matured, cities began to spring up in new locations and for reasons other than those that dominated the nation's early history.

On July 24, 1882, Theodore Lyman, a wealthy Boston Brahmin, addressed a crowd in Dedham, Massachusetts. In answer to a question about protective tariffs, Lyman quoted his father, the ex-mayor of Boston, as saying "as long as New England was overlaid by ten feet of gravel, she would have to manufacture or starve." (*Boston Herald,* July, 22, 1882). This was a lesson New Englanders, with their poor, acidic, and rocky soil, learned early. Within a couple of generations, New England farmers gave up growing grain and looked for other ways to earn their livelihoods. Key to that new livelihood lay the region's rapidly flowing waters. As New England falls off the upper reaches of the Appalachian mountain range, its rivers and streams flow out of the mountains and into the sea. Early millers captured those flowing streams' energy, which could amount to four times the power of a hefty horse and forty times the power of a strong man, to turn their saws and millstones. By the early nineteenth century, the energy of those streams was turning spindles, pressing paper, bellowing fire, and hammering iron. Timothy Dwight, the president of Yale College, noted at the turn of the century that "works erected for various manufacturing purposes" were springing up everywhere along New England streams.

Often housed in converted gristmills or sawmills, these manufacturing enterprises were small affairs, typically employing fewer than twelve employees and using smaller dammed streams for their power. The mill workers lived nearby in the local village or on neighboring farms. They fed themselves from their own gardens or farms, and they relieved themselves in outhouses behind their houses. Wastes from household chores were merely thrown on the ground out the back door.

In 1808 the federal government, in response to the Napoleonic Wars, enacted a trade embargo followed by the Non-Intercourse Act of 1809. Under these acts, the importation of manufactured goods into the United States was radically curtailed, a move that would eventually help spur the development of domestic industries to meet the national demand. The War of 1812 also greatly increased the nation's demand for goods. Ships needed sails, and soldiers needed uniforms and guns. In response, dozens of small mills, adopting the spinning technology brought to this country by Samuel Slater in 1789, sprang up throughout the Northeast. In 1808 there were fifteen powered spindles; one year later there were eighty-nine (Cumbler 2001, 33–35). When peace came, most of these mills failed, and those who worked in them—mostly women and children—returned to the familiar routines of country life. But for some manufacturers, particularly those of southern New England and southeastern Pennsylvania, manufacturing was not something so easily abandoned.

While visiting England in the early nineteenth century, Francis Cabot Lowell, a Boston merchant, asked his British hosts to show him their factory. Assuming Lowell to be a simple trader, the English let him see their impressive new powered looms. Lowell carefully noted the design of the loom and, upon his return to Boston, persuaded two fellow merchants, Nathan Appleton and Patrick Tracy Jackson, to join him in building a textile factory to run not only powered spinning machines, but power looms and carders as well. The partners acquired a mill site on the Charles River in Waltham, hired a talented mechanic, Paul Moody, to design the looms and carders, obtained a far-reaching corporate charter from the state legislature, and, in 1813, began building a textile-manufacturing operation on a new and unimagined scale. The new association, the Boston Manufacturing Company, purchased and transformed the site of an old mill. The old milldam had flooded nearby fields, whose owners were paid for the flooding of their land as prescribed by the mill acts. The Boston Manufacturing Company had far grander visions of waterpower. Taking advantage of the same mill acts' provisions, the company rebuilt the dam significantly higher than the old paper mill's dam. As a result, water began to spill out into far more fields, eventually flooding 200 acres belonging to farmers in four different towns. The farmers complained that they could no longer get to their hay fields and sued the company to have the dam lowered. Unfortunately for the farmers, the courts allowed that the mill acts, originally passed for the common good of local farmers, also protected the dams of the new corporations. With the courts' protection assured, the Boston Manufacturing Company next built huge overflow waterwheels to harness the water's power and bring it into the multistory mill building. Recruiting women from the surrounding area, the Boston Manufacturing Company packed not dozens, but hundreds, of workers into its new mill to tend thousands of spindles and produce over a million yards of cloth a year.

More than just profits and cloth flowed from the new mill on the Charles River. Seeing the success of their bigger neighbor, the Waltham Cotton and Wool Factory Company downriver from the Boston Manufacturing Company attempted to increase production by raising the height of their dam. The rising water behind the heightened dam backed up into the Boston Manufacturing Company's waterwheel. In 1815 the Boston Manufacturing Company sued and won, forcing the Waltham Cotton and Wool Factory Company to lower its dam. Although the court found for the new company, the waterpower of the Charles River, although utilized as never before imagined, was simply insufficient for satisfying the imaginations and ambitions of the new manufacturing entrepreneurs. By the early 1820s, the leaders of the Boston enterprises were looking for new locations where they could capture even more waterpower and build not just bigger mills, but whole mill towns.

Mills, Towns, and Cities 119

Mills and smokestacks line the Merrimack River in Lowell, Massachusetts. (Corbis)

From its marshy headwaters thirty miles southwest of Boston, the Charles River meanders slowly through the flatlands of eastern Massachusetts. To its north runs the Merrimack, a mighty river rushing 116 miles down the slope of the White Mountains, carrying with it eighteen times the waterpower of the Charles. In the early 1820s, Boston textile investors began buying land along the three major drops on the Merrimack, first at Pawtucket Falls where the Concord River joins the Merrimack. At this site, the investors began developing a major industrial enterprise that involved several mill sites, dams, canals, and boarding houses. In 1820, 1,500 people lived scattered about the town of East Chelmsford. With the coming of the mills, the town—newly renamed Lowell—grew to become the second-largest city in New England, with over 33,000 inhabitants by 1850 (U.S. Bureau of the Census of Population 1820–1850). Downriver at Amestig Falls, the town of Lawrence also rose to become a major textile center; while in New Hampshire, people gathered together into boarding houses and the new factory buildings to build Manchester.

New Englanders were taking control over water and harnessing it to their industrial needs. The initial mill towns' tremendous financial success led to other mill towns, some spun off by the original investors, others by imitators. To the south, Providence investors harnessed waterpower in Fall River and along the Blackstone River and began to build mills and hire workers. Water cascading over the Passaic River's falls brought investors eager to build a New Jersey equivalent of the Boston industrial model. South of Philadelphia along the Brandywine River, gristmill and sawmill dams were raised to increase waterfall over them, and spinning machines turned out yarn where previously farmers had brought their wheat, while the Dupont family used the waterpower to make gunpowder.

The return on the investments in Lowell, Manchester, Lawrence, Fall River, Providence, and Paterson, New Jersey, convinced investors that waterpower was the cornucopia from the gods. It needed only to be harnessed to turbines to produce wealth. Soon most of the rivers and streams of the Northeast were enlisted in this great enterprise of wealth generation. In 1847 investors from Boston, after noting that every falls on the major and minor rivers in New England "had been seized upon for the use of some mill or factory," turned their attention to the still-untapped but mighty Connecticut River as a source of power and therefore wealth. The investors felt that with its flow of 6,980 cubic feet per second in the driest month and its 59.9-foot drop at Hadley Falls, the river, once dammed, could be capable of delivering 30,000 horsepower. Merely half of the Hadley Falls site's potential power could maintain a hundred factories the size of those at Lowell with a million spindles. But to do so the Connecticut River would have to be dammed, and that would be no easy feat, for it involved a bigger and far more challenging engineering project than had been completed before.

Optimistic that they could harness the mighty river, the investors began placing orders for lumber and hiring on workers. The dam was completed in 1848, and, as the press looked on, water backed up behind it—but only for three hours. With a huge rush, the rising water overwhelmed the dam, washing a massive amount of northern forest hemlock downstream. The determined engineers and Boston investors did not plan to be beaten by nature, however. They felt destined to control nature and make it do their bidding, which in this case meant producing power for their machines. Construction began immediately on a new, substantially stronger dam. By the fall of 1849, the new dam was completed. This time it held, and builders began to erect textile mills and boardinghouses. By 1852 there were two mills in Holyoke, Massachusetts, each 268 feet long, 68 feet wide, and five stories high, and the town began filling up with operatives (those who worked the machines), mechanics, clerks, and tradespeople. By the end of the 1850s, 4,632 people occupied land where ten years earlier only a few dozen farm families eked out a living (Cumbler 2001, 43–44).

Although Holyoke grew, it did not generate the profits foreseen by its original investors. The cornucopia of waterpower turned out to have limits after all.

A market glut of textiles discouraged new investments in the factories. Having built not only the dam, but also canals, machine shops, and boardinghouses, but unable to sell factory sites, the company was forced into receivership in 1859.

The company's new management began to look for nontextile companies that needed mill sites. Paper companies, which needed extensive waterpower to run their huge Fourdrinier machines, began buying up sites and building mills. By 1880, Holyoke's twenty-three paper companies were producing over 150 tons of paper daily (Cumbler 2001, 44–45). At the same time, the development of a standard-gauge national rail network allowed Holyoke's paper to reach a nationwide market. The mills producing all this paper needed clean water that, after it was used, was simply dumped back into the Connecticut River.

Holyoke in 1880, with a population well over 20,000, had become one of the towns Theodore Lyman believed would save New England. Where once there had been farms, there was now industry. Lyman noted that the manufacturing towns, "with their humming machinery, their prosaic faces, their long hard streets, ugly brick buildings and sickly smells" were the "foundation of national power." (Cumbler 2001, 5). And around the mills gathered laborers, operatives, clerks, and overseers. They lived in boarding houses, tenements, and small residences. They ate, washed, and excreted wastes into the waters of the Connecticut River.

As the canals, and later railroads, opened in the west, corn, wheat, beef, and pork moved east. Except for those in the fertile valleys of eastern Pennsylvania and central New York, most of the farms east of the Appalachian Mountains could not compete with the farms of the tramontane region. Rural New England, particularly the hill country, lost population, allowing farms to revert to forests. But the growing cities—whether inland and industrial like Lowell and Holyoke, or on the coast like Boston, New York, and Philadelphia—not only absorbed people from the surrounding countryside, they also took perishable goods.

Although the soils of eastern farmland were not as rich as those of the interior, the farmers themselves were closer to the growing cities and towns. Townspeople worked in the new mills, factories, shops, and offices and lived in the towns and cities. They needed places to sleep, warmth at night, and food. Local quarries cut slices out of mountains to build factories and mansions for factory owners. Local clay pits supplied the brick kilns that produced the bricks for office buildings, courthouses, tenements, and mills. The forests of the upriver hills and mountains furnished the lumber that built homes and dams. And until the late nineteenth century when coal became the dominant source of energy for heating and cooking, those forests also supplied the firewood. The food city and townspeople also needed came from surrounding farms. Farmers, who in the early nineteenth century grew crops for their own use and sent their surplus into

Farm abandonment, 1850. (Harvard Forest Dioramas/Fisher Museum/Harvard Forest, Petersham, MA)

the larger West Indian and European markets, now found new markets in the local industrial towns and coastal cities for vegetables, fresh fruits, eggs, chickens, milk, butter, and cheese.

In response, farmers reorganized their land. Earlier, the vegetable garden and milking shed, usually the province of the female members of the household, had been small affairs located by the side of the house and focused on meeting just the family's needs, while grain fields dominated the farm. In the second half of the nineteenth century, however, area farmers began to specialize in growing vegetables, raising chickens, or dairy farming. Farmers converted cereal fields into pastures for cows, cornfields for silage, or acres of fruit trees, potatoes, or green crops. By the end of the nineteenth century, as mild weather arrived each spring, it brought with it hundreds of early morning wagons laden with farm produce plying town roads on their way toward urban farmers' markets or train depots where the goods would be sent on to larger coastal city centers. These farms took on the name "truck farms" based on the habit of farmers to truck their goods to local markets. Farms became gigantic feeding machines, sending perishable foods to the hungry cities.

Though the new industrial economy first eclipsed and then transformed the older, rural economy, it was also transformed by it. Farmers redirected their efforts toward feeding the very industrial world that was overshadowing them. As the new centers grew, they took power and influence away from rural communities. Yet the new centers also provided the salvation to the older rural society.

Producing food for the new urban centers was not the only way the rural countryside benefited from the growth of cities. Manufacturing, as Theodore Lyman reminded his audience in 1882, brought the nation great wealth, but wealth unevenly distributed. For most of the nineteenth century, operatives who crowded into tenements and boardinghouses worked ten to fourteen hours a day for between ten and twenty dollars a week. Families could survive only by putting several members of the household into the labor force, and then only barely. Infant mortality rates grew throughout the nineteenth century, eventually reaching appalling levels. In Holyoke, mortality rates ranged above twenty per thousand for most of the nineteenth century (Cumbler 2001, 47–48). Diphtheria, dysentery, typhoid fever, measles, scarlet fever, tuberculosis, and cholera took a heavy toll on urban residents. Poor quality and adulterated food (e.g., watered-down milk) lowered the average height reached by urban residents by several inches and contributed to the toll disease took on the cities' poor. While little of the wealth generated by the new corporations filtered down to the working classes, it did accumulate handsomely for the owners of the new means of production. Some wealth also accrued to the growing population of the middling sorts—tradespeople, supervisors, and professionals.

Many of the wealthy, like Theodore Lyman, "detest[ed] manufacturing towns," but were dependent on manufacturing for maintaining their livelihoods. For the well-off who lived in these industrial centers and in the crowded commercial cities, the surrounding countryside became more than simply the source of food. It became a place of escape, a haven from the tensions, smoke, soot, noise, smells, and bustle of the city. Cities in the summer months were considered disease incubators. Fleeing from the very source of their affluence, the wealthy built their own homes along the coast or in the country. Land long considered wasteland along the coasts of Long Island, New Jersey, Rhode Island, Massachusetts, New Hampshire, and Maine now gained value as locales for the wealthy's summer estates. Magnificent homes sprang up on marginal hill-country farmland, characterized by poor soil but spectacular views and fresh country breezes. Local residents looked after these homes in the winter months and tended them and served in them in the summer. Warmer weather brought the city families out to these more remote areas, where the household would be reconstructed. Because farming required such varied talents as equipment repair, building construction, well digging, fence mending, and foundation laying, local farmers could handle maintenance and repairs on the estates that were beyond the skills of the urban merchant, banker, lawyer, or property speculator. In addition, once the urban family arrived at their country home, meals needed to be cooked, and cleaning and laundry done. Again, rural families supplied the necessary labor.

The very wealthy were not alone in looking to the rural countryside for escape and in depending upon the rural labor force to sustain their recreational retreats. By the end of the nineteenth century, it became part of the ritual of summer for middle- and upper-middle-class urban residents to escape to the hundreds of resort hotels, country inns, and guest houses that had sprung up or expanded to take in seasonal visitors. Farm families that were no longer capable of sustaining themselves on their market produce alone began opening up their homes and marketing themselves to seasonal visitors. Women and children of urban families often spent entire months in these rural retreats, while male members of the household stayed for only a couple of weeks before returning to the cities and coming back to the country for weekends. These seasonal visitors ate local farm produce and bought local crafts. They stayed in homes and inns maintained by the labor of local farm families as illustrated in William Dean Howell's, *A Traveler from Altruria* (1894).

The urban interlopers who looked to these country retreats as a connection to a more tranquil, romantic past seldom saw the deep connections between the urban future and the very existence of these arcadian reprieves. And oftentimes they were just as oblivious to their dependence upon the rural folk whose labors were so necessary for the retreats' upkeep. In recreation, as well as in work, the countryside sustained the city as much as the city sustained the countryside. The old nineteenth-century country homes and resorts one sees today in the Poconos, Catskills, Adirondacks, and the Green and White Mountains are as much a product of urban America as the rural countryside they inhabit.

Lowell's textile mills were dramatic harbingers of the future. Other new enterprises began to gather about the fault lines of the region's major waterways as well. Mills needed machinery, and soon machine shops and ironworks were competing for space along river and canal banks. Just as Francis Cabot Lowell and his investors needed Paul Moody to build their machines, so, too, did the investors in Manchester, Lawrence, Fall River, Rockdale, Paterson, and Holyoke need the skill and labor of the artisans who built the machines that converted their raw material into finished goods. Large mills may have dominated the urban landscape, but most industrial cities included a number of machine shops feeding the mills the tools used to convert waterpower or steam power into the movement that produced finished products. Machine shops not only turned out machines, they also turned out machinists. The shops were schools of mathe-

Industrial town located near Bellows Falls, Vermont, on the Connecticut River. (Library of Congress)

matics, geometry, hydraulics, mechanics, and engineering. By the end of the nineteenth century, many of these cities had also added schools of technology and engineering, funded by the companies and their owners that needed the machines and tools to run their industries.

Boston, New York, and Philadelphia arose as trade centers. They built their commercial success from their links to goods produced in their immediate hinterlands (and seacoasts). Goods of the immediate surrounding areas were of small importance to the success of industrial cities like Lowell or Fall River. The raw cotton they processed into cloth was grown hundreds of miles away to the south in a different ecosystem entirely. The mills' existence owed little to local products of nature. Instead, they built their success around the energy flows of falling water, the entrepreneurial vision and risk-taking of the investors, and the labor and skill of those who came to work for the investors. The falling water located the cities; the entrepreneurship, skill, and labor of the peoples of the region converted that energy into a thriving urban community.

ECONOMIC DEVELOPMENT OF THE INTERIOR

Other cities of the nation also grew through a combination of nearby power sources or transportation routes and local products. When the Erie Canal opened, it not only linked New York to a significantly larger interior, it also linked regional cities of the interior to New York. The Erie Canal flowed through the Mohawk Valley. West of Oneida, the valley opens up into a broad plain some fifty miles wide along the southern shore of Lake Ontario. Birch-maple forests grew out of the rich alfisol soil, which was far more fertile than the soil of New England. A long growing season and a moderate twenty to forty inches of annual rain and snowfall helped contribute to the region's appeal to those looking for a place to farm. In the early nineteenth century, settlers poured into the region and, with the coming of the canal, forests were soon replaced by wheat fields, farmhouses, and bustling villages. The Genesee River, 230 miles west of the Hudson River, originates in the Allegheny Mountains to the south and flows north through the broad plain before spilling into Lake Ontario at Rochester, New York. Before merging into Lake Ontario, the Genesee River drops 200 feet at Rochester Falls.

Early-nineteenth-century farmers from the surrounding countryside grew wheat and brought it to Rochester to be ground into flour by mills located along the falls of the Genesee River. Wheat grew well in the inland plain's rich soil and cool air. With the Erie Canal, farmers gained a more convenient means of transporting their grain to market. Between 1821 and 1835, plowed acreage doubled as more and more fields were planted in wheat, which came to Rochester to be ground and then shipped east through the canal. Once in Rochester, farmers bought the goods they needed for farming—plows, shovels, rakes, scythes, axes, and wagons; those they needed to maintain themselves and their households—boots, shoes, cloth, jars, pots, pans, and stoves; and those that reflected their newfound agricultural prosperity—pianos, lace curtains, books, and mirrors. The goods they bought were brought in along the same canal that carried out their wheat, or they were made in Rochester itself.

Between 1820 and 1830, Rochester became the fastest-growing town in America. By the end of the 1830s, Rochester had a population of 25,000 people and was shipping out 200,000 barrels of flour annually (Johnson 1978, 18). Rochester's flour-milling enterprises soon became elaborate integrated systems. Farmers' wagons and barge boats full of wheat pulled up to the gigantic mills, where workers shoveled the wheat into buckets attached to conveyor belts that took the wheat up to the top floor where it was mechanically sorted and cleaned. The wheat was then ground into flour, spread out on the floor to cool, then swept up and poured into barrels. Most of this work was done by water-powered machinery.

The machines of Rochester's mills, like the machines of Holyoke or Lawrence, were built in local machine shops by skilled machinists. These machinists soon branched out to other activities so that by the early twentieth century, Rochester's skilled artisans were turning out a variety of machines and optical instruments. Prosperous farmers' demand for finished boots, shoes, clothing, and furniture helped expand local production of these items. By the end of the century, after wheat and flour production had shifted west, Rochester's economy moved to more general manufacturing, particularly of furniture, men's clothing, boots and shoes, optical instruments, and machine tools. Instead of growing wheat, local farmers supplied urban Rochester as well as more distant New York with fruits, vegetables, and dairy products.

As in Rochester, location and local produce combined to contribute to Cincinnati's urban expansion. Cincinnati is located on the Ohio River, the major artery of the central region of the nation. Stretching 981 miles from Pittsburgh to the Mississippi River with only one fault line (at Louisville), the Ohio River follows a portion of the southern edge of continental glaciation. Glacial melt carved out the river's original channel. At their confluence in Cairo, Illinois, the Ohio more than doubles the volume of water in the Mississippi. North of the Ohio River lies a rich, humid, temperate zone of the warm continental province covered in broadleaf deciduous forest.

At the turn of the nineteenth century, settlers floated down the Ohio River from Pittsburgh and began clearing land for farms. The pigs they brought along thrived in the forest regions foraging for nuts (particularly acorns), fruits, roots, or small animals. Pigs fare poorly both in hot sunny regions (because direct sunlight burns their skin) and in very cold regions. Compared to other animals their size, they also need a great deal of water to survive. The lower Midwest was not too cold, contained an abundance of water, and its forests offered plenty of shade. By the early nineteenth century, the region was overrun with swine. As farmers cleared forests for farms, they harvested their pigs. Each fall, farmers would go into the woods to try to round up their hogs and drive them to the nearest market. In southern Ohio that market was Cincinnati. There the hogs were slaughtered, and the meat was salted down, packed in barrels, and shipped out on the Ohio River. Butchering, particularly hog butchering, produces excess fat (lard) and leather. Soap and candles were made with lard as their base. Soon soap and candle shops sprung up alongside Cincinnati's slaughterhouses, as did tanneries and boot and shoe shops. By the second half of the nineteenth century, Chicago had grown to overshadow Cincinnati as the nation's slaughterhouse, but the soap and related products of the Proctor and Gamble Company continued to give the city its economic vitality, as well as make it a target for the social critic Sinclair Lewis in his famous novel, *Babbitt* (1924), whose fictional town of Zenith was based on Cincinnati.

Other cities on the Ohio River besides Cincinnati also prospered from advantaged locations and the resources of the surrounding countryside. Pittsburgh's valuable location, at the Ohio's origin where the Monongahela and Allegheny rivers converge, was noted before the city gained its current name. In the 1750s, the French located Fort Duquesne at the present site of Pittsburgh because they believed it would give them control over the whole of the Ohio River valley. American colonials, under the command of a young George Washington, tried to seize the fort from the French in 1754. Failing in this attempt, the colonial and British troops attacked again in 1755, only to be soundly defeated once more. In 1758, however, the French abandoned the fort, giving the British control over this strategic location. With the coming of American independence, settlers began to pour across the Appalachians into the cheap lands of the west. Pittsburgh, at the beginning of the wide Ohio, provided these settlers with a natural embarkation point. Coming over the mountains in wagons and on horseback with their pockets full of savings, the settlers arrived, sold their wagons, stocked up on supplies, and headed into the interior on flatboats. Supplying settlers became a major Pittsburgh activity, helping the city's population grow from 1,565 to 8,000 between 1800 and 1815. Nature had endowed the Pittsburgh area with other advantages besides the Ohio River. From the north, the Allegheny River flowed down through the heart of the fertile Allegheny valley, while the Monongahela River flowed northward through the iron- and coal-rich mountains of Pennsylvania and West Virginia. Both rivers also flowed out of mountains thick with forest hardwood.

The migrants heading west needed farm tools. These were heavy items to lug over the mountains. With iron and charcoal readily available, Pittsburgh entrepreneurs soon invested in furnaces to produce iron. They also converted wood, cut from upriver forests, into charcoal. Large mounds of wood interspersed with kindling would be covered with a thin layer of soil, and then the wood would be slow burned into charcoal. The charcoal was then hauled to local foundries where it was used to convert iron ore into pig iron. Local ironmasters melted iron ore in furnaces burning charcoal and limestone. Pummeling hammers powered by waterwheels removed impurities from the iron, which was then sold to Pittsburgh smiths. Blacksmiths using charcoal heated the iron and then beat it into items needed by those traveling west, or remolded it in a second furnace and poured the molten metal into castings. In 1811 a number of Pittsburgh's merchants invested in a rolling mill. From then on, large slabs of iron from surrounding furnaces could be reduced to bars by steam-powered rolling, which eliminated the need for the rural hammer mills. Within four years, Pittsburgh was turning out over $750,000 worth of iron products (Wade 1971, 47). By 1820, Pittsburgh began puddling (a process whereby a skilled puddler stirs the molten iron to increase the oxygen in the caldron and burn off impurities) at its furnaces. Increasingly, Pittsburgh took control over all aspects of the production of iron and iron implements. The countryside furnished chunks of iron ore, limestone, and charcoal, and Pittsburgh ironworkers produced finished wrought-iron bars and iron products. By 1830, Pittsburgh was a veritable workshop, and its air hung heavy with charcoal smoke and rang with the sound of hammers shaping iron on anvils. Pittsburgh boasted nail factories and steam engine workshops, as well as factories manufacturing pots, pans, shovels, rakes, plows, and scythes.

As steam power brought iron making into the city of Pittsburgh, it also increased the demand for fuel. Charcoal, already in high demand for iron making, increased in price. Forest cutting moved farther away from available transportation, mountainsides became stripped of trees, and wood fuel of any sort grew more expensive. Still, manufacturers needed cheap fuel for their steam engines, and residents needed inexpensive fuel for heating and cooking. The solution lay embedded in the hills of the surrounding countryside. Coal, particularly bituminous coal, could be dug out of the hills and shipped by barge up the Monongahela to Pittsburgh's steam engines. Coal allowed ironmongers to turn to a new fuel perfected in England by George Darby and his son in the eighteenth century: coke. Coke was produced by slow roasting coal to burn out the impurities. Coal converted to coke could then be used in the production of iron without risk of contamination by the sulfur and other impurities in the coal.

Readily available crude iron, inexpensive fuel, and the high cost of shipping finished iron products over the mountains all contributed to turning Pittsburgh into "Iron City" by midcentury. Cheap fuel, materials, and the fragility of glass also encouraged growth in the city's glassworks. At the turn of the nineteenth century, Pittsburgh's entrepreneurs began making glass products from local silica and coal. By the end of the War of 1812, "Pittsburgh Glass" was known throughout the west, and within another two decades, the city's glassworks would rival its iron foundries as centers of economic activity. Both depended upon skilled workers, a ready market, and an abundance of local raw materials—especially coal. Cheap local coal fueled Pittsburgh's industry, heated its homes, cooked its meals, and filled its air with soot. Bituminous coal contains high levels of impurities, but even without those impurities, a good amount of coal particulates become airborne when coal is burned. Dirty, pungent coal smoke engulfed Pittsburgh and the surrounding valleys. A visitor in the 1840s commented that, "a

View of oil wells at Church Run near Titusville, Pennsylvania, 1868. (Bettmann/Corbis)

dense cloud of darkness and smoke visible for some distance . . . hides the city . . . coal being the only fuel of the place." (Nye 1999, 75).

Local coal and iron maintained Pittsburgh's thriving economy. Cheap labor, innovative entrepreneurs, and the railroad's steady demand for iron—and later steel—rails kept Pittsburgh growing. By the 1860s, railroaders had become dissatisfied with iron rails, which wore out quickly under the strain of carrying heavier, higher-speed trains. The adoption of first the Bessemer and then the open-pit furnace allowed Pittsburgh manufacturers to produce relatively inexpensive steel and steel rails. By 1880 production of iron rails had virtually ceased, and Pittsburgh manufacturers dominated the steel rail market. But like its iron mills, Pittsburgh's steel mills consumed massive amounts of coal, in the form of coke, and iron ore that layered the city with soot.

If Pittsburgh was hidden by a dense shroud of smoke in the first half of the nineteenth century, by the second half glowing molten iron and steel and shooting flames could be seen from miles away twenty-four hours a day. Steel making moved to a two-shift, twelve-hour-day routine in which workers kept the furnaces burning continuously. Pittsburgh became a city of night and day. Residents no longer worked with the rhythms of nature but with the rhythms of the clock. And the clock did not stop at sunset. Others besides steelworkers also found their workdays stretching into the evening. Clerks, tradespeople, artisans, and housewives worked long past the setting of the sun—and not just in Pittsburgh.

Night work required lighting. Prior to the nineteenth century, people read and worked by candlelight. Coal gas lit city streets and larger establishments. For small shops and work at home, lamps burning coal oil provided light for late workers and readers. The wealthy preferred cleaner-burning whale oil, which did not leave the same lingering odor as coal oil.

In 1859 petroleum was discovered in Titusville, in northwestern Pennsylvania. Believing that petroleum could be refined into a burnable fuel to compete with coal oil, speculators rushed into the area, drilling wells and barreling oil. The oil was then sent west to Cleveland, where small refiners distilled the oil into kerosene for the home-lighting market. Sensing that the market for refined kerosene was ever growing, John D. Rockefeller began consolidating refiners into his new Standard Oil Company. If Pittsburgh glowed at night with molten steel, Cleveland illuminated the night with the shooting flames of burning gases from the refineries. Petroleum emerged easily and in abundance from the ground of northwestern Pennsylvania—in fact, too easily. When drillers struck oil, the gusher would pour out hundreds of barrels before it could be effectively tapped. Excess oil spilled on the ground and into local streams, discoloring the water for miles. In Cleveland, Rockefeller's refineries burnt off wastes into the air, but thousands of gallons of additional wastes and spills flowed into canals and ditches, eventually draining into the Cuyahoga River and then on to Lake Erie, the shallowest of the Great Lakes.

Rockefeller was right that cheap refined petroleum would beat out coal oil as a home-lighting fuel; what he could not have anticipated when he entered the oil business in 1863 was that a Detroit mechanic would later produce a cheap and efficient automobile powered by an internal combustion engine and fueled by refined petroleum. Mechanics had fantasized about horseless carriages for years. By the turn of the twentieth century, three types had emerged on the market. All three were very expensive. Using railroad locomotion as a model, one vehicle ran on steam. Another used the new electric technology being developed by Edison. The third burned petroleum in an internal combustion engine. By 1900 manufacturers were producing all three models. Of the 8,000 automobiles in the country at that time, those powered by gasoline were the least common

Rows of completed Model T cars roll off the Ford Motor Company assembly line, c. 1917. (Library of Congress)

(Nye 1999, 175–176). The more popular Stanley Steamer was easy to start, ran on kerosene (which it burned to boil the water for the steam), and had fewer moveable parts than the internal combustion engine. But it was also heavy and easily became bogged down on the dirt roads that predominated at the time. By contrast, internal combustion engines were lighter, and gasoline delivered more energy for its weight. But the key to the success of the internal combustion automobile was Henry Ford. By simplifying the automobile and rationalizing and standardizing production, Ford produced a car that people other than the very rich could afford.

Ford revolutionized the automobile industry and the nature of private transportation in America. In 1910 just under a million people owned automobiles; by 1925 that number surpassed 10 million. Commuters like Sinclair Lewis's Babbitt prided themselves on driving to work in their own private transportation vehicle. Car owners who lived in the country could drive to the city, while those in the city could visit the country for day trips. Motor touring became a major leisure-time activity for middle-class Americans. Private, state, and national parks began to cater to campers and visitors arriving by car for vacation camping. Ford trucks allowed farmers to bring produce to market more readily than before. With all the industry growth, Fort Detroit leaped from a small lake-boat producing center of 45,619 people in 1860 to a city of over a million by 1920.

Refined petroleum was not the only new energy source found in the ground of northwestern Pennsylvania. As the wildcat drillers hunting for oil punched holes in the earth, they often hit pockets of natural gas. To them, the gas was simply an impediment to retrieving oil. They burned it off or just discharged it into the air. But to others, who were using it as an alternative to coal gas, natural gas was a valuable fuel source. In 1821, Fredonia, New York, lit its streets and buildings with local natural gas. By the 1870s, technology was developed to control and transport natural gas through pipes and tanks. Soon natural gas was outselling coal gas. Towns near natural gas reserves blossomed as industry using the cheap local fuel expanded. Natural gas transformed the sleepy little midcentury farm town of Muncie, Indiana, into a significant manufacturing center of glass jars, with cheap natural gas fueling its glassworks.

Located on the Chicago River, the city that took its name offered a convenient port and haven for early-nineteenth-century sailing boats and canoes. The Chicago River originated in a swampy high ground several miles southwest of the city. Although the eastern shore of this marsh fed the stream that eventually flowed into Lake Michigan, the other side of this high ground drained west into the Illinois River and then into the Mississippi. These rivers cut through the rich, fertile prairie lands of southern Minnesota and Wisconsin, as well as Illinois, Iowa, and Missouri. Chicago rests on the southern border of Lake Michigan. Michigan is the longest of the Great Lakes, stretching some 380 miles north to south. In the nineteenth century, huge forests of white pine interspersed with maple, yellow birch, and hemlock rose up along its northern shores. Under the forests lay rich deposits of iron ore, as well as tin and copper.

Before railroads, one reached Chicago by boat or by horse, foot, or wagon. The overland trip was difficult. Although much of the midwestern lands were covered by rolling forests or prairie, the glaciers had also left behind huge tracts of flat, poorly drained soil. These wetlands—such as the 1,500-square-mile Black Swamp in northwestern Ohio, the swamps of northwestern Indiana that covered hundreds of thousands of acres, or the Wet Prairie of east-central Illinois—confronted farmers with land that was difficult to manage and that they believed to be unhealthy. With plenty of arable land available nearby, settlers passed these regions in favor of drier land more amenable to the farmer's plow. The wetlands these settlers avoided accounted for some 64 million acres. Nearly a quarter of the state of Illinois, 8.3 million acres was wetland. Federal surveyors had noted that with drainage the wetlands would provide good farming land, but draining marshland was no simple task.

For over a hundred years, farmers had understood that drained marshland was highly productive land. The draining itself was difficult, time-consuming, and labor intensive, however, and required building drainage ditches feeding into lower-lying creeks and streams. Only farmers with property bordering such creeks and streams could drain their property without crossing over another's land. In 1832 the state of Indiana, in an attempt to encourage wetland draining, passed a ditch law giving farmers the right of way to build ditches across other farmers' land. Illinois passed a drainage law in 1845 that allowed for the development of cooperative committees to oversee drainage-ditch construction. In 1850, Congress passed the Swamp Land Act, giving Indiana 1.3 million acres and Illinois 1.5 million acres of formerly government-owned wetland. The states could then sell off the swamps and use the revenue to initiate reclamation projects. Despite these encouragements, the abundance of arable land nearby meant that little actual drainage occurred.

New technology and demand for land increased interest in draining wetlands. In the late 1820s, John Johnston, a Scottish immigrant to western New York's Lake Erie plain, bought wetlands most others avoided. In 1835, after struggling with the problem of poor drainage, Johnston read about horseshoe-shaped tiles that were being used in England and Scotland to drain fields. After securing the drainage tile pattern from Scotland, Johnston had a local earthenware manufacturer fashion a tile-molding machine and begin manufacturing drainage tiles that Johnston buried in long trenches in his fields. The underground clay tiles increased the movement of water from the marshland into adjacent drainage ditches. Johnston's project worked. Productivity on his farm dramatically increased. Johnston's success led other farmers to buy and bury drainage tiles in their wet fields. By the late 1850s, dozens of tile potteries sprung up in western New York to meet the needs of farmers plagued with the poorly drained fields of this glacial plain. In 1864 there were some 6,060 miles of tiles in New York, along with 13,000 miles of stone drains and over 7,000 miles of open ditches (Vileisis 1997, 124).

As word of the success of tile drainage spread west, farmers in Ohio, Indiana, and Illinois began digging ditches and laying drainage tiles, while drainage tile manufacturers moved west as well. By the 1880s over a thousand drainage tile factories were operating in Ohio, Illinois, and Indiana, and horse-drawn machines such as the Blickernsdorfer Tile Ditching Machine, were digging four-foot ditches across the Midwest. In 1892 the Buckeye Trencher, a steam-powered ditch-digging machine, became the new favorite of swampland owners. Continued innovation in ditch digging and drainage encouraged the conversion of wetlands to farmlands throughout the Midwest. Drainage brought more oxygen into the soil and increased the breakdown of organics, increasing fertility. Drained land produced significantly more crops per acre than traditionally dry land. Corn production on drained Illinois wetlands was 50 percent higher than on traditional farmland (Vileisis 1997, 124). Dried land also reduced the incidence of malaria.

Wetlands may have made for highly productive farmland when drained, but left undisturbed they functioned as water reservoirs, fowl habitat, and fish-breeding grounds. Diminished wetlands increased flooding. Water that was previously held in marshes and swamps to seep into the water table and slowly drain out into streams and creeks with the new drainage tiles and ditches flowed off of the land in a rush, filling streams, creeks, and rivers to overflowing. Water tables dropped without the continual, steady infusion of water from the wetlands, forcing midwesterners to dig deeper wells. With wetlands drained, fowl populations declined, and fish habitat shrank. Of course, these effects were of little concern to land speculators, who saw the value of drained land increase 500 percent.

Around the original marshes east of Chicago lay deciduous forests of broadleafed oak and hickory stretching east into the Appalachian Mountains. Across this region warm, moisture-laden winds blow up from the Gulf of Mexico and meet drier, colder air coming down from the northwest. As these weather patterns collide, they drop rain over the broad midsection of the American Midwest.

To the west of Chicago, the climate changes from wet to dry. Here the drier, colder winds from the northwest predominate. Fires sweep across the dry rolling land, burning everything in their path. Trees have a hard time holding their own in this fire ecology, but grasses thrive. The soil below the tall grasses is rich in calcium and well drained. As the glaciers that scraped across Wisconsin and Minnesota retreated, they left behind a till that washed and was blown south across the broad, flat plain. The generations of grasses that grew on this deposit in turn left rich deposits of humus. By the time white settlers moved into the region, several feet of well-drained mollisol matted with grass roots lay below the flowing prairie grasses. To the south lay rich forests sitting atop deposits of coal and lead.

By the middle of the nineteenth century, Chicagoans had begun taking advantage of the city's unique location in the center of three ecosystems—dry mollisol prairie to the west, damp alfisol hardwood forests to the east, and spodosol conifer forest to the north. Pine lumber from the north came down through the Great Lakes to wait in Chicago wood yards until it could be shipped west to waiting farmers breaking the prairies with John Deere plows. Wheat, pork, and beef arrived by train from the plains and filled the city's grain elevators and stockyards before being shipped east to hungry residents, both in American and European cities. Chicago's population jumped from 4,470 in 1840 to 503,000 in

1880. By 1900, Chicago had over 1,699,000 people, making it the nation's second-largest city, trailing only New York (U.S. Bureau of the Census, 1920 Population, 62–75). After hosting the Columbian Exposition in 1893, Chicago could reasonably claim the mantle of the ultimate American city.

Although Chicago served as an important transfer point, it also processed many of the goods brought in from the different ecosystems. Pigs and cows were not simply shuttled east; they were slaughtered in Chicago. Gustavus Swift's development of refrigerated railcars, combined with assembly-line techniques for slaughtering or disassembling animals made Chicago ideally situated to achieve economies of scale for the mass slaughter of animals. As a result, it became cheaper in New York City to buy midwestern beef already dressed than to buy beef from cows raised on Long Island farms and slaughtered in Brooklyn slaughterhouses.

As America prospered and its middle class grew in size so, too, did its taste in meat become more demanding. Previously, tough, lean Spanish long-horn cattle flowed into the stockyards of the growing midwestern cities. But with the growth of a national market for dressed meat, Americans began to look more critically at their meat. Marbled beef of English Angus cows became the meat of preference for the middle and upper classes. By the late nineteenth century, short-horned cattle from English stock raised on the southern and high plains ranges were driven to feedlots in Nebraska, Iowa, and northern Illinois; fed on corn from Illinois, Indiana, Iowa, northern Missouri, and southern Minnesota; then shipped to Chicago, where they were slaughtered and dressed, and the meat shipped to waiting tables around the nation. Beef and pork soon surpassed lumber as Chicago's primary import-export item. Originally, the convenience of the Great Lakes system and the Erie Canal persuaded farmers to send their surpluses to Chicago rather than other interior cities. Carried on steamboats across the Great Lakes and cheap barges moving through the Erie Canal, manufactured goods could be sent to Chicago cheaper than to St. Louis via the Atlantic, Gulf of Mexico, and the Mississippi by steamboat. Likewise, it was cheaper to send out grain from Chicago. As a result, farmers received a better price for their produce and brought home more goods with their earnings. By 1850, Chicago was the destination of choice for farmers of the central region.

The Illinois-Michigan Canal, completed in 1848, linked Chicago to the Mississippi through the Illinois River. The canal followed the glacier channel that millennia earlier had allowed the lake to drain into the Mississippi watershed. Once the canal was finished, farmers from the rich Illinois River valley began sending corn to Chicago and receiving lumber in return carried on canal barges. But even as the Illinois-Michigan Canal was completed, a revolution in transportation had already begun to sweep across the nation's midsection. In the late 1840s, Chicago

Farmer selling his roadside produce. (Library of Congress)

boosters fanned out northwest of the city selling bonds for a railroad to connect the rich farming regions of northern Illinois and southern Wisconsin. In the fall of 1848, construction began on rails stretching northwest from Chicago. Within a week of the opening of the first ten miles, farmers sent over thirty carloads of wheat by rail to Chicago. By 1852 half the city's wheat arrived by rail. The initial railroad's success encouraged the building of others, and construction took off. By 1870, Chicago had railroads radiating northward through Wisconsin to southern Minnesota, west through northern Illinois into Iowa, southwest into Missouri, and south and southeast into southern Illinois and Indiana. From the east, rails brought in finished goods from the manufacturing districts of New York, Pennsylvania, and Massachusetts and shipped out dressed meat and grains in return. Chicago provided an ideal terminus for railroad builders.

Chicago flourished with the railroads. Railroads cut the cost and time needed to ship bulk agricultural goods across the western prairie. Although St. Louis enjoyed a natural transportation link north into the rich prairie lands provided by the Mississippi, the Rock, the Illinois, the Wisconsin, the Cedar, the

Des Moines, and the Missouri rivers, river transport was always problematic. Rivers froze in the winter; ran perilously low in the dry months; and were plagued with snags, floods, and disruptions. The railroad suffered none of these complications. It cut across the countryside, it put down terminals in the richest farmland, and it ran with more regularity than river transport. It also ran anywhere its directors opted to go. For many farmers, however, access to markets and rich prairie farmland proved a mixed blessing. It took a full generation for farm families to turn 140 acres of forest into a farm in full production. Growing urban populations and the hungry Europeans eagerly absorbed American produce as these farms increased production. With the introduction of John Deere's steel plow, prairie farmers found they could bring 140 acres into production in less than a decade. Farm families spread out across the prairies, plowing under bluestem and Indian grasses and planting wheat and corn in their stead. With new equipment, new agricultural techniques, and new farmland, production increased dramatically. Increased production held down prices, however. Banks supplied farmers with loans to buy land; lumber to build homes, barns, and fences (or for barbed wire after its invention by Joseph Glidden in 1873); McCormick harvesters; and Deere plows. Following the brutal labor of the harvest, farmers often found to their dismay that the price they received for what they produced barely covered their costs and loans. This occurred with particular frequency after the economic collapse of 1873. Farmers who transformed the land from prairie to fields of grain through considerable sweat and labor came to believe others were benefiting from the wealth this labor created. The railroads carried the farmer's goods to market, but the railroads seemed to be carrying off the wealth the farmers were creating along with those goods.

At the same time, railroad rates were determined by competition rather than actual costs. Operators running rail lines to rural areas without alternative transportation routes—known as short hauls—charged high rates to move produce to market centers. Transportation routes from market centers like Chicago to distant markets, called long hauls, were more competitive. Several railroad lines, as well as the river, lake, and canal transportation network, moved goods between Chicago and east coast cities. Accordingly, short-haul rates for moving bulk goods were usually substantially higher per mile than those charged to move goods over the long haul. Until their goods arrived at the major market centers, farmers bore the burden of this rate differential. Additionally, farmers felt merchants and bankers were overcharging them for manufactured goods and the money they needed to borrow. Farmers responded to these conditions by forming the National Patrons of Husbandry or the Grange in the 1870s and 1880s and demanding state laws to regulate railroad rates. In the 1890s, they protested by participating in the Populist political upsurge.

Prairie farmers needed more than markets for their produce—they also needed agricultural machinery to help them work the land. By the 1850s, Chicago was already a major center for manufacturing agricultural implements. By that time, Cyrus McCormick's Chicago factory was producing over a thousand reapers a year, and he was just one among many. Agricultural implements of all kinds came out of Chicago factories and workshops. Railroads linked the city to the rich Illinois coalfields that stretched south from the Illinois River all the way to the intersection of Illinois, Kentucky, and Indiana. Cheap coal brought in by the Illinois Central Railroad from Centralia to Chicago fueled the furnaces and foundries of the city's factories, while powering the trains along the way. Low-cost transportation and Chicago's central location guaranteed that the city would become a major metropolitan center, but the cheap coal of central Illinois helped make Chicago the nation's second-largest city. During the late 1870s, midwestern coal companies began using strip-mining techniques whereby the land overlaying the coal seam was literally scraped off to expose the seam. This new technique, which utilized powerful steam shovels, doubled the recovery rate of shaft mining and drove down coal prices—particularly in Chicago with its easy access to the coalfields by either rail or river routes. In addition to agricultural implements, Chicago's factories also manufactured cooking utensils and stoves for the prairie farm families and mining families in the coal-producing region. The industrial expansion in the later nineteenth century involved large-scale heavy mechanization. Industries capturing economies of scale demanded unprecedented amounts of cheap coal. In 1870, Chicago, with a population of 299,000, consumed 777,000 tons of coal. Twenty years later, with a population of 1,099,800, it consumed over 4,013,000 tons of coal. By 1900, Chicago had 1,698,600 people and gorged itself on 7,373,900 tons of coal (Platt 1991, 8). All that coal produced heat and energy for homes and factories, and from those factories flowed a vast quantity of goods.

Having sold their wheat, corn, pigs, or cattle, farmers with money in their pockets went to Chicago's stores to load up on items not easily procured at home. As the iron fields of northern Wisconsin and Michigan's northern peninsula opened up, iron ore began coming through the Great Lakes on lake freighters to be smelted in Chicago foundries and then converted into shovels, harvesters, plows, stoves, and iron kettles. In the beginning of the twentieth century, newly discovered iron reserves were exploited in Minnesota, and the U.S. Steel Company built a new satellite steel city on Chicago's eastern edge in Gary, Indiana. Gary used Chicago's labor supply and rail connections, but built its own docks to service the ships that brought the iron ore down from the Upper Great Lakes. Gary's wastes, like Chicago's, flowed back into Lake Michigan.

Despite Chicago's prodigious manufacturing, the demands of the region's growing populations always outpaced what Chicago could provide. Goods from eastern factories predominated on the store shelves of Chicago, as well as in the outlying stores that obtained their goods from Chicago's wholesalers. In 1870, Aaron Montgomery Ward, who had spent the previous three years as a traveling salesman, realized that despite the rail connections between most rural and small-town communities and the larger metropolitan areas, the selection of goods offered in country stores remained limited and expensive. In Chicago, by contrast, large volume and centralized rail connections offered a wide selection of goods and low prices. In an effort to take advantage of this discovery, Ward moved to Chicago and set up a company to sell directly to farmers through the mail. He offered rural families wholesale prices and won the endorsement of the Grange organization. Ward could offer farmers low prices because he bought in large volume, sold directly to the consumer, and took advantage of Chicago's central location and cheap rail costs to bring into his warehouse the items that he then sent out to the customers.

Ward met a need, and his enterprise grew accordingly. By 1880 his catalogue contained 540 pages, boasting more than 24,000 items for sale. Ward's success led Sears, Roebuck, and Company, which was also located in Chicago, to launch a catalogue business in the 1880s. At the turn of the century, Ward's company moved into a new twenty-story building, with each floor packed with the items offered in the company's 1,200-page catalogue. The two lower floors housed an army of clerks opening mail and routing the orders to the appropriate departments. Electric generators in the basement supplied light to keep the building working deep into the night. Montgomery Ward's catalogue sat in the kitchens and parlors of nearly every farmhouse across the nation. More than just a means to cheaper commodities, the Montgomery Ward catalogue brought, as the historian William Cronon noted, the city to the country.

Before becoming a bustling metropolis at the turn of the century, Chicago had to address its location's shortcomings. Adjoining the southern tip of Lake Michigan at that part of the lake's only natural port offered definite advantages, but those benefits were countered by the location's poorly drained, swampy lowland setting. Chicago rested just a few feet above lake level. Heavy rains failed to drain adequately off of streets and away from homes and shops. Spring showers inevitably produced mud that bogged down travel. Rather than move the city to higher ground, city leaders opted to bring higher ground to the city. Beginning in 1849, the city passed a series of ordinances to raise the grade level of streets. Over the next twenty years, Chicago literally rose several feet above the wet and mud. Hotels and department stores were raised up as well. By the end of the century, few visitors realized that the city had lifted itself out of the mud.

The American continental economy, created in the nineteenth century, depended on more than just the movement of goods and people; it also required the movement of information. Knowing where goods were, their prices, and when they could be moved was as important as being able to move goods. The development of electrically transmitted information allowed for almost spontaneous communication. The telegraph allowed information to travel at the speed of electric current. As the nation's hub and primary transportation junction, Chicago by the 1880s grew to become the nation's second-largest information carrier. In this capacity, Chicago's information-transmitting enterprises also grew. Western Electric Company produced telegraph machinery for Western Union and, with the development of Bell telephones, phone equipment as well.

Farmers worked the land according to the rhythms of the seasons and the days. They rose to work as the sun broke the horizon and continued until dusk called them inside. City workers labored to a different pattern, however. Neither the seasons nor sunlight dictated their workdays. Urban residents continued to labor, shop, prepare meals, clean up, and recreate long past sunset, especially in the winter. Indeed, for many, the city's late-night activity constituted the very essence of its appeal. But working, shopping, and playing after dark still required light.

As in most nineteenth-century cities, Chicago's main streets, stores, and factories were illuminated by gaslight and had been since the 1830s—in Chicago's case by the Chicago Gas and Light and Coke Company. Gaslight was twelve times brighter than the light from candles or oil lamps and became the preferred light for those needing light for more than simply reading. By the start of the Civil War, factories, department stores, and the homes of the wealthy were signing up for gaslight. Soon gas companies began laying pipes into middle- and upper-class residential areas as well.

Although gaslight was brighter and more consistent than oil lamplight, gas did produce soot, and the dangers of using gas were omnipresent. Though brighter than candles, for some tasks gaslight still represented insufficient illumination. Experiments in electric light soon resulted in a brighter, cleaner alternative. Following the example of eastern cities, Chicago first experimented with arc lamps—electric powered, glowing carbon tips. These lamps gave off significantly more light, but were also fire hazards, and the light they produced far exceeded the needs for most interior spaces. In 1881, Thomas Edison developed the self-contained glass bulb, which provided a safe, practical electric light source.

By the 1880s, urban residents were experimenting with a number of lighting systems: gas lights, electric lights powered by basement generators like the one in Montgomery Ward's, or electric lights powered by central dynamos generating electricity that was then shipped out to several locations. Within a decade, electric lighting had won the day, and hotels, offices, and department stores all chose the "modern and progressive" solution by illuminating their establishments with electric lights. Electric lights proved so successful that some of the wealthy were demanding service in their homes.

By the end of the nineteenth century, electric power lines radiated out from huge central power stations to factories, stores, streets, and homes. Meanwhile, those homes were moving ever farther from the central business district.

By definition, a city is a place where people cluster. Cities of the first half of the nineteenth century resembled cities of the seventeenth and sixteenth centuries. Jumbles of stores, shops, taverns, warehouses, wharfs, homes, offices, and artisans' production centers were jammed together to be readily accessible on foot. People walked to work and to shop, and they lived accordingly. Omnibuses, carriages, and the early railroad were too slow, uncomfortable, and irregular to allow for anything but a bare minimum of commuting. The larger cities did spread out, but only in nodules, for residents still had to live within walking distance of work, stores, and services. Most cities' residential neighborhoods were confined to within a two- or three-mile radius of downtown or an equivalent nodule. Two significant innovations transformed the walking city. The most visible was the rail streetcar. On rails, large carriages capable of holding a dozen or more passengers could be pulled by a single horse or a team of horses. As a result, the streetcars could travel significantly faster, farther, and offer a much smoother ride. By the post–Civil War period, cities throughout the nation had put down streetcar rails radiating out four or five miles from the city center. The streetcars allowed more people to move away from their places of work. With the development of the horse-drawn (and later cable and electric) streetcar, middle-class families sought out newly accessible residential areas where they could own detached or semidetached houses or triple-deckers with small yards. New suburban neighborhoods sprang up around the old cities. The suburbs concentrated residences along the new streetcar lines. The investment in laying track encouraged companies to maintain the routes that had been established. People could buy homes with an assurance that they could catch a nearby streetcar to work. Within these neighborhoods, stores specializing in household goods, drugs, groceries, and hardware emerged, usually at junctions along the streetcar lines. The arrival of electric lights and municipal water and sewage services linked these neighborhoods to the larger urban community.

With the development of the railroads, Chicago began to spread out across the flat countryside as the more well-to-do moved north along the lake into Evanston, Winnetka, Willamette, and Highland Park to avoid the crowds and smoke of the city. The development of streetcars accelerated Chicago's advancing march outward.

The suburbs did not function as intermediaries between the city and the country. The country was a center of production that sent its produce to urban population centers and took back manufactured goods. The city and country were tied together through the cash nexus.

But the suburbs were appendages of the city. They grew out from and were part of the city—an outer edge of it—but part of it. Taking advantage of cheap land, developers laid out roads, promised services, and sold lots. The flat landscape gave rise to a particular midwestern suburban residential home, a low one- or one-and-a-half-story detached home sitting on a single lot. By the early years of the twentieth century, a prospective homeowner could buy a lot with the promise of electricity and water and sewage hookups, and then purchase a home from Sears and have it delivered to the lot in pieces. These suburban communities were removed from the old city's crowds, dirt, and ethnic and racial diversity. People lived in the suburbs but did not work there.

Residential suburban communities represented just one kind of appendage the growing city extended across the countryside. As manufacturing expanded, space became a cost variable. Because urban land was already built up, the expansion-minded increasingly left the central city for the land beyond, particularly areas along the railroad lines. By the beginning of the twentieth century, cities like Chicago had industrial districts radiating out along the rail lines surrounded by crowded residential working-class neighborhoods and satellite industrial cities just beyond the borders of the central cities.

Beyond creating a new urban ecology, the turn-of-the-century city also created several independent urban ecosystems. The paper-work industry (law offices, insurance companies, and corporate offices) as well as high-end commerce offices, fancy department stores, specialty shops, hotels, and restaurants were concentrated in the central business district. Land costs combined with the need of those in the paper-work industry for face-to-face contact in a centralized location drove office buildings, department stores, and hotels skyward, restrained only by gravity and the difficulty of moving body weight up several stories of stairs. The Otis Elevator Company developed a moveable container that could be hoisted by using steel cable wound about a rotating drum. Steam power could then lift people up as many stories as the builder chose to erect the building. Because the new steel beams and reinforced concrete were strong but lightweight, buildings could go up without massive foundations. These offices concentrated thousands of people in a relatively small ground-floor area by piling them on top of one another. But these were not the concentrations of residences, they were the concentrations of workplaces. Surrounded by concrete sidewalks and paved roads, the buildings created a region not of greenery, but of concrete, brick, stone, glass, and steel. In the summer these districts stored heat, while in winter snow turned to slush and mixed with the dirt and grime of the city.

Elisha Graves Otis shows his first elevator in the Crystal Palace, New York City, 1853. (Bettmann/Corbis)

Around these new downtown office skyscrapers arose poorly capitalized but labor-intensive industries. These businesses rented space in manufacturing buildings, offered low wages, and experienced high employee turnover. To reach the largest possible labor pool, they needed to be centrally located. Mixed in with these industries were artisan and craft shops, bars, taverns, tenements, and boarding houses. Population densities jumped dramatically just outside the immediate central business district. Crowded into these areas were thousands of the city's working poor. These dense communities generated wastes, ashes, garbage, and trash that piled up in alleys and on streets, while dust, dirt, and smoke filled the air. Populations of animal species that were adept at living among people, including mice, rats, starlings, cats, dogs, and horses, found a home in this very human environment. By the turn of the century, most human waste was being funneled into the sewer systems to be flushed downstream, but the animal populations' wastes continued to plague these dense neighborhoods.

Contrasted with their urban counterparts, residential suburban neighborhoods specialized in greenery. Trees and shrubs identified a neighborhood as suburban. This flora theme was reflected in the localities' names: Gardenlawn, Mapleview, and Highland Park. Green lawns, which imitated the grounds of landed estates in miniature, reflected home pride, while tree-lined streets created a parklike setting. Suburban tree planting on former farm fields to the city's north side brought shade to areas that hadn't seen shade for decades. To the west of Chicago, treed suburbs built on the old prairie shaded the ground that hadn't been arbored in millennia.

In the nineteenth century, cereals supplanted bluestem, side oats, grama, and Indian grass on the prairies west of Chicago. In the twentieth century, fescue, bluegrass, azaleas, elms, oaks, dogwoods, and crabgrass came to dominate the land. Cats and dogs, robins and jays, doves and finches, and squirrels and moles all found food and places to nest in the diverse—if artificial—suburban ecosystem. By the twentieth century, Americans had radically altered the natural world around them. They created what the environmental historian William Cronon calls a "second nature." More than just one second nature, Americans were busy creating multiple second natures, both in rural and urban settings.

6

POLLUTION AND HEALTH

COASTAL CITIES

In the summer of 1973, my mother visited me when I lived in Cambridge, Massachusetts. It was hot that summer day. Garbage from the surrounding triple-deckers, originally packed in plastic bags, gathered on the street waiting for trash pickup. Rats, feral cats, and dogs had ripped open and picked through the garbage the night before. The stench of the garbage mingled with that of dog and cat feces that lay abundantly and hazardously about the sidewalk. My mother, familiar with the different odors of rural Wisconsin, was not accustomed to such urban sights and smells. "You get used to it," I suggested when she expressed concern. Getting used to urban smells and sights was part of the American experience.

In 1810, William Levi and Joseph Chapin opened a Massachusetts cotton mill on the Chicopee River. It was a typical mill of its time, consisting of two carding machines and two frame spinners. The mill employed six to eight people. Millwork was not hard, but it was different from the agricultural labor the workers were used to. Agricultural work involved movement across fields and into and out of barns. Farm workers rose early to a hearty breakfast, then toiled all day in fields, woods, and meadows. Behind every farmhouse sat a privy. Every few years, farmers might dig a new hole and move the privy. The privy or a pause in the field, meadow, or woods met the human waste disposal needs of these farmers and their families.

The hands in Levi's and Chapin's mill could not pause in their work, unbutton their pants or lift their skirts, and relieve themselves as they had in the fields. Work in the mill confined them to one place, and that confinement required a set place for human waste disposal. For only six or eight workers, the waste disposal solution was easy. Behind the mill an outhouse was constructed, one much like those already familiar to the former farm workers. To reduce the need to move the privy, the privy builders dug a ditch, rather than a hole, which ran from the outhouse to the nearest stream.

Industrial discharge into Fox River, Wisconsin. (Wisconsin Historical Society. Image 28350)

The wastes from the Chapin and Levi mill degraded the nearest stream, but with a continuous flow and rain inundation, the only noticeable effect was an unpleasant odor in the immediate vicinity of the privy, especially during hot, humid summer days. People were used to the smells around privies, however, and expected such unpleasant odors nearby. What people were not used to was the volume of wastes generated by the type of large-scale industrial mills that replaced the Chapin and Levi mill and came to dominate the New England countryside wherever a falling river provided enough power to operate large numbers of machines. The huge mills changed the nature of the sewage problem. Residents of the small farming village of Irish Parish, Massachusetts, were scattered about a rural and semirural countryside. The town was healthy and "favored by nature." Its residents drew water from streams, springs, and wells. They fed food scraps to their pigs and used outhouses for their wastes. They threw wash water

Paper mills, Holyoke, Massachusetts. (Library of Congress)

and slops out on to the yard to be absorbed into the ground. When the industrial investors hired workers to dam the river at nearby Hadley Falls, renamed the town Holyoke, and built gigantic brick mills, they changed more than just the town's name, the falls of the river, and the landscape of the surrounding countryside; they also changed the quality of the river water and the air.

In 1872 the Massachusetts Board of Health noted that the region's inland streams had initially been dammed to "saw the lumber and grind the corn." In doing this, they argued, the "benefits have been great and their evils have been infinitely small. But our rural saw-mills and grist-mills are by and by turned to other uses" as the "village becomes a factory town." Wastes from all the people who collected in these factory towns and cities still needed to be disposed of, and the volume of those wastes grew exponentially.

Naturally, inadequate waste disposal made cities more noisome and unsightly. But poor waste disposal combined with population densities also contributed to disease. Disease microbes need live hosts. To survive as a strain, microbes need to send out colonies to new hosts. When I'm struck with a cold, for example, I contribute to the process of maintaining the cold virus by coughing, sneezing, and touching my mouth and nose and then leaving behind germs on whatever I touch. If I lived in isolation, the virus would eventually die out as my antibodies killed it off. To survive, the virus needs to move to another host before its first host kills it. Because I live with others, the virus that attacks me is well positioned to perpetuate itself.

Some microbes, like those responsible for the common cold, are mostly innocuous. Others, however, are deadly. The need to move to new hosts is particularly strong among the deadlier microbes. As they move from host to host, some diseases depend upon other species to act as intermediary vectors. Malaria, yellow fever, and West Nile disease use mosquitoes to get from host to host, while typhus relies upon lice. Some of these microbes move between different types of hosts. Most contemporary diseases arise within animal populations before spreading to humans. Although this was not the sole reason, this trend accelerated once humans domesticated animals and began living in close proximity with them (although some diseases such as yellow fever or more recently AIDS migrated from wild animals to people). For millennia domesticated animals shared living quarters with their herders or owners. This facilitated the ability of disease agents to cross species and infect humans. Measles originated as rinderpest in cattle. Smallpox evolved from cowpox or some other related livestock poxvirus. The Spanish influenza, which killed some 21 million people at the end of World War I, came from pigs. The appearance of avian flu in Asia in 2004 was the result of a virus jumping from chickens to people. The SARS (Severe Acute Respiratory Syndrome) outbreak in the early twenty-first century spread from wild civet cats kept in markets and sold to people for meat in China.

During the millennia that humans have wandered the earth, several diseases have emerged that move between people. Those infected either get better or die. Survivors usually carry immunity to the microbes. To continue to exist, the microbes must find new hosts who lack immunity. This can be accomplished two ways: first, by moving across populations in search of ever-new hosts or, second, by attacking a new generation born without the immunity of the original hosts. Generally speaking, to sustain themselves infectious diseases need a human population sufficiently numerous and dense to provide a constant source of new hosts. The population of fifteenth-century Europe fit these criteria perfectly, as Columbus pointed his boats west to the New World and its unsuspecting inhabitants.

European cities were crowded places where diseases spread quickly from person to person, and continuous trading across the continent into Asia transported these diseases over great distances. Cholera, which arrived in Europe in the nineteenth century, spread to the United States in 1832. Dysentery, measles, rubella, smallpox, polio, typhoid, tuberculosis, mumps, diphtheria, pertussis (whooping cough), and typhus were commonplace in European cities, and they took their toll on city populations, usually among the young. Until the second half of the nineteenth century, European cities needed a constant influx of new migrants from the countryside just to maintain population levels, let alone to grow in size. The diseases that need dense populations to thrive are known as crowd diseases.

When the colonists came to America, they brought their crowd diseases with them, inflicting the catastrophic suffering and death on Native Americans, as noted in earlier chapters. Settlers spread out across the countryside. Within a couple of generations, Americans had become a rural people, with less than 6 percent of the population living in towns or cities by 1790 (U.S. Bureau of the Census, Historical Census of the United States, Bicentennial ed. 1976). One result of this population dispersal was that Americans were healthier and lived longer. Indeed, by the late colonial period, Americans visited Europe with trepidation because of its reputation for disease. But, because the very prosperity of this new land encouraged more trade and more densely settled communities, it also increased potential exposure to infectious diseases.

Boston, Philadelphia, and New York all claimed populations exceeding 10,000 in 1790 (U.S. Bureau of the Census, 14th Census of the United States 1920, 62–75). The Americans gathered in these population centers lived along coastal bays or tidal rivers. They lived and worked in locations scattered about the urban space. Privies certainly contributed to the noxious smells of the early American city and compounded the problems of dysentery, but as long as urban densities remained small, the privies sufficed for waste disposal. In the early years of the nineteenth century, however, city populations began to overwhelm nature's ability to dispose of their wastes. Smells from manure, rotting garbage, and night soil could be detected from on board ships coming into harbors, providing visitors to these coastal centers with an unexpectedly pungent greeting. Roadside ditches that gathered rain runoff and household slops, but did not drain, and dams, streets, and poorly built bridges that cut down or cut off stream outlets to larger rivers created large pools of stagnant water for mosquito larvae without accompanying larvae-eating aquatic animals. Cisterns, shallow wells, and discard barrels and buckets also created a thriving urban habitat for mosquitoes. Wells became dangerously polluted and the incidence of epidemics grew more common.

In 1793 a yellow fever epidemic struck Philadelphia, killing one in twelve of the city's inhabitants (Bass Warner 1972, 25). More than 23,000 people fled the city. African monkeys originally spread yellow fever to humans through mosquitoes, and African slave traders carried the disease to the New World. In the Americas, yellow fever traveled with mosquitoes and trade. Eighteenth-century physicians had little knowledge of germs, microbes, and mosquitoes as disease vectors. But what they did know was that the air smelled far worse in the city than in the country, and the city was where more peopled died. For this reason, they associated the disease with the city. Those who tasted well water from the various sections of town knew that the contents of privies had penetrated into wells, particularly in the poor districts. The poor, who lived in filthy, foul-smelling neighborhoods, were hit particularly hard. The air in poor neighborhoods reeked of rotting garbage, left in the streets for roving pigs to scavenge, and wastes from poorly maintained overcrowded privies. The city's poor bore the brunt of the disease because they lacked the resources to leave town during the epidemic. They did not enjoy the more nutritive and sustaining diets of their better-off neighbors and hence were more liable to die once they contracted the disease. Their homes also tended to be in back alleys with standing water and overflowing privies and on swampy or marshy less-desirable building sites where mosquitoes tended to breed and congregate. Although doctors did not understand the cause of the epidemic, city leaders came to believe that by cleansing the streets and yards of filth and providing clear water for drinking and washing the streets, they could help prevent the disease from reoccurring. Yellow fever's return in 1797 and 1798 finally convinced the city leaders to develop a waterworks that tapped the Schuylkill River northwest of the city and pumped the water into the city. The system did deliver clear water to the city (or at least to the city's hydrants), until it became overwhelmed by city growth twenty years later.

In the 1820s, when more than 114,000 people lived in Philadelphia and its immediate environs, it became clear to city leaders that a new water-delivery system was needed. Concerned about maintaining a continuous supply of clear water for drinking, cleaning, washing down streets, and fighting fires, the city water committee proposed damming the Schuylkill and using waterpower (waterwheels at the dam) to pump the water to large reservoirs on Fairmount Hill, from whence a gravity-fed system would deliver the water to the city. In the 1830s, the system was completed, just as middle-class Americans were experiencing a revolution in water usage. By 1837 almost 1.5 percent of Philadelphians had installed bathrooms with running water. As a result, average water consumption doubled from the city's 1823 levels to 20 gallons per person per day. By 1850, when almost 10 percent of city residents had water closets and another 2 percent baths, Philadelphians during the summer months were using 44 gallons of water per person per day (Bass Warner 1968, 107).

The surfeit of clear, fresh water relieved the minds of city leaders. They surrounded their waterworks with the nation's earliest large, beautiful urban parks. The new influx of clean water also deflected attention from the other half of the urban environmental health equation: wastes. Wastewater from Philadelphia and its surrounding communities first flowed into privies, then ditches, and ultimately into streams that drained into the Schuylkill and Delaware, polluting the city's once clean water. By the end of the century, Philadelphia's death rate from typhoid, a disease contracted by drinking water contaminated by human waste, was higher than New York's or Boston's.

Philadelphia's water system became a model for others. But because it was first, Philadelphia also had to solve the problems of its system on its own. At the turn of the nineteenth century, few Americans knew much about engineering or hydrologic design. Millwrights had a practical knowledge of dam and race construction, but there were no schools, college engineering courses, or even centers for apprentice training in engineering. When confronted with its yellow fever crisis, Philadelphians looked to a visiting British engineer and architect, Benjamin Latrobe, for a plan to provide water.

With fewer than thirty civil engineers prior to 1816, America suffered from a shortage of trained experts. At the same time, America embarked on an ambitious canal- and dam-building campaign. Engineers for these projects were either hired from Europe or trained on the spot. The canals' construction launched the careers of many of the nation's early civil engineers.

John Jervis, who designed New York City's water system and helped Boston rebuild its system, was a veteran of the "Erie School of Engineering." He went on to become chief engineer of the Chenango Canal Company, for which he designed a number of artificial reservoirs to supply water for the canal. Jervis tapped into that knowledge when he was hired by the New York City Water Board to construct a system to bring clean water to the city in 1836. For the next six years, Jervis oversaw the building of the Croton aqueducts, dam, and distributing reservoir system that helped relieve New Yorkers of the horrors of drinking the polluted waters of the East River, local wells, or the arbitrary actions of the private Manhattan Water Company. Although the system was hailed as a success in 1842, the city's growing population forced a radical rebuilding of it in 1858.

Like Philadelphia, Boston confronted the problem of polluted water supplies in the 1790s. From Boston's settlement until 1796, city residents drew their water from underground wells or cisterns that collected rainwater. By the last decade of the eighteenth century, Bostonians' water quality appalled even the stiffest Puritan. The water was hard, discolored, odorous, saline, and—not surprisingly—bad tasting. In 1796 the Aqueduct Corporation received a charter from the General Court to supply Boston with water from Jamaica Pond, but the corporation was never able to supply the whole city. The city's rapid expansion after the War of 1812 exacerbated the problem. Concern for an adequate supply of water for fire protection added to the city's fears. By the 1830s, the condition of Boston's water reached crisis levels. An 1834 survey found that polluted surface and groundwater had contaminated wells throughout the city—a quarter of the city's 2,767 wells either failed regularly or were so polluted as to be classified nuisances. Clean water became an obsession among civic leaders. Finally, after beating back opposition by private water companies and well owners wary of higher taxes, Mayor Josiah Quincy Jr. in 1846 pushed through an "Act for Supplying the City of Boston with Pure Water." By this time Bostonians were not only concerned about odorous and salty water, but they had become convinced by city doctors that noxious, polluted water contributed to disease. Temperance activists contributed the idea that clean water would offer the city's working population an alternative to alcohol for drinking. Under the direction of Jervis, Boston's water committee began construction on a major waterworks to bring the clear and abundant water of Long Pond (renamed Cochituate after the building of the waterworks) to Boston's residents. The water board ran and expanded the system so that the city no longer suffered water shortages. This was never an easy task, however. By 1860 the average Bostonian was using 100 gallons of water a day for a total daily outflow of 17 million gallons. In 1878, Boston completed the Sudbury River and Mystic Lake connection to its water system, adding an additional capacity of 62 million gallons per day (Nesson 1983, 10–11).

Clean water also improved citizens' health. Although it was not until the late 1870s and 1880s that American scientists became convinced of the germ theory, they did believe that malodorous water contributed to disease. After the introduction of readily available clean, clear water, Boston never again faced the large-scale epidemics of typhoid, cholera, and typhus that plagued crowded nineteenth-century cities. Even so, late-nineteenth-century Boston could not be considered a healthy city.

For early-nineteenth-century cities, rainwater posed a serious problem. Because streets were either dirt, brick, or cobblestone and laid down with little thought for grading or elevation, rain quickly turned roads into impassable streams and puddles. Garbage, trash, manure, ashes, and human wastes mixed with rainwater to create foul, mosquito-breeding slurry. Those making the unpleasant trip across the street, let alone across town, risked, at the very least, ruining their clothes. And rain came often to northern cities. Boston responded with a campaign of storm drain construction. These drains channeled water off the streets and into the nearest waterways—the Back Bay along the Charles River, the Mystic River to the north, and Boston Harbor.

Storm drains built by New York, Philadelphia, Boston, Pittsburgh, Cincinnati, and Chicago removed rainwater, but they did little to remove human and animal wastes. Storm drains were meant to clear the streets of surplus rainwater, but increasingly more than just rainwater plagued urban communities. Increased populations compounded the human waste problem. Most waste was directed into privies and privy vaults. In the country, when a privy hole filled, the family simply filled over the hole and dug another. In the city, shortage of space precluded such tactics, so privies had to be emptied. In better-off neighborhoods privy vaults were built to hold the waste until it could be hauled away by "night soil" scavengers who sold it to surrounding farmers for fertilizer or simply dumped it in local streams or rural areas.

Increasingly, privies and privy vaults overflowed into cellars, yards, and streets. Even well-maintained privies saturated the surrounding soil with wastes. Scavengers hired to remove night soil from vaults often left behind a stream of waste that sloshed out of the back of their wagons. Rules passed by concerned city councils regulating privy vault cleaning were seldom enforced and completely ignored in poorer neighborhoods, where landlords had little interest in cleaning privy vaults and tenants lacked the necessary resources. In 1833, in an attempt to get a handle on the problem of privy overflow, Boston passed an ordinance allowing landowners to connect privy vaults to the city storm sewers. Other cities also provided landowners the opportunity to connect to storm sewers. As a solution to the problem of waste, however, this practice not only failed in the short term—it also proved disastrous over the long run.

Wealthier neighborhoods and the commercial districts were first to have storm sewers, built primarily in response to landowners' requests. The wealthy were also the first to connect their privy vaults to the new sewers. In most cities, abutting landowners were responsible for the cost of laying the sewers. Renters in poor neighborhoods and back alleys seldom had even storm sewers laid down to which to connect. Even for the poor who did have access to a storm sewer, their landlords were typically reluctant to pay to connect to the line. As late as the 1870s, most urban residents still used privies and cesspools. By 1870, Philadelphia still had 88,000 vaults and cesspools, while close to three-quarters of New York's residents had no access to city sewers (Tarr 1996, 114–116). Pittsburgh's board of health warned the city in 1875 that overflowing privy vaults threatened the city's health. In 1880 the Allegany county engineer noted that overflowing cesspools were what was available for the waste needs of the city's southside. Even a connection to a sewer line did not free a neighborhood from sewage problems, however. Human wastes now flowed into storm pipes that were often either blocked or leaky and that inevitably dumped the wastes untreated into the nearest body of water. Wastes built up around sewer outflow pipes, blocking the pipes and fouling the surrounding areas. But the failure of the storm sewers ultimately came home to city leaders when urban Americans began taking advantage of cities' new water systems.

As in Philadelphia, the availability of water flowing under gravity pressure into residents' homes in other cities encouraged additional consumption there as well. Running faucets and flush toilets became a necessity, not only for the wealthy, but, increasingly, for all but the poor. With flowing clear water, city residents bathed more, washed more dishes, and flushed their wastes down toilets. Cesspools and privy vaults were quickly overwhelmed. Wastewater flowed into yards and from there into streets and alleys. Cities began passing ordinances like Boston's 1871 law requiring landlords and property owners to connect to city sewers. Pittsburgh outlawed cesspools where sewer service existed and ordered the removal or cleaning of the thousands of privies scattered about the city. Such ordinances failed to solve the waste problem; inspectors were in short supply and landlords and poor property owners resisted the costs of connecting to sewers if they were even built in the area, and poor areas of the city were seldom serviced by sewer lines.

Storm sewers were designed to carry rainwater off of city streets. Even in this function they were plagued with blockages. Mud built up at every turn. Poorly built lines failed to maintain continuous downward grades. Outflow pipes—particularly in low-lying areas—often dumped their waters below the high-water line, leading to backflow at high tides or floods. The tidal basins that originally received rainwater and waste runoff were regularly flushed with tidal and river currents. Just as these wastes were being dumped into basins such as Boston's South Bay, Back Bay, and Roxbury Canal, landfill, bridges, and street construction were effectively filling in much of the basins, serving to limit or cut off the larger water currents. Bays and basins became shallower, and wastes began to build up. In the age of the "walking city," land near the urban center was in short supply. In response, owners of marginal marshes and tidal flats filled in marshy areas and constructed homes, warehouses, and wharves. As development spread across low-lying areas, sewer lines were extended, but without the proper grade to guarantee that sewage would continue running out. The problems of tidal in-wash (tidal water pushing back into the sewer lines and flooding basements) became even more acute.

Greater water use increased these problems exponentially. Outflow pipes fouled ever-larger areas. Wastes pilled up around and over these outflow pipes. As accumulated wastes filled the harbor bottom and reduced clearance, harbor wharves had to extend docks or dredge channels to accommodate boats and ships. Responding to complaints, sewer departments laid more and more pipes and dredged canals and riverfronts to reduce the buildup of wastes and the accompanying odors. More homes were connected to the pipes, but the wastes continued to flow into the nearest stream or tidal outlet, and the odor problem remained. Although they failed to understand why poor sanitation caused disease, the public and civic leaders recognized that there was a link. Unaware of the germ theory of disease causation, they simply assumed that foul-smelling air carried disease. Believing that the "poisoned atmosphere"—miasma—caused typhoid, diphtheria, meningitis, cholera, and pneumonia, doctors and laypeople alike called for sanitary reform.

Boston's Back Bay suffered particularly from poor drainage. As early as the 1850s, reformers were calling for tide gates and standard grades for sewage pipes, but new ordinances and new sewer mains failed to remedy the problems. By 1870 the Back Bay, on the very doorstep of the state capitol and fashionable Beacon Hill, was avoided not only because of foul smells emanating from its stagnant waters, but also because residents feared disease. Other sections of Boston besides the Back Bay also suffered from nauseating odors and festering wastes and garbage. Road construction along the waterfront had cut the flow of deeper harbor water from the South Bay. Sewage, garbage, and wastes that had initially been flushed with harbor tides and storms now lay stagnant and odorous. The Roxbury Canal built at the end of the eighteenth century to move goods in and out of Roxbury, increasingly became the dumping ground for wastes from South End industries, particularly soap-making and tannery factories.

In 1875, Boston appointed a committee to find a solution to the city's sanitation problems. The committee reported back a year later that the problems rested on old tide-locked sewers, badly laid pipes clogged with debris and lacking proper down-flowing grades, and leaking cesspools. The committee also argued that the problems were regionwide and needed a comprehensive solution. The committee members suggested building a massive sewer main that would intercept wastes before they left existing outflow pipes. This new drain, the Main Drain, would then carry the wastes out to Boston Harbor to be released with the outgoing tides.

The Main Drain and the intercepting sewer lines captured a majority of Boston's wastes and dumped them far enough out in the harbor to avoid waste accumulation in stagnant pools and basins. What the Main Drain could not accommodate was storm water, the originating reason for city sewers.

Boston's Main Drain, like others built in New York and Philadelphia, was designed to accommodate average waste flows. Summer rainstorms filled city streets with millions of gallons of water an hour that needed to be removed from the city quickly. Designing and building a system to accommodate the maximum amount of water that might flow through it would have been extremely expensive. City leaders instead opted to build a system to accommodate wastes based upon normal human use. For larger storms, the system relied on special outflows that mixed excessive rain with sewage and channeled it all directly into the closest waterway. It was assumed that the storm water would dilute the sewage and the combined high volume would be enough to wash the wastes away from settled areas.

Fear of disease and oppressive foul odors mobilized city residents in the larger urban communities elsewhere along the East Coast as well. By the 1880s, patterns for sewage and waste pollution removal in coastal cities were firmly established. Unpleasant odors, foul-tasting water, the disgusting sights of rotting garbage and festering human wastes, and epidemic diseases drove public action. Public waterworks that captured and conveniently brought clearer upstream water to urban households through gravity-fed systems solved some of these urban sanitary problems, particularly for the better-off urban residents. But while clear water under pressure could travel through hoses to clean streets and through pipes to bring clean water into the home, it also exacerbated the problems of water wastes. Water consumption increased dramatically, especially with flush toilets, and with increased consumption came increased wastewater. As private privies and cesspools became overwhelmed, citizens began to link waste pipes to city storm sewers, effectively dumping the wastes into the nearest waterway. For coastal cities, those waterways were often harbors and tidal basins. When shorelines became unbearably polluted with out-flowing wastes, coastal cities responded by laying longer sewer lines that reached deeper into coastal waters.

Although they never completely solved the problem of diseases from wastes, Boston, New York, Philadelphia, and the other port cities did manage to find supplies of clear water and eventually locate their sewer outflow pipes far enough out into the tidal wash so as to mitigate the worst of their sewage waste problems. But not all the nation's cities had the advantage of tidal waters to carry away their unwanted by-products.

INLAND CITIES

In 1882, Theodore Lyman, whose father as mayor of Boston in the 1830s had advocated a municipal water system, noted that in his youth there were few mills, but by the 1880s, there was "an extraordinary amount of manufactures and of the greatest variety" (Cumbler 2001, 182). Waterpower had originally moved the machines of those manufactures. Sawmills, fulling mills, gristmills, and even the early carding and spinning mills such as those of William Levi and Joseph Chapin needed relatively little waterpower to turn their stones, saws, carding machines, or spinners. But when Francis Cabot Lowell and Nathan Appleton hired Paul Moody to build looms, they were not thinking of a small dam. The Boston Associates' mills at Waltham were successful, but could be even more so if not limited by the Charles River's sluggish pace. In 1822 the company looked for a site for an even larger enterprise. Rivers along the northeastern coast crossed several fault lines as they descended out of the Appalachian mountain chain. The Pawtucket Falls was the product of one such fault line. The Boston Associates created the Merrimack Company to take control and ownership of the waterpower at the site. They built a dam across the river, flooding some 1,000 acres of land and creating a drop of over thirty feet. The water that built up behind the dam was channeled into canals and from there into waterwheels that powered the machinery inside the mills that had been built along the canals.

Hundreds of workers came to work in these mills and their supporting factories. Boarding houses and homes were built, and shops and stores went up along side streets. By 1860, Lowell boasted over 36,000 inhabitants; by 1880 that number had reached almost 60,000 (Steinberg 1991, 216–217).

Their success in capturing waterpower and turning it into profits encouraged the original Boston Associates to seek out new sites for manufacturing centers and spurred others to imitate them, and not only in New England. West of New York City, where the Passaic River crosses a major fault line in Paterson, New Jersey, mills went up and workers crowded into tenements and boarding houses. Similarly, on the Brandywine River southwest of Philadelphia and on the Delaware River north of the city, mills and factories produced dense human settlement around the falls at Lambertville and Trenton.

While waterpower drew the investors inland to dam the faults along the nation's major waterways, it was the mills that went up beside these dams that drew the people. And these people arrived in unprecedented numbers. Throughout the first half of the nineteenth century, most of the nation's urban growth hugged the coast. The only notable inland population growth was concentrated among the growing industrial cities of the interior that tapped inland waterpower. Lowell reached a density of 2,334 persons per square mile by 1850, while Lawrence, which grew 113 percent between 1850 and 1860, in that same period jumped from 1,150 persons per square mile to 2,450 by 1860 (Steinberg 1991, 216–217).

These new and rapidly growing inland industrial cities approached the problems of wastes much as the coastal cities had. Originally, landlords and homeowners used privies and cesspools to dispose of human wastes, while horse manure and garbage was left on the streets to be dealt with by scavengers and concerned property owners. Factory and tenement buildings presented a more difficult problem. While a privy hole might suffice for a family or even a small shop of a half dozen hands, the new industrial mills of five and six stories employed hundreds of workers tightly packed together for thirteen or fourteen hours a day. The human sewage alone from that many people was immense. Companies built privies and later water closets behind the mills that emptied into ditches that ran either into sewer lines or directly into the nearest stream. Except for periods of heavy rain, however, the streams were seldom capable of carrying away the volume of wastes that flowed out of mill privies, much to the chagrin of any who lived near the streams. Health inspectors found that one stream running down from a series of mills on the Connecticut received a daily average of 2,500 pounds of feces and 3,500 gallons of urine (Massachusetts State Board of Health Fourth Annual Report 1873, 40). Though the stench from these waste streams was unimaginable, little was done to ameliorate the condition.

Besides the mills, the growing industrial cities' tenements and boarding houses also demanded some means of disposing of wastes other than the traditional privy. Residential densities grew with the increased density of workplaces. Several thousand people working in a single mill needed housing within walking (or later trolley car service) distance from their workplaces. Five- and six-story tenements soon replaced two- and three-story boarding houses. Tenement water closets emptied into sewer lines that in turn emptied into the nearest body of water. The city of Lowell's first storm sewers flowed into the Middlesex Company's millpond. Until the 1870s, most of the runoff into the pond was rainwater, but increasingly sewage from tenement houses and boarding houses came to predominate. The sewage began to build up until the Middlesex Corporation in 1884 sued to stop the city from discharging into the pond and to remove the deposit of sludge.

Most urban residents accepted the foul odors emanating from local streams and brooks as a condition of modern life. Some towns and cities even renamed local streams to reflect their new role as waste receptacles. Town Brook in the small manufacturing town of Keene, New Hampshire, was renamed Town Sewer. When odors became overpowering, many towns, in an effort to hide the smell, built culverts and diverted local streams into covered pipes. But hiding sewage did nothing to solve the greater problem of waste disposal.

When the local sewage-filled streams became too offensive, mills and mill towns looked to follow the model of larger cities like Boston, New York, and Philadelphia for solutions. As in Boston, the mill towns built sewer systems connecting privies and cesspools to sewer lines and then directed the sewage into the larger rivers and streams, in hopes that the sewage would be diluted and flushed away. As the Massachusetts Board of Health noted in 1873, "the temptation to cast into the moving waters every form of portable refuse and filth to be borne out of sight is too great to be resisted" (New Hampshire State Board of Health 3rd Annual Report 1884).

Rivers and streams absorbed ever-increasing volumes of sewage. A New Hampshire doctor, A. H. Crosby, estimated the average person's daily amount of solid and fluid excretal sewage was approximately equal to three cubic feet per day with another three cubic feet of water from daily water usage. Multiply those figures by the tens of thousands of people who crowded into the inland cities and it becomes clear why torrents of wastes soon overwhelmed natural water systems. In 1888 the Connecticut Board of Health noted that an industrial village of only 2,000 people dumped a half-ton of fecal matter and 500 gallons of urine into a local stream daily (Connecticut State Board of Health 10th Annual Report 1888, 203).

Unlike Boston, New York, or Philadelphia, these inland cities did not have tidal water to wash away their wastes. They instead dumped their wastes into the rivers and streams that ran through their communities. At a time before the scientific works of Lister, Pasteur, Henle, and Koch that linked disease to microscopic germs were commonly accepted in the United States, most people believed that water naturally cleaned itself through dilution and flow. Nineteenth-century Americans believed that diseases stemmed from foul odors, nasty tastes, and visible dirt. Reasonably enough, clear water without an unpleasant odor or taste was assumed to be clean.

Large bodies of running water did seem to clean the water of sewage. Although the shore immediately adjoining the sewage outflow pipes might have looked filthy while giving off nasty odors during periods of low water, the same river a couple of miles downstream might typically look and smell clean. Firm in the belief that running water cleaned itself, cities responded to growing wastes by extending their outflow pipes farther out into rivers or diverting the sewage into interceptor lines that dumped it downriver. But unlike the coastal cities' wastes, which were dumped into harbors and washed by tidal flows, the wastes inland cities dumped into rivers and streams flowed downriver past other towns and communities. Those towns and communities used the water for drinking and cleaning.

The practice of dumping ever-greater amounts of sewage into the nation's waterways led to a general outcry over pollution. Downstream farmers became concerned that their animals might grow sick from drinking polluted water. People did believe that malodorous water caused disease, and as water systems grew more polluted, general concern also grew that even the larger bodies of water like the Delaware, Passaic, Hudson, Connecticut, Merrimack, Ohio, or Mississippi rivers might become too polluted to clean themselves. Health investigators reported in 1888 that the Connecticut River was receiving over 42.25 tons of fecal matter and 45,900 gallons of urine from the large industrial cities that lined its banks (Connecticut State Board of Health 10th Annual Report 1888, 260).

The Passaic River, which in the early nineteenth century had enjoyed a flourishing fishing industry and was a major recreational area, became so polluted by the second half of the century that Newark was forced to abandon it as a water supply. The fishing industry was destroyed, and homes along the river were abandoned because of the stench that rose from the Passaic, especially in hot weather.

Epidemics of typhoid, meningitis, diphtheria, dysentery, diarrhea, and cholera became common in the industrial cities of the interior, particularly in cities that took their drinking water from bordering rivers. Farmers began to sue upstream towns for polluting the water and harming their livestock. In 1906, Missouri, on behalf of the city of St. Louis, sued Illinois and the Sanitary District of Chicago over Chicago's dumping of sewage into the Illinois River that in turn ran into the Mississippi. Chicago's wastes, including excessive waste effluent from the city's meatpacking industries, flowed into the Chicago River. Odor from the river permeated the whole south-central part of the city. The river emptied into Lake Michigan, from which Chicago took its drinking water. Fear of cholera and typhoid epidemics, combined with what the Chicago *Times* referred to as the river's "stench," led city leaders to finance a major engineering project to redirect its sewage away from Lake Michigan and into the Illinois River. The river was diverted, directing Chicago's sewage southwest through the sanitary canal into the Illinois River and away from the city's water supply intake pipe. Chicago's victory over sewage pollution became a problem for St. Louis, which, in drawing water from the Mississippi River, found itself on the receiving end of wastes from the nation's second-largest city. The case ultimately went to the Supreme Court as *Missouri v. Illinois and the Sanitary District of Chicago* (1900) (180 U.S. 208; Ct.331; 45L. Ed. 497; 1901 U.S.) where Justice Oliver Wendell Holmes wrote, reflecting the still-confused state of science concerning pollution, that it was not clear that it was Chicago's pollution that was harming the health of St. Louis residents. Holmes also argued that Missouri was not in a good position to sue Illinois for harm because other Missouri towns north of St. Louis also dumped raw sewage into the Mississippi. Moreover, St. Louis's own wastes also flowed downstream past other towns and cities taking in the river water. Holmes said that St. Louis did seem to experience a dramatic increase in typhoid after Chicago's wastes were redirected, yet he also noted that the sanitary district's scientists had evidence that the Illinois River had become clearer because the dumped sewage was mixed with an influx of clear Lake Michigan water. Holmes accepted Chicago's argument that St. Louis should solve its own problems with waterborne disease by cleaning river water before using it, rather than trying to prevent Chicago from following the time-honored practice of sending wastes downstream.

Pollution and Health 163

Homestead Steel Works Plant in Pittsburgh, Pennsylvania, 1907. (Corbis)

St. Louis responded that it wasn't downstream from Chicago, but that only through the reengineering of the Chicago River and the drainage canal had the city managed to move its wastes into the Mississippi watershed. Chicago's natural watershed was Lake Michigan, St. Louis argued, and that is where it should dump its wastes. Holmes disagreed, noting that the Court "c[ould] not but be struck by the consideration that if this suit had been brought 50 years ago it almost necessarily would have failed. There is no pretence that there is a nuisance of the simple kind that was known to the older common law. There is nothing which can be detected by the unassisted senses—no visible increase of filth, no new smell. On the contrary, it is proved that the great volume of pure water from Lake Michigan which is mixed with the sewage at the start has improved the Illinois River. . . . The plaintiff's case depends upon an inference of the unseen" (*Missouri v. Illinois and the Sanitary District of Chicago* 1906). Holmes realized, however, that just as the science of the present made the Missouri case untenable, science of the future might indicate a different outcome for such cases, saying "what the future may develop of course we cannot tell."

As inland cities like Lowell, Holyoke, Hartford, Chicago, and St. Louis dumped their sewage into the nearest down-flowing outlet, Pittsburgh and Cincinnati similarly emptied their wastes into the Ohio River, with smaller cities along the river following suit. And from the Ohio River they all drew drinking water. Pittsburgh's sewage flowed into the Ohio River, but in 1900, 350,000 people dumped their untreated sewage into the Allegheny River, from which Pittsburgh drew its water. While water from the nearest river or lake provided cities with adequate supplies of pressurized water for fire protection and cleaning city streets, it did not necessarily provide safe drinking water. In its case against the Sanitary District of Chicago, Missouri argued that Cincinnati was plagued with disease from Pittsburgh's sewers 200 miles upstream, and that Buffalo suffered from high rates of typhoid caused by the sewage Cleveland discharged into Lake Erie.

New York, Philadelphia, and Boston reached beyond their borders to capture water upstream, before sewage wastes polluted it. Boston built its Main Drain sewer lines and ultimately formed the Metropolitan Water District to keep pollution out of its water system. New York built suburban waterworks as early as the 1840s, and by the turn of the century was capturing pure water as far away as the Catskills. Philadelphia protected its water supply by purchasing outright much of its upstream watershed. But few cities had the political or economic muscle of New York, Philadelphia, or even Boston to take control over pure water far to the interior. Most cities depended on the nearest water supply, whether the local stream, river, or lake, and difficulties arose when others used the same water system to dispose of wastes.

Nineteenth-century mortality data testify to the crisis of waste-filled water supplies. The numbers of infant deaths from diarrhea and dysentery were appallingly high—especially so in interior industrial cities. Cholera and typhoid took a particular toll on urban residents. In 1863 the mortality rate in Holyoke, Massachusetts, was 33 per 1,000 live births, the majority of deaths resulting from intestinal diseases. The story was the same in other U.S. cities. Philadelphia, Newark, Louisville, and Pittsburgh all had yearly death rates of over 50 per 100,000 inhabitants from typhoid. From 1873 to 1907 Pittsburgh had over 100 yearly typhoid deaths per 100,000 people, and in 1882 the number reached 150.

In 1870, New York City's infant mortality rate was 65 percent higher than it was in 1810 (Melosi 1980, 16–17).

Concern over urban deaths and the link between poor living conditions and high mortality had been growing since the middle of the nineteenth century. In 1850, Massachusetts established a special commission under Lemuel Shattuck to investigate sanitary conditions. In his *Report of the Sanitary Commission*, Shattuck demonstrated that mortality was significantly higher in dense urban communities than in sparsely populated areas. Similar studies were conducted in New York and Pennsylvania. By the 1880s, concern reached a fevered pitch. Legislatures began considering antipollution legislation, towns were suing towns, and health officials and reformers were lobbying for laws to provide "clean air, clean water, and clean soil."

In the late 1870s, the work of European scientists, particularly Robert Koch, Joseph Lister, Louis Pasteur, and Jacob Henle, pointing to germs as the cause of disease began to influence Americans concerned about public health. These Europeans found that disease spread when microscopic germs were ingested by a victim who touched a contaminated person, ate contaminated food, or drank water containing germs. Although reluctant to abandon the belief that disease arose from general filth and miasma, Americans increasingly recognized the likelihood that germs in contaminated water were the true culprits. With death rates from waterborne diseases such as typhoid climbing, city leaders and health officials became increasingly concerned about germ-contaminated water supplies.

In 1887, Massachusetts empowered its state board of health to conduct a new investigation of the state's waters. Helen Swallow Richards, a laboratory-trained biological chemist from the Massachusetts Institute of Technology (MIT) who understood chemistry, biology, and germs, conducted the study. In 1887 the state board of health authorized William Sedgwick, the new professor of biology at MIT, and Hiram Mills, a hydraulic engineer from Lowell, to set up an experimental station in Lawrence to investigate water treatment. Reflecting the optimism and confidence in technology and science peculiar to late-nineteenth-century America, Mills, particularly, believed that the solution to the problem of pollution lay in engineering. He was confident that with the proper application of science and technology, America's water could be rendered safe.

In the winter of 1890–1891, typhoid struck the Merrimack River cities of Lowell and Lawrence, and the following year the epidemic hit cities along the Connecticut River. Deaths from typhoid rose from the midteens per 100,000 to over 40 per 100,000. Lowell reported 171 typhoid deaths in November alone (Steinberg 1991, 232–233).

Spurred to action, the Lawrence experimental station scientists worked to develop a method to purify the water of disease-causing germs. They discovered that filtering water through sand and exposing it to air caused a biological decomposition of the harmful bacteria. Their experiments' success led to the development of a filtration system for the water supplies serving Lowell and Lawrence. News of this success quickly spread throughout the nation. Cities sent agents to inspect the work at Lawrence, and filtration soon became a must for water systems drawing their supplies from nearby rivers.

The experiments in Lawrence also underscored that rivers did not necessarily clean themselves of pollutants. While dilution may have reduced the odds of contracting disease, the scientists realized that the rivers did not always eliminate pathogens and certainly not the hearty typhoid bacillus. Americans increasingly became aware that actions upstream could prove harmful for those who drank that water downstream. St. Louis had been a decade too early in its court case against Chicago.

Understanding the interdependence of water systems did not immediately move Americans to clean their waters. Economics and technology still played heavily in the game of urban politics. With the availability of running water under pressure and the establishment of flush toilets as a necessity rather than luxury, urban water use rose dramatically. Hundreds of thousands of gallons of wastewater poured out of city sewer lines daily. City interceptor lines took these wastes downstream or further out into rivers and harbors. The slogan during the nineteenth century for water engineers might have been "the farther away the better." The work at the Lawrence experimental station suggested that this water could be filtered to remove most of its harmful matter. The notion of filtering millions of gallons of water did not appeal to most budget-conscious urban leaders, however. Yet, city leaders were forced to confront the reality that human sewage pollution in their water systems caused disease. From there it was no great leap for them to figure that it was easier and cheaper to clean their drinking water and then simply dump the used, waste-filled water downstream, rather than clean the water after use and expect others upstream to do the same. Cleaning the water after use would have produced clean safe water, but at a much greater expense, and only if everyone was on board. It was also more difficult to convince ratepayers to foot the bill for an expensive water-filtering system for the benefit of downstream towns. The more expensive filtering also required regional water-systemwide cooperation and trust, something in short supply in nineteenth-century America. This plan would have worked to clean waters a generation or more sooner than was the case, but such a solution required too much of nineteenth-century leaders. Instead, reflecting the values and priorities of the time, they chose to clean their water before using it and to let the downstream communities look out for themselves.

Having made the decision to clean their drinking water, communities continued to look to science and technology to protect their water supplies. In the

1890s, George Fuller and George Whipple discovered that coliform bacteria in water indicated human and animal feces contamination. Thereafter water boards could test their water for fecal pollutants as well as pathogenic organisms. Cities and states across the Northeast began establishing biological stations and laboratories to investigate the quality of urban drinking water and recommending filtration and later chemical cleansing additives where tests indicated a need.

Much as cities and towns adopted filtration and chemical cleansing additives to clean their water, at the same time many continued to resist cleaning up their own pollution. Court challenges, however, kept these communities on the defensive. Experiments in 1890 in Brockton, Massachusetts, proved that sewage treatment was viable, but the key issue of cost remained, especially for cities much larger than Brockton, like Chicago or Pittsburgh, where dumping untreated water still seemed both more sensible and economical. Certainly downstream towns, especially those without elaborate water purification systems, would have preferred that the larger upstream cities move to the Brockton model. Several took their cases to the courts, achieving mixed results. In some cases, the courts ruled in favor of downstream towns, encouraging those upstream to install sewage treatment facilities. But in other cases, the courts used a balancing doctrine in deciding against downstream litigants. The balancing doctrine encouraged courts to decide cases on standard cost-benefit terms. If building a sewage treatment facility would not prove prohibitively expensive, and the benefits clearly outweighed the costs, then the courts tended to find the polluting community at fault in harming the downstream riparian user. If, on the other hand, the courts found the costs of building a treatment facility far outweighed the benefit to a downstream riparian user, they seldom found for the litigant. Pittsburgh's wastes may have overwhelmed the farmers and farm communities downstream, but in the eyes of the courts, the costs of cleaning those wastes outweighed the downstream farmers' and small communities' need for clean water.

Human wastes represented just one of the pollutants plaguing nineteenth-century American waters. In 1886, James Olcott warned Connecticut's agricultural board that the region's once clear and clean water supply was becoming "one of the worst in the world." Although Olcott gave due credit to human wastes as a polluting factor, his message was primarily directed at industrial wastes. Industrial wastes concerned Olcott because the mills that brought large numbers of machine operatives to the inland industrial cities generated massive amounts of industrial wastes that were also dumped into the nearest water system. Besides needing water to power their machines, industrial enterprises also needed water in the industrial processes themselves. In textile mills, water washed away dirt, packing salts, bleaches, and surplus dyes. Leather producers likewise dumped organics and tanning acids into the closest running water.

Paper production needed water to process rags and pulp, the wastes of which were unloaded into streams and rivers. In the metal industry water washed and cleaned parts and equipment of caustic chemicals used in the manufacture of tools and metals.

The volume of industrial wastes dumped into the nation's waters was staggering. During a single year a medium-sized cotton textile mill employing 250 workers would on average dump 600 million gallons of polluted water contaminated with arsenic, bleach, acid, starch, dung, dyes, and other toxic matter into the nearest moving water. Water flowing downstream from even a small mill ran black with a frothy mixture of deadly wastes that flowed, in the case of one mill in Lawrence, Massachusetts, for more than a mile downstream (Cumbler 2001, 57).

Industrial furnaces, steam-powered engines, and coal-burning fireplaces also produced prodigious amounts of coal ash. Most manufacturers and cities conveniently dealt with coal ash by depositing it into the local stream, adding yet another pollutant to an already toxic cocktail of arsenic, bleaches, acids, and organic wastes.

For most industries, local rivers and streams provided power and were the most convenient place to dump waste products and used processing fluids. As the Massachusetts Board of Health noted in 1872, "Manufactories are located on river-banks, particularly for the sake of water-power, particularly on account of the desire to use the water in various manufacturing operations, and particularly because the running stream affords the opportunity of readily disposing of waste liquors and other refuse" (Massachusetts State Board of Health Fourth Annual Report 1872, 21). As the industries grew and the processes became more complex, the volume of waste products also grew and became more diverse and deadly.

Wastes, whether human or industrial, affect water systems in several ways. Live water systems contain dissolved oxygen (DO). Dissolved oxygen supports marine life. It also functions as the water system's cleanup crew. By interacting with the oxygen and small microbes in the water, organic matter breaks down in a process that takes oxygen out of the water. Live plants and moving water capture oxygen from the air, contributing to the flow of oxygen into the system. Thus, within the ecology of a water system there is a constant interaction between organic matter, oxygen, and plant life. This process produces nitrates that enrich aquatic plant life, which in turn provide food for fish and smaller marine life. A healthy water system has a high level of dissolved oxygen. Fish need relatively high levels of dissolved oxygen to live. High levels of nitrates poured into a water system generate plant growth that, when it dies, can overwhelm the oxygen level in the system. Less dissolved oxygen means less aquatic animal life. Fish die out. Sewage and other organic industrial wastes absorb DO as oxygen is taken up in breaking down the organic matter dumped into the water. Wastes represent a biochemical oxygen demand (BOD) because they reduce the amount of dissolved oxygen in the water.

In addition to high BOD wastes, the waters of the Northeast were also plagued by the waste chemicals themselves dumped into the waters, which directly killed aquatic life. Solids flushed from industry vats that settled on river and stream bottoms exacerbated the problem by smothering marine life. Nineteenth-century Americans did not understand precisely how wastes disrupted and destroyed water systems, but they did recognize foul water, an all too common sight in the nineteenth and early twentieth centuries.

More difficult than recognizing pollution was controlling it. The use of running water in disposing of wastes was deeply entrenched in habit, custom, and even law. Although common law recognized the rights of downstream riparian owners to water "uncorrupted" by upstream users, in practice the courts used the principle of "reasonable use" in determining whether the "corrupted" water did, in fact, represent a harm or "tort" to downstream riparian owners. But the courts were reluctant to shut down prosperous industries. In an 1872 case, the courts argued that reasonable use necessarily entailed some degradation of water quality. In a fashion typical of many such cases, the courts ruled that any use "will tend to render the water more or less impure," and hence downstream riparian owners should not expect to find their water pure and uncorrupted, providing upstream users were using the water reasonably (*Merrifield v. City of Worchester* (1872), 216).

In 1871, Judge Gray, a Massachusetts superior court judge, noted that "one great natural office of the sea and all running water is to carry off and dissipate . . . impurities" (*Haskell v. City of New Bedford* (1871), 208). By referring to the idea that running water's natural role was to take away wastes, Gray's opinion reflected an important nineteenth-century utilitarian understanding that a resource's role in nature was to serve man. In this case, running water served being a sink to accept human wastes. By its reasoning, if one of water's natural roles was to carry off wastes, then disposing of wastes in local streams and rivers represented a natural use. And dispose of them manufacturers did.

By the early twentieth century, wastes in the nation's major waterways had become genuinely hazardous. The chemical industry and developments in older industries added new toxins to the water mix. By the 1920s, heavy metals, such as lead, and organic chemicals, such as benzene and naphtha, flowed freely into the nation's waters. Coke—cooked coal—that was necessary for steel production was another addition to the wastes polluting American waters. Before 1910 coke was produced in beehive ovens that sent wastes spewing into the atmosphere. Looking to capture tar and other coke by-products and placate citizens irate over the visible air pollution, producers shifted to by-product coke ovens. While these ovens produced significantly less air pollution, they also generated larger quantities of liquid effluents high in phenols and ammonia. City leaders were reluctant to challenge manufacturers' dumping policies because to do so risked alienating those who provided jobs and tax revenue. Pollution's costs were in turn shifted to the environment and to the communities that lived with polluted streams and foul-smelling neighborhoods. Heavy metal and chemical pollution particularly burdened midwestern river systems. Phenols, ammonias, caustic solutions from refinery wash waters, and mine acids flowed freely into the Ohio River basin as well as the lower Great Lakes basin.

Industrial wastes killed marine life and changed the taste and smell of water. When cities added chlorine to the water supply to control pathogenic organisms, the chlorine's reaction to phenols made the water undrinkable. Acid discharges made river water harder. An investigation of the Ohio River beginning in 1913 found the river seriously compromised by industrial, as well as human, wastes. And the Ohio River wasn't the only major midwestern river system compromised by industrial wastes. The Mississippi and the Illinois River basins were so badly polluted that early twentieth-century investigators doubted that they could ever be cleaned up.

Water pollution was not the only pollution that plagued American cities. Although campaigns to clean water supplies dominated nineteenth-century discussion of the environment, citizens also mounted campaigns against smoke particulates in the air. Protection from winter cold always occupied the attention of northerners. European settlers cut and burned cordwood to keep warm. Smoke from these wood fires spewed out of chimneys across the North, mingling with fall brush fires to give the countryside its distinctive fall-winter smoky smell. Fireplace smoke in cities was more of an irritant. Concentrated populations meant concentrated fires for cooking and heating. Wood ash blew about the urban landscape and settled down on sidewalks and streets, soiling pant legs and skirt hems and frustrating those hanging laundry. Improvements in stoves decreased the volume of wood burned and the attending smoke. But as urban areas burned off the cordwood available from the surrounding countryside, more and more homes turned to coal for their source of heat. High-density coal provided more warmth per volume of burnable material than wood.

It wasn't only homes that burned coal. Operators of steam engines, which increasingly replaced waterpower as the source of energy to power machinery, found coal a more convenient and cheaper fuel for their machines than cordwood. Coal also powered the railroads that spread rapidly across the nation. Although Henry David Thoreau, residing at Walden Pond in Massachusetts, complained of the smoke from the railroad engines, it was in the cities that railroad smoke and ash caused the most problems. Adding soot to air already overburdened with household smoke made breathing city air a difficult and halting effort.

Industrial steam plants, coal-burning boilers in tenements and office buildings, railroad engines, and industrial furnaces sent hundreds of tons of particulate pollutants into the urban atmosphere in the nineteenth and early twentieth centuries. Anthracite from the fields of central and northeastern Pennsylvania was the preferred coal for heating homes and for fuel in many industries. Hard anthracite coal was high in carbon and, when properly tended (a practice far too often breached), burned relatively cleanly. When the vast soft bituminous coalfields of western Pennsylvania, southern Ohio, West Virginia, Kentucky, Tennessee, western Indiana, and Illinois opened up, steam boiler operators, railroads, and—with the development of coke—the iron and steel industry switched to bituminous coal. Bituminous coal did not burn as cleanly as anthracite and it contained high levels of sulfur, which added a particularly acidic pungency to its smoke. But if bituminous coal was smoky and acidic, it was also cheap. Bituminous coal outcroppings were scattered across western Pennsylvania, Ohio, West Virginia, western Indiana, and Illinois. Many of these mines were within easy access to river transportation. Although these shallow mines were dangerous to work and plagued with flooding, they were easy to open up and exploit. Cheap bituminous coal flooded into coal markets, pushing out anthracite and even cordwood as the fuel of choice. In most contexts, only 30 percent of bituminous coal was actually consumed in producing heat. The rest went up smokestacks as residue, filling the air with soot. With the electrification of cities in the second half of the nineteenth century, new electric power plants that were initially located within the cities added their coal smoke to the already dense urban atmosphere.

All major inland urban centers found their air thick with smoke and soot, but it was the valley cities like Pittsburgh, Cincinnati, Louisville, and St. Louis that particularly suffered from air pollution, especially in winter months when inversions held the foul air in the valleys. Although east coast cities continued to rely on anthracite and thus avoided the heavy sulfur- and particulate-polluted smoke that plagued inland cities, some nevertheless found cheap bituminous coal tempting. New York attempted to control its smoke nuisance with ordinances against the burning of soft coal and fines for excessive smoke. The city's Edison Electric Company, repeatedly cited for violating the city's smoke ordinance when it turned to cheaper soft coal, hired spotters placed on the company rooftop to warn the workers to stop feeding soft coal to the furnaces when sanitary inspectors approached.

Smoke was both a sign of progress and industry and the bane of urban living. Urban residents suffered from a variety of respiratory ailments, while the sick and aging faced a shortened life expectancy. Dr. Charles Reed, a respected ex-president of the American Medical Association, told nineteenth-century Cincinnatians that the city's smoke was exacerbating consumption. A health official in Cleveland told audiences that smoke was so contaminating city's residents' lungs that in autopsies the exposed lungs were black. Health investigations in Pittsburgh found strong correlations between soot-fall and pneumonia rates. Although city boosters saw smoke-spewing stacks as signs of prosperity, other urban residents viewed smoke as a health risk and social burden. Even some businessmen, particularly in the retail trades, saw smoke as having an economic cost. Acid in the smoke ate away at buildings and bridges, pocked storefronts, and deteriorated merchandise within stores. Smoke not only dirtied clothes, buildings, and indeed everything it settled on, it also killed urban greenery. Citizens' groups and foresters noticed high levels of smoke pollution killing all but the hardiest trees. Urban women's clubs campaigned to clean city air of unnecessary smoke. Agitating among politicians, civic leaders, and leading families, these women got smoke abatement ordinances and legislation debated and, in some cases, passed. Although a long way from producing clean air, the smoke abatement campaigns and measures did draw attention to the physical, social, and even economic costs of heavily polluted air. Despite these campaigns and the resulting ordinances, coal-based businesses, the railroads, electric-power generators, and processing industries that used large amounts of coal fought smoke restrictions. This business intransigence and a fear among civic leaders that smoke abatement would harm economic development meant that in most cases smoke continued to spew from the nation's urban smokestacks until the switch to cleaner-burning natural gas and oil in the post–World War II period. Indeed, the discovery and subsequent utilization of nearby natural gas in the 1880s drastically reduced Pittsburgh's smoke pollution, though the city's supply was soon depleted and, by the 1890s, it was back to coal and smoky skies for Pittsburgh residents.

In addition to air and water pollution, urban residents gathered around them the more general wastes of living. To get about the city before the age of the streetcar and automobile and to move goods before the truck, people either relied on walking, carrying, pushing, and wheeling, or they relied on horsepower. Until the beginning of the twentieth century, horses were as common to urban areas as to the countryside. Horses pulled wagons, carriages, carts, and streetcars and carried people. Horses also urinated and defecated. By the end of the nineteenth century, despite the market for manure among surrounding farms, accumulating horse manure had created a major urban crisis. One horse could produce as much as twenty-five pounds of manure and several gallons of urine a day (Tarr 1996, 323–393). Multiply that horse by the thousands occupying the dense urban set-

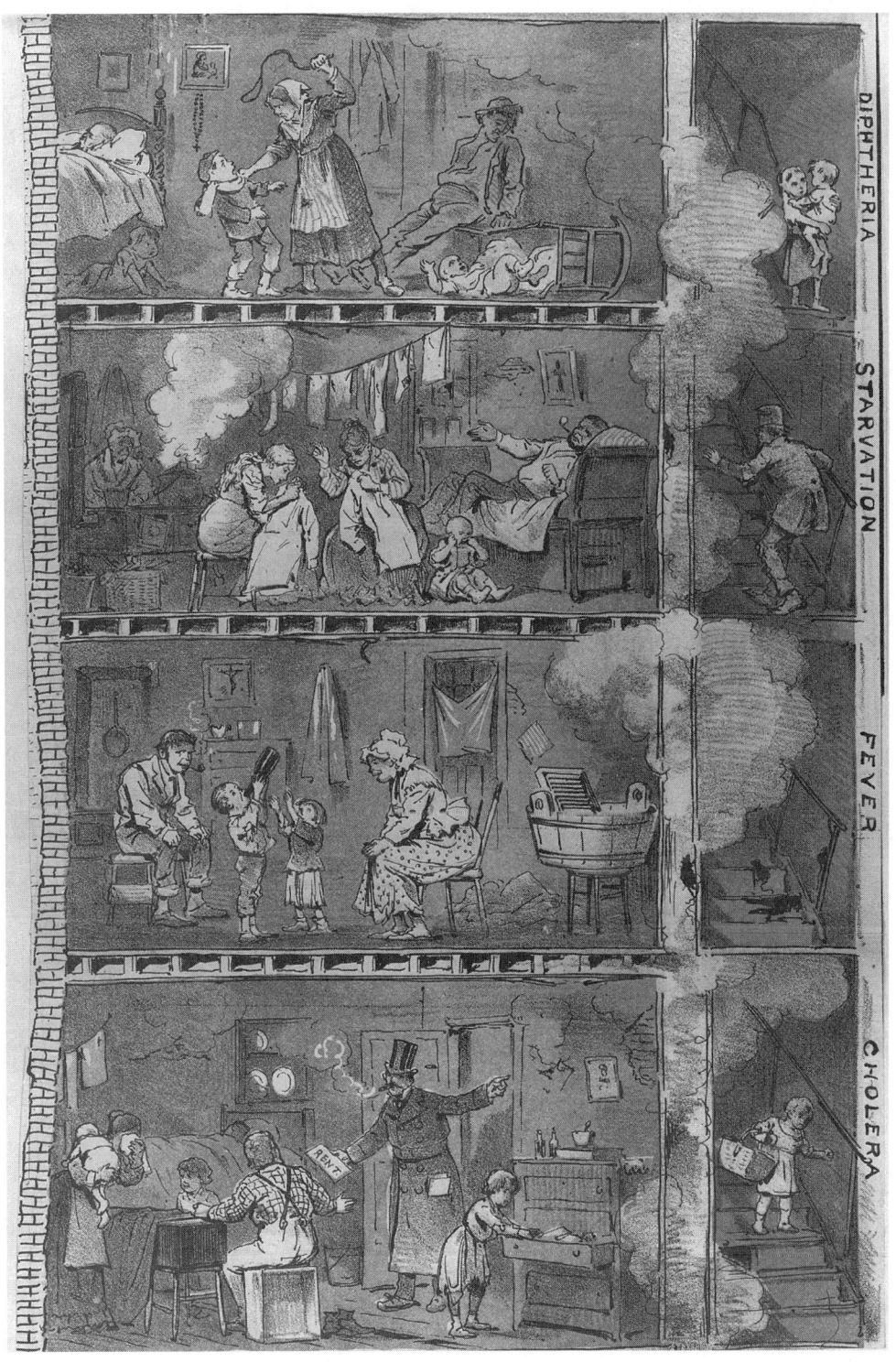

Cross section of slum dwelling in lower New York showing poverty and disease. Lithograph created in 1880. (Bettmann/Corbis)

tings and one can imagine why urban leaders labeled the horse manure problem a crisis. Horses, mules, and cows in Chicago produced more than 600,000 tons of manure daily. It filled the streets, which as a result, required constant washing. Horses even proved problematic in death. Disposing of a dead horse was difficult and expensive. Often a wagon owner would simply unharness a dead horse and pull his wagon away, leaving the carcass in the street. In commercial districts and wealthier neighborhoods, dead animals were removed quickly, but in poorer communities they all too often were left to rot.

Americans ate prodigious amounts of meat. In the country, animals slaughtered for meat were kept scattered about the farm and their smells were mingled with that of the rural countryside. That was not the case for those in the city. Cities concentrated meat-eating people. Pigs, cows, and sheep were driven into the city and gathered at slaughterhouses to wait for their ultimate fate. Animal wastes and waste products from the slaughtering accumulated about these slaughterhouses. Slaughterhouse stench was a notorious nineteenth-century problem. Court records are full of complaints against the smell of these operations. Increasingly, as the slaughter of animals became concentrated in Chicago, the slaughterhouse smell declined in other American cities as it in turn rose in Chicago.

Alongside the manure and dead animals, food scraps, broken furniture, old newspapers, and coal ash were all thrown into city streets or back alleys. Urban congestion contributed to accumulating garbage. Increased prosperity and industrial production encouraged consumption. Similarly, the integrated national economy made possible by the national rail system allowed companies to reach out to national markets. Increasingly, shipping to those mass markets also meant packing goods. By the 1870s, paper bags were increasingly common for shoppers. In 1879 the machine-made folding box made its appearance, and by the end of the century, the collapsible metal tube housing everything from toothpaste to glue became a fixture in people's waste bins. The developing consumer market encouraged manufacturers to develop product loyalty through elaborate packaging. By the second decade of the twentieth century, cities generated thousands of tons of garbage—much of it discarded packaging. Historians don't know exactly how much garbage was generated in American cities, but we do know that Boston, with 750,000 residents, collected over 52,000 tons of garbage a year in 1920. Philadelphia, with 1.75 million people collected over 101,000 tons of garbage. New York, with over 5.3 million people, collected almost half a million tons of garbage, most of which was dumped at sea (Melosi 1981, 23). For those neighborhoods with garbage collection, bins held wastes, but seldom contained them. Urban garbage collection was spotty at best. Some cities contracted the job to private companies, but more often than not these companies failed to adequately collect and haul away the wastes. Poor neighborhoods were particularly plagued with piles of uncollected garbage. The smell of garbage became the signal of urban America. It was said that what first greeted arriving immigrants to America was not the Statue of Liberty, but the smell of New York City's garbage.

In New York City, the stench and filth of the city's wastes motivated city women to form the Ladies' Health Protective Association and later the Women's Municipal League, which led the campaign to clean the city of garbage and improve overall health. Under the leadership of George Waring, the city's commissioner of sanitation in the 1890s, New York created a public system of garbage collection and street cleaning that became a model of efficiency. Other cities tried, with mixed results, to copy New York's success. Settlement house leader Jane Addams of Hull House was so outraged at the failure of Chicago's private contract method of waste removal that she made the garbage issue a major project of the settlement house and even got herself appointed garbage inspector for her ward. City leaders knew that the growing mounds of garbage and ash created breeding and feeding grounds for rats and other pests. In response, they appointed sanitary commissions, the members of which faced both the problem of efficient collection of the growing mountains of waste and the issue of what to do with the garbage once collected. New York's method of dumping refuse at sea was a plan neither well received by those who saw the garbage floating about along the shore, nor widely adaptable. Other methods adopted included burning, dumping in marshes and the open countryside, and—for edible wastes—turning it into slop for pig farms.

Reformers' attempts to protect drinking water and to clean up smoke and garbage were part of a general campaign for a cleaner environment begun during the decades following the Civil War. In the 1870s, New England health reformers claimed that citizens had a right to "clean air, clean water, and clean soil." Water filtration, chlorination, and the germ theory temporarily put aside public health concerns about water pollution, but with increased recreational use of rivers and lakes the issue resurfaced in the early twentieth century.

In 1912, Congress authorized the Public Health Service (PHS) to investigate how the condition of the nation's rivers and streams might affect health. In 1913 the PHS established the Center for Pollution Studies in Cincinnati, Ohio. Under Earle Phelps, a veteran of the Lawrence experimental station, the center began looking into water pollution. Phelps and his team investigated both sewage and industrial wastes. The PHS studies did much to increase the scientific understanding of pollution and its impact on water systems, but its work produced little direct improvement of the nation's waters. In 1924, Congress passed the Oil Pollution Control Act, reducing the flow of petroleum products into the nation's waterways, but other than this act, no other federal environmental legislation

Early twentieth-century recreational water use, Holyoke Canoe Club. (Library of Congress)

was passed. Some communities, responding to state legislation or threats of lawsuits, did work to limit severe industrial and sewage nuisances. But concern for cleaner water in general was drowned out, first by the drumbeat for the war effort of 1917–1919 and then by the call for progress and prosperity during the 1920s.

The 1920s brought the nation not only prosperity but also a new array of goods. Research and development laboratories established or expanded during the war years continued to bring forth a variety of products for the American consumer. In 1924, Dupont, interested in getting out of the munitions industry after bad press about war profiteering, brought cellophane on to the market. New plastics and chemical products came flowing out of American manufacturing plants, and new wastes poured from smoke stacks and outflow pipes. Americans had new things to fill up their homes and new wastes to fill their garbage receptacles, and new toxins to fill their lungs.

Not all Americans fell under the sway of business, however. In his popular 1922 novel, *Babbitt,* Sinclair Lewis paints his title character in a negative light for putting sewers in his real estate development that "had insufficient outlet" for the wastes. But lingering doubts and concerns did not translate into significant ameliorative legislation, leaving the nation's waters and air abysmally filthy. What little relief came resulted from the general economic downturn during the Great Depression rather than from serious legislative action.

7

PROTECTING THE PLACE

My sister's job regularly takes her to Boston's medical complex, located off the Fenway Green Belt. Theodore Lyman, whose father as mayor in the mid 1830s worked to get Boston's waterworks, was part of a committee to create the Fenway. A leader of that committee was Frederick Law Olmsted, the landscape architect who designed, among others, Central Park in New York, Belair Park in Detroit, and the fairgrounds of the Chicago World's Fair of 1893. Olmsted's parks, designed to create a planned but naturalized place, reflected urban Americans' concern for the transformation of their physical world and an urge to protect something of the natural world even in urban settings.

By the second half of the nineteenth century, that natural place did indeed seem threatened. In 1886 when James Olcott warned Connecticut farmers about pollution, he reminded his audience that only a generation or two earlier the region's waterways were places where one found not factories dumping industrial wastes but "barefoot boys, fishing these same brooks in the most enterprising manner" (Cumbler 2001, 49). As Americans began to take stock of their newfound prosperity, Olcott's call to a time of a purer environment resonated for a large number of them. By the end of the nineteenth century, smoke pouring out of the nation's smokestacks and wastes pouring from factory outflow pipes were the by-products of industrial activity that produced millions of yards of cloth, gallons of refined petroleum, and 29.5 million tons of iron and steel—far outstripping our rivals across the Atlantic. Smoke and pollution might have made life more difficult, but they also were seen as signs of progress and industrial production. America's vast natural resources, combined with aggressive manufacturers, a relatively well-educated population, a continuous flow of new labor, and limited governmental restrictions created the conditions generating wealth. A stroll down any number of poor urban neighborhoods was evidence that this wealth was not evenly distributed. In these residences newly arrived immigrants crowded together in appalling conditions while laboring long hours for limited pay. New York City had districts with significantly greater crowding than anything found in Europe. An 1894 investigation of the city found some 30,000 people living in a five- or six-block radius (Chudacoff 1988, 124).

Although poverty and disease were common conditions for poor city neighborhoods in a nation where, by 1920, over 50 percent of the population lived in cities and towns, not all Americans suffered the ravages of poverty. Industrial output produced wealth, and increasing numbers of Americans came to share in some of that wealth. The rapid expansion of industry required ever more workers. Immigrants who managed to stay healthy, marry late, limit their family size, and learn English could expect to gain skills and move, if not into the comfortable middle class, at least out of poverty. As long as the economy continued to grow, they could expect even better for their children. Rural white American migrants to the city with the right religious affiliation and having the advantage of knowing the native tongue could expect an even faster track into economic security.

Industrial expansion also spawned a proliferation of middle-class jobs and professions. Factories needed managers, bookkeepers, salesmen, chemists, scientists, directors, and supervisors. Banks needed managers, clerks, and tellers. Stores needed salespeople, insurance companies needed agents to hawk and handle policies, the sick needed doctors, and the young needed teachers. By the early years of the twentieth century, over 20 percent of the labor force was employed in education, the professions, personal services, trade, and government. Not all these workers enjoyed the comforts of the middle class, but many now had the resources to buy a home and the leisure time to enjoy recreation (U.S. Bureau of the Census, Historical Statistics of the United States, Bicentennial ed. 1996).

The wealthy historically enjoyed leisure and recreation. Owners of mills, mines, and factories were long accustomed to traveling in Europe or to the country or seacoast. The huge nineteenth-century "cottages" in Newport, Rhode Island, or on Long Island are testaments to that activity. For males in the post–Civil War era, leisure activity increasingly centered on the sports of field and stream. Hunting and fishing became obsessions of the well-to-do, especially for those whose workdays tied them to city offices. Removed by a generation or more from the countryside, these men now looked to rural America as a source of renewal and a place to practice the "manly art" of chasing game. But they no longer lived in rural America. The super-rich, the robber barons, built country estates where they could live in comfort and still wander the countryside killing animals and fishing streams and lakes. Others of the upper classes were forced to join hunting or fishing clubs that collectively owned lands and lodges. Less fortunate aspirants to the good life were forced to rent the right to hunt and fish from others.

As the nineteenth century progressed, many of these vacationing outdoorsmen came to realize that the rural countryside they looked to for renewal was itself in need of renewal. Game was depleted from fields and forests and fish from

Early twentieth-century sportsmen. (Courtesy of John Cumbler)

rivers and streams. Pollution, habitat destruction, and overharvesting were taking their toll on the rural environment. In an attempt to protect both the rural ideal and the game they made their targets, many of these sportsmen joined together. In 1886, under the leadership of Theodore Roosevelt and George Bird Grinnell, the editor of *Forest and Stream* magazine, these elites formed the Boone and Crockett Club to protect wildlife game habitat and espouse the value of "good sportsmanship" in the field. Among its members were the "who's who" of American business, including Henry Cabot Lodge, J. Pierpont Morgan, and Elihu Root. While the Boone and Crocket Club worked to promote sportsmen's values, state Audubon groups joined together in 1905 under the leadership of George Bird Grinnell to form a national lobbying organization to protect songbirds from hunters and prevent the hunting of birds for profit made by selling their feathers to hatmakers.

Concern for the loss of the older and presumed "purer" America incited these elites to support the creation of city, state, and national parks, and the creation of state fish and game commissions to protect habitat, stock game, and exert control over hunting and fishing. Control over hunting and fishing usually meant limiting the wildlife harvesting of poorer Americans, particularly immigrants and subsistence farmers. Out of these concerns came not only fish and game hatcheries and nurseries, but also laws regulating who could hunt and fish and when and for what game. The new laws required hunters and fishers to purchase costly licenses, limited how one could fish and hunt, and established game or bag limits. These regulations, as well as stocking programs, did help stabilize the game populations, but they also worked against traditional hunters and fishers. At the same time, the practice of placing bounties on predators like wolves, mountain lions, and coyotes and other policies privileging certain game animals within a habitat encouraged by these sportsmen wreaked havoc on the habitat and upset traditional predator-game balances.

The Boone and Crocket Clubs appealed to wealthy Americans with its espousal of rugged sportsman values and its support of policies that facilitated the leisure-time activity of the wealthy classes. But American elites were not the only ones with growing concerns about the quality of the physical world.

For urban Americans, work was removed from the household by the second half of the nineteenth century. By the beginning of the next century, Americans increasingly attempted to divide their work life from their leisure activity as they looked to take their play and relaxation away from any reminder of their workplace. For the American middle class and skilled working class, the teens and twenties brought comfort and increased leisure time, and the opportunity for that separation of leisure from work. Unable to afford a home at the beach or rural retreat, these Americans looked to local lakes, ponds, and rivers for recreation. Canoeing became a popular pastime, and swimming moved from the play of rough youth to a family activity. Fishing for recreation gained popularity among the urban middling classes. Automobiles and improved roadways brought the countryside closer to the city for the growing middle class, while at the same time moving its occupants farther from their places of work. In 1916, Congress passed the first Federal Aid Road Act that sent money to states for highway construction. And more Americans had the means to take advantage of these new highways. By 1922 there were 10 million automobiles available to drive into the country. Seven years later there were over 23 million—one car for every five Americans. When the anthropologists Robert and Helen Lynd studied the people of Muncie, Indiana, in their famous 1929 *Middletown* study, one of the most significant changes affecting people's recreational activity was automobile ownership (Sutter 2002, 24).

The nineteenth-century creation of a national transportation network that brought agricultural produce to the doors of urban Americans also pushed marginal farmers to bankruptcy. Throughout the hill regions of the Northeast and the upper regions of Wisconsin and Michigan, farmers abandoned their property. Fields returned to forests, and farmhouses and barns collapsed, leaving behind forlorn foundations and stone walls wandering through thick woods. On just such a farm the naturalist Aldo Leopold found a weekend retreat from his hectic university life in the same county where John Muir was put to work by his father to help cut out the family farm. Fortunate farmers working the rich prairie soil of the Midwest prospered growing corn, milking cows, and raising beef. Farms closer to towns and cities survived by selling perishable goods to urban markets, but much of rural America struggled to survive on the farm.

While these farm families labored long and hard under the continual burden of debt, to many middle-class Americans the rural farmer represented an ideal of purity and virtue. From the roadside or country inn, rural farmland was the essence of idyllic beauty. To appreciate that beauty, families took up auto touring, packing their cars for weekend or vacation trips up the nation's back roads. For those not fortunate enough to afford a weekend or longer vacation, a day trip to the country to picnic in some farmer's field was the reward for a week of hard labor breathing dirty city air and struggling with urban filth. Farmers catered to these interlopers by setting up roadside stands and selling local produce, such as maple syrup, fruits, vegetables, and cider. The more enterprising opened their homes to guests, who lugged along fishing poles and picnic baskets. In the 1920s, states began to acquire abandoned and unwanted land for state parks particularly catering to the auto-tourist. The cut-over districts of northern Michigan and Wisconsin were abandoned by lumber companies after the removal of profitable lumber. These companies either abandoned the land to have the state seize it for taxes, or sold it cheaply to poor farmers. After a few decades of struggle, many of these farmers gave up—particularly in the hard agricultural times of the 1920s. With large amounts of unwanted land on their hands, and in some cases with railroad lines and rough lumber roads built into the region to take out lumber, states converted the land into state parks catering to the newly emergent middle-class interest in rural recreation.

As more middle-class Americans looked to their environment for recreation—whether the local lake, pond, river, or stream, or even the neighboring countryside or state park—they confronted the disquieting reality of the environmental costs of the nation's increased prosperity. Country roadsides were littered with discarded tires, bottles, and cans. Lakes and rivers were not swimmable or fishable. Canoeing through wastes and filth had little appeal. Those who could not afford automobile ownership were confined even closer to home for recreation. And the rivers, lakes, ponds, and green spaces closer to home were in even worse shape.

The development of chlorination and filtration at the end of the nineteenth century had alleviated much of the health concern regarding the nation's drinking water. With the growing interest in waters for recreational purposes, the nation once again became interested in the quality of water, not only for drinking, but also for swimming, bathing, canoeing, and fishing. This increased interest led political leaders, letter writers to local newspapers, and civic leaders to begin pressuring boards of health and fish and game commissions to consider conservation measures for local waters and environments. It was in response to this public concern that New Jersey and Pennsylvania created the Delaware River Basin in 1922, adding New York in 1925. In 1928 all nine states within the Ohio River Basin formed the Board of Public Health Engineers of the Ohio River Basin. The states along the Great Lakes joined together in an organization to work toward protecting the Great Lakes.

Yet despite this flurry of activity, little actual legislation resulted. In 1899, Congress passed the Rivers and Harbors Act, banning the dumping of garbage and ash into navigable waters. However, the act explicitly exempted sewage discharge, a major cause of pollution in the waters of the rivers and on the beaches of urban areas. In 1924, Congress enacted the Oil Pollution Act to protect coastal water fish life, but left rivers and streams untouched, and even so the act was seldom enforced. Increased public awareness, as well as legal nuisance court cases, encouraged the building of sewage treatment facilities through the teens and twenties. Yet, by the end of the decade of prosperity, out of a total urban population of over 70 million, only 18 million urban Americans were served by a water treatment facility (Melosi 1980, 77). Unless it was profitable to process the wastes, industrial wastes continued to be dumped in the nearest stream. And most of the rivers and lakes located in urban areas were far too polluted for healthy swimming, bathing, or fishing, and too foul for sailing or canoeing. And urban green spaces continued to be plagued by smoke and soot.

Concern over the failure of the Forest Service, created in 1905, the National Park Service, created in 1916, and state fish and wildlife commissions to protect habitat and waterways for fish and game led in 1922 to the formation of Izaak Walton Leagues in Chicago to push for the protection of outdoor America. Will H. Dig, a public relations expert from Chicago, headed the organization. He used his professional skills to draw in members, and before the end of the decade, the Izaak Walton Leagues signed up over 100,000 members, especially in the Midwest. Larger by over a factor of ten than either the Audubon Society or the Sierra Club, the Izaak Walton League became the nation's first mass-based environmental organization. The Izaak Walton League published the popular *Outdoor*

America magazine that both popularized fishing and hunting and functioned to recruit new members. The league drew to it professionals and outdoor enthusiasts. Its president from 1928 to 1930 was Henry Ward, a University of Illinois professor of zoology. The league's popularity also drew political figures. Herbert Hoover was a supporter, and the league used its influence on fly fisher Calvin Coolidge to get the huge Upper Mississippi River Wild Life and Fish Refuge established. The league demanded reform of the management of public resources. After it organized a national water quality survey in 1927, the league used its data to push for the passage of federal water-pollution control bills. Congress finally passed the league-sponsored water pollution bill in 1939, only to have Roosevelt veto it as too costly. The rapid growth of the leagues and their popularity reflected an increasing anxiety about the American commons, but effective remediation remained an illusive goal.

In 1929 the bubble of prosperity burst. Americans stopped worrying about canoeing, sailing, and bathing and began worrying about mortgages, rents, food, and clothing. Malnutrition became a national concern, as city families lacked the funds to purchase adequate and nutritional food, and farmers lacked the selling prices necessary to transport goods to market. Failing markets closed down American industry. Between 1929 and 1932, industrial production fell over 50 percent. Steel plants ran at 12 percent capacity (Galbraith 1999, 142). With the country economically crippled, the nation rejected Herbert Hoover and the Republican Party and elected Franklin Delano Roosevelt to the White House.

Needing to put Americans to work, Roosevelt initiated a number of government employment programs that directly affected the nation's environmental quality. To help farmers, Roosevelt's New Deal initiated the Agricultural Adjustment Act (AAA). The AAA attempted to fight dropping farm prices by cutting production. Farmers were paid to produce less corn and milk and fewer pigs, and thus to take land out of production. They retired their most marginal land. Less farmland in production, coupled with programs for soil conservation, helped mitigate erosion, reduced silt wash in rivers and streams, and protected land. Taking young men between the ages of eighteen and twenty-five, the Civilian Conservation Corps (CCC) sent them to work in the countryside, protecting watersheds and working on park and forest conservation projects. In state and national parks and forests the CCC created hiking paths and campgrounds, many specifically designed for automobile campers.

Unemployment haunted not only the young. To put people to work, the New Deal launched a number of projects through the Works Progress Administration (WPA), the Civilian Works Administration (CWA), and the Public Works Administration (PWA) that were necessary for the nation's well-being. Parks, sewer lines, and waste treatment plants were favorite New Deal projects. The number of Americans served by sewage treatment facilities rose from 18 million in 1930 to over 40 million in 1940 (Melosi 1980, 77). New Deal workers also built urban parks and cleaned up abandoned land, making urban life a little more habitable for those without cars to get to the country. Road building, even more than parks and sewer systems, enthused New Deal officials looking to put people to work. Roads in parks, such as Skyline Drive along the Appalachian ridgeline, put people to work and made the national parks more accessible.

Building roads in parks put people to work and brought the parks closer to automobile-owning Americans, but it also opened wilderness areas up to motorized travel. Naturalists who treasured wilderness areas removed from even the noise of cars or motorboats began to fear that America's remaining wilderness areas would be overrun with roads, cars, and modern conveniences. Responding to that threat, leading conservationists like environmentalist Aldo Leopold, forester and wilderness advocate Robert Marshall (who had spent his youth in camps in the Adirondacks), canoe enthusiast Ernest Oberholtzer, parks advocate Robert Sterling Yard, and others came together to found the Wilderness Society to advocate for the protection of areas reserved for hikers and removed from roads, cars, and motorboats.

The Second World War finally brought an end to the deprivations of the Great Depression as the nation shifted its concern to winning the war. Industrial production increased dramatically, as did research and development. Limits of raw materials such as rubber led to development of coal- or petroleum-based synthetic materials. Concerned about the potential loss of life and fighting ability from the inadequate hygiene that had plagued previous wars, military leaders pressed for protection against disease. The need for food production at home and troop safety abroad encouraged the development of new petroleum-based fertilizers, cleansers, and pesticides—particularly chlorinated hydrocarbons such as DDT and phosphorous ones such as malathion and parathion. American farmers had long been accustomed to using pesticides. Prior to World War II, the costs and dangers (including popular concerns over possible food poisoning) of the most commonly used pesticides, such as lead arsenate, limited their use. The new hydrocarbon-based insecticides were cheaper and easier to dispense. Crop dusters could spray them out by the ton. Farmers could easily disperse them from the back of their tractors, and homeowners could spray them from handheld canisters. Cheap chemical ammonia-based fertilizers boosted agricultural production, while pesticides seemed to reduce crop loss.

Synthetic pesticides, herbicides, and cleaning materials were sold as harmless. The pesticide DDT (dichlordiphenyltrichloroethane) was dusted on American troops as if it were talcum powder. The chemical industry touted its claim for "better living through chemistry," as it produced ever-greater quantities of organic chemicals for cleaning cars, homes, and laundry, and dispensing with farm, garden, and yard pests. Fertilizers were so cheap that they were applied with abandon. The consequences of dumping massive amounts of fertilizers, pesticides, and herbicides on fields and lawns, a good bit of which washed off into river and streams, seemed of little concern for those who saw only a rapidly growing bounty of goods. Americans came to expect spotless and bugless produce, and farms delivered it with the aid of the multibillion-dollar American chemical industry.

But all was not as rosy as the "better living" campaign implied. In 1958, Duxbury, Massachusetts, a quiet, scenic town located along a marshy coastal area south of Boston, sprayed DDT as part of a state mosquito-eradication program. A Duxbury bird-watcher, Olga Owens Huckins, noticed an increased number of dead birds and a dramatic decline in songbirds following the spraying. Concerned, Huckins turned to her close friend, Rachel Carson, a western Pennsylvania-born, professional naturalist writer, who worked for the national Fish and Wildlife Service. Carson was a trained biologist who wrote several successful books on nature and the sea. Huckins suggested that Carson look into the bird deaths. Carson put aside other projects and focused her attention on synthetic pesticides. After careful research, Carson wrote her wide-ranging critique of chemical pesticides. First published serially in the *New Yorker* in 1962, then later that same year as a book, *Silent Spring* captured the attention of the nation. Carson was attacked by the petrochemical industry, but her careful research and direct style convinced many Americans that chemicals could also poison their fields and streams. Following the publication of *Silent Spring*, new information about declining numbers of shorebirds, especially pelicans, and the discovery of pesticides in remote areas of the globe increased public concern and led ultimately after a long and protracted political and legal battle to the banning of DDT in the United States in 1972.

But the battle to stop widespread pesticide spraying was not the only environmental concern that awakened Americans in the prosperous postwar period. Smokey air, sore throats, and sooty clothes were for years seen as the costs of prosperity. Yet Americans also understood that smoke was bad for one's health. In 1948 an inversion, in which a layer of warm air traps colder air below it, occurred over Donora, Pennsylvania, a steel town near Pittsburgh. The inversion lasted for an extended period, concentrating air pollution in the city. Hundreds of residents were hospitalized with lung ailments, and twenty died as a direct result of the pollution. These deaths, along with word of London's killer fog of 1952, which contributed to several thousand deaths, were national news. Citizens of other industrial cities such as Gary, Indiana, or Youngstown or Cleveland, Ohio, became concerned that a similar tragedy was just waiting to happen to them. In Pittsburgh, the smoky city, a wealthy banking family, the Mellons, had already backed a thirty-year smoke-abatement campaign that finally produced an air pollution control ordinance in 1941, although it was not effectively enforced until the late 1940s, when cheaper, cleaner fuels began to replace coal, and industrial decline reduced pollutants flowing out of the city's smokestacks. Nonetheless, ordinances enacted in cities across the northern states against backyard waste burning, changes in fuel sources, and newer technologies encouraged Americans in the belief that they could avoid such a large-scale tragedy as had happened in Donora. Burning in public dumps was drastically curtailed. Coal-fired steam engines were replaced with diesel engines, especially for locomotives. Urban air became cleaner as cities switched from coal to petroleum or natural gas for heating fuel, as large tankers filled with cheap, Middle Eastern oil sailed into American ports, and new pipelines brought gas produced in Texas and Louisiana to consumers in the Midwest and Northeast.

Clean air was not the only worry of the newly prosperous postwar Americans. The development of penicillin and the successful campaign to develop a vaccine against polio convinced many Americans that, if they put their minds to it, they could conquer whatever problems nature threw at them. But nature seemed to be throwing Americans ever more problems. In the late 1950s, untreated sewage began to congeal off of Staten Island into a huge mass known as the Arthur Kill blob. Waters off the beaches along Boston Harbor, long a working-class and lower-middle-class respite from summer heat, were found unfit to swim in, as were the waters along many Great Lakes beaches. In 1969 the Cuyahoga River, heavily loaded with industrial wastes from Cleveland refineries, dramatically caught fire. The 1969 fire was not the Cuyahoga's first. In 1952 the Cuyahoga exploded in flames that swept downriver and through a shipyard, burning three tugboats and causing thousands of dollars in damage. Not only were rivers and streams that flowed through industrial cities like Cleveland, Detroit, Pittsburgh, Boston, New York, and Philadelphia unfit for fish and unswimmable, some were now fire hazards. Continued pollution and especially the added impact of phosphorus-based laundry detergents and increased use of household bleach overloaded water systems' ability to process pollutants. Oxygen levels in streams and rivers dropped dramatically with the increase in biological oxygen demand (BOD). Pollution loads threatened major water systems with eutrophication. Lake Erie became known as the dying Great Lake.

Lake Erie had long been on the receiving end of untreated sewage from Detroit and Buffalo. The postwar years saw an increased use of phosphate-based detergents. Households dumped used wash water, rich in phosphates, down the drain and into the nearest water system. Phosphates act as a nutrient, encouraging algae blooms. Phosphates in Lake Erie caused huge blooms that died and drained oxygen from the water. Huge blobs of dead matter, sewage, and garbage, like floating islands, began drifting about the lake, and hills of foamy soapsuds clung to the shore. Citizens living along the lake organized a Save Lake Erie campaign and demanded action.

Lake Erie was not the only American waterway calling out for concern. The Susquehanna River, the main feeder into the Chesapeake Bay, one of the nation's richest estuaries, had suffered severe degradation due to mine waste, particularly acid mine drainage, as had the Ohio River's feeder streams and rivers. Although Pennsylvania had passed a Clean Streams Act in 1937, the act excluded coal-mine drainage from enforcement. It was a major omission. Drainage from mounds of highly acidic coal waste rendered much of the upper part of the Susquehanna and the Monongahela River orange or black in color and devoid of life. But it wasn't what was happening above ground that created the most dramatic event in the Susquehanna's modern history. For over a century, miners had dug coal from under the river basin. For months, the Knox Coal Company encouraged their miners to dig coal dangerously close to the river's bed. On January 22, 1959, the consequence of that decision became apparent when the ceiling supports gave way and the mine's roof collapsed. The river itself poured down into the mine, creating a huge whirlpool. Over 100,000 gallons of water per minute swirled down into the mineshafts below. It took over four months and thousands of tons of cement and sand, dirt, mine waste, and railroad cars dumped into the hole before it was plugged enough to reduce the flow of water to only 400 gallons per minute. During the four months before the hole was partially plugged over 10 billion gallons of water rushed into the mine shafts under the Wyoming Valley, the major coal region of the Susquehanna, killing three miners (Brubaker 2002, 54).

In 1960 and 1965 concern over water encouraged Congress to pass clean water legislation. This legislation led to the gathering of data on water pollution, including industrial wastes. Americans became worried that a compromised environment would undermine their newly won prosperity. Walter Ruether, the president of the United Auto Workers Union, caught the sentiment of many Americans when he asked, "What good does it do to win vacation time, if working Americans can't enjoy it because the lake is polluted" (Cumbler 2005).

Concern over the loss of recreational space drew people into conservation actions. In 1952, when several steel companies were looking to build mills along Lake Michigan's southeastern shore, twenty-one women under the leadership of Dorothy Buell came together to form the Save the Dunes Council. The council argued that the dunes represented a delicate and unique landscape and should be protected as a national park. By 1960 the council had over 2,000 members and, after several years of fighting the steel industry, it managed to win a compromise that allowed Bethlehem Steel to build a mill, but also set aside 8,600 acres for what became the Indiana Dunes National Lakeshore in 1966. Save the Dunes continued to push for legislation to protect the lakeshore, joining up with a Michigan group headed by Genevieve Gillette.

The 1960s were a time not only of growing prosperity and concern over pollution, but also a time of increased political involvement and rebellion. Students across the country joined with minorities in demanding political participation and accountability—what the Students for a Democratic Society called "participatory democracy." One of the demands raised by this deepening democratic involvement was for a cleaner environment. Students, activists, and environmentalists joined together to demand legislation for cleaner air and water. In September 1969 Senator Gaylord Nelson from Wisconsin called attention to the environmental crisis facing the nation. Nelson challenged Americans, especially young Americans, to mobilize to protect the environment as they had mobilized for civil rights and peace. Particularly, he called for a national teach-in on the environment. Activists began mobilizing for a 1970 Earth Day to draw attention to the critical condition of the nation's environment. Maine's Edmund Muskie, chair of the Senate Committee on Air and Water Pollution, told the nation that the issue wasn't about national parks, but the quality of life of average Americans. He reminded people that the issue of clean air and clean water affected all Americans, those living in inner cities as well as those in the suburbs.

Muskie's comment that environmental degradation affected people in the ghettoes as well as the suburbs was not news for African Americans, nor should it have been a surprise to white Americans. America had always distributed her goods unequally, and as W. E. B. Dubois noted in 1902, the color line was the most pronounced measure of that inequality. Environmental goods—public space and healthy and clean air, water, and soil—were more available to white Americans than black Americans. What might have surprised Americans was that environmental racial inequality had grown over the last half century. Certainly, the black citizens of Pittsburgh lived in the most polluted neighborhoods, but poor white residents of Steel City did not fare much better in terms of air and water quality. Although in the city of Gary, Indiana, African Americans found themselves working in the most polluted and hazardous departments in the mills, until the 1960s their neighborhoods did not have greater concentrations of air pollution than those of Gary's white working class. Before the great black urban migrations of the 1950s and 1960s, most African Americans lived in the rural South. Although they often lived in abject poverty, in the presynthetic-pesticide era, their physical environment was more healthful than those of poor urbanized whites.

The environmental color line began establishing itself in the second half of the twentieth century. Unionization and the general prosperity of the postwar

People gather for the first Earth Day celebration on April 22, 1970. (Hulton Archive/ Getty Images)

era enabled working-class whites to move out of older residential areas bearing heavy burdens of air and water pollution. Because of patterns of residential discrimination and overt discrimination by the Federal Housing Authority's lending policy, African Americans took over the residences abandoned by whites. From New York to Chicago, African Americans found the homes available to them were those downwind from power plants, next to urban dumps, alongside polluted streams, and among chemical plants. Because these neighborhoods lacked effective political clout, they were seldom cleaned or protected from wastes. Concern by African Americans over this environmental disparity gave rise to the environmental justice movement of the late twentieth century. Working mostly out of community groups such as South Chicago's People for Community Recovery, minority communities began challenging long-standing patterns of environmental neglect of their neighborhoods. But the long history of the color line contributed to the fact that it took several years for the concerns of white environmentalists to find common cause with people of color, whose environmental issues were usually more immediate than aesthetic.

Although direct, active political coalitions made up of white environmentalists and African Americans were few and tenuous, white activists owed a tremendous debt to the black community. Black Americans taught white America the lesson of democracy in the 1950s and 1960s. From the activism, energy, and political demands of the civil rights movement, white American activists learned political activism and the importance of demanding a voice in the political system. They also learned democracy and carried that lesson into action. The environmental movement learned from the civil rights movement about mobilizing people and pressuring the political system for results. They took those tactics and energy into the national political arena, ultimately bringing hundreds of thousands of Americans into the streets of cities across the nation on April 22, 1970, for the nation's first Earth Day. And the Black Political Caucus provided a solid block of votes for environmental issues.

Feeling growing political pressure for action on the environment, President Richard Nixon signed the National Environmental Protection Act (NEPA) in 1970. In December of that year, the NEPA created the Environmental Protection Agency (EPA) to set standards and enforce environmental legislation.

Under the leadership of Edmund Muskie, Congress passed the Clean Air Act of 1970. And in 1972, Congress passed the Clean Water Act. The Clean Air Act of 1970 set up the National Ambient Air Quality Standards to protect air quality for health. It directed the EPA to lower carbon dioxide (CO_2) and hydrocarbons emissions into the air by 90 percent and nitrogen oxide emissions by 80 percent. The Clean Air Act was reauthorized in 1977 and again after a difficult political battle in 1999. Unfortunately, although urban air quality has improved, the increased number of cars on the road, an increase of over 300 percent from the mid-1950s to the mid-1990s, thwarted the goal of making our urban air healthful (Opie 1998, 456–457). Since the late 1990s, the popularity of large, less-gas-efficient sport-utility vehicles (SUVs) has also worked to undermine attempts to lower air pollution. Recent health studies have suggested that acceptable levels of air pollution need to be lowered, but with ozone pollution alerts that declined in the 1980s on the rise again, cities are facing increased challenges to meet air pollution standards.

The Clean Water Act set dates for cleaning the nation's waters. Sewage dumping was drastically reduced. Billions of dollars in federal funds became available for municipal water treatment plants. Rivers and lakes once thought unswimmable or unfishable increasingly sported public beaches where swimmers and fishers flocked on weekends. Although a long way from its initial purity, America's environment made improvement toward the old nineteenth-century reformers' goal of clean water and clean air. But it still had a long way to travel before reaching it. In 1993 the city of Milwaukee, which proudly announced itself as one of America's healthiest cities, realized that decades of ignoring the declining quality of the regional water basin and years of ignoring water-system infrastructure could have disastrous consequences. Snow lay thick on the ground until late March in the winter of 1992–1993, when a warm snap sent snowmelt rushing into the city's gutters and sewers. Local streams overflowed their banks and sewers backed up. Eventually, the overflow spilled into Lake Michigan. Warm weather did not bring joy to Milwaukee families. Unusually high incidents of diarrhea illnesses plagued the city. Children stayed home in such numbers that some schools were closed. On April 8, 1993, the city announced that it was suffering from an epidemic of *Cryptosporidium parvum*, and the probable source of the contamination was the municipal water supply. Eventually, some 400,000 people suffered the consequences of the contamination (Foss-Mollan 2001, 162).

The battle to control the ever-mounting volume of wastes continued to frustrate environmentalists and mobilize local action. Neighbors and community groups established recycling centers and initiated educational programs to encourage and extend efforts to reduce waste through reuse. In 1971, Vermonters succeeded in their "bottle bill" campaign, forcing bottling companies to charge and redeem deposit fees for drinking containers. Several other states quickly took up the idea. After long and protracted battles with the bottling industry, which claimed the bills would hurt the economy and limit jobs, most northern states passed versions of the bottle bill. Bottle bills have reduced litter, but with beer and soda companies using nonrenewable containers of metal, thin glass, or plastic, the bottle bills have done little to slow the growth of waste. When neigh-

Gases rise from a vent on high ground in the town of Centralia, Pennsylvania, 1981. A coal-buring fire started in 1962 as a result of people discarding garbage into an abandoned mine shaft. The fire continues to burn underground as of 2005. (Bettmann/Corbis)

borhood complaints and air pollution laws limited or shut down old waste incinerators, cities and towns turned to landfills to dispose of their trash. But with limits on available land for dumping and landfills reaching their capacities, cities and towns have increasingly looked to recycling to decrease the waste flow. Expanded municipal-sponsored recycling programs have vastly increased the volume of waste redirected back into the production process over the limited progress of citizen-sponsored volunteer recycling efforts.

But bottles and trash were not Americans' only concern. In 1962 the small Pennsylvania mining town of Centralia still practiced open burning in the town dump. Unfortunately, below the dump lay an abandoned coal mine. Town workers expected the fire at the dump to burn itself out in a couple of days, but it unexpectedly burned ever hotter for over a month. When workers finally bulldozed off the burning wastes, they discovered that the fire had ignited an underground coal seam. The fire burned downward into the mines below. Carbon monoxide filled abandoned mines and closed active ones. For the next sixteen years, state and federal agencies attempted to extinguish the fire—to no avail. Finally, in 1983 the Department of the Interior decided to leave the fire and buy out the town. To this day in 2005, vents from the still-burning coal break through the ground to release sulfurous gases that kill trees and wildlife. Smoke rises from cracks in highways, and the ground collapses. Centralia is not Pennsylvania's only underground coal fire burning out of control; there are dozens. Not all underground-burning coal seams are products of human action. But mine tunnels help bring air to the flames and make extinction more difficult. News of out-of-control underground flames, along with persistent and growing reports of water pollution, gave fuel to the idea that more needed to be done to protect the environment.

In late 1976, Congress passed two important environmental acts, the Resource Conservation and Recovery Act and the Toxic Substances Control Act. Although bitterly fought by the chemical industry, these bills passed with strong public support. The importance of controlling toxic waste became apparent shortly after the passage of these bills, when the community of Love Canal, New York, became news across America. For decades the Hooker Chemical Company had dumped toxic industrial chemicals on a waste site in Love Canal. Later, after the waste dump was covered over, a school was built on the site. Residents of Love Canal began noticing abnormal levels of health problems. Led by Lois Gibbs and the Love Canal Homeowners' Association, local activists, mostly women, mobilized. They conducted health surveys and contacted and confronted state and national officials—including holding EPA officials captive until they addressed the concerns of Love Canal residents. The women found high levels of epilepsy, liver disease, rectal bleeding, miscarriages, and birth defects. The surveys showed that the number children in the area born with birth defects was twice the national average. Although some experts claimed that a direct connection between the chemicals dumped and the increased health risks could not be proven, the state and federal governments spent close to 2 million dollars on study and cleanup of the site. Eventually, the activists forced the government to buy their homes, and Love Canal became a national symbol of both the carelessness of industry and the power of a mobilized local community.

The events in Love Canal signaled to the nation the potential dangers of toxic wastes that might lie just below the surface of the ground. Concerned over the possibility of future Love Canals and aware that other communities might mobilize, Congress rushed through the Comprehensive Emergency Response Compensation and Liability Act, or Superfund Act, in 1980. This act taxed toxic-waste generators to create funds for cleaning up heavily contaminated sites. At its inception, the Superfund was originally allocated 1.6 billion dollars, but even with additional funding added over the years, the Superfund cannot come close to covering today's costs of cleaning up the many toxic sites scattered across the country. This is mostly because the money necessary to deal with the problems has not kept pace with the rising number of sites and their requisite costs for cleanup.

Love Canal was also very much an American story because it wasn't an old, decaying industrial city. Love Canal was a new community. It was a community of the newly emerging middle and lower-middle classes, a community of well-paid unionized workers, skilled mechanics, insurance salespeople, teachers, and accountants. In the nineteenth century, workers and the lower-middle class found their homes and their jobs in the cities. They walked or took the streetcars or trolleys to work. Their homes and their jobs were centrally located.

In the postwar period, that pattern began to change. Change had already begun for the prosperous middle class, but after 1945 more lower-middle- and working-class Americans participated in this change. It was change fueled by funds guaranteed by new government-loan programs of the Federal Housing Authority and the Veterans Administration. It was change facilitated by innovations in the construction industry, particularly tract housing. It was change sped along on the vehicles pouring out of the nation's automobile plants. Americans called it suburbanization.

Using heavy bulldozers that cleared hundreds of acres of land, leveled hills, and removed trees, brush, and grass, developers brought mass-production techniques to the home-building industry, confident that federal loans programs would guarantee them buyers for what they built. By 1950 there were over 1.9 million new housing starts (Rome 2001, 15–16). Leading this new home construction was William Levitt. Levitt bought up old potato fields on Long Island, twenty miles from New York City, and threw up hundreds of standardized homes, economizing in design and scale. With block upon block of similar and simple homes, "Levittowns" became both a source of ridicule, as in the Malvina Reynolds 1962 song "Little Boxes," as well as the symbol of American prosperity.

Although the suburbs offered Americans home ownership, often at a cost lower than urban rents, these communities also promised a cleaner, more healthful, and—for many—"whiter" environment. But under the promises of the good life offered in promotional literature lay a darker environmental story. The new suburb builders tore through the landscape. Forests, streams, and hills fell before the power of profit that the construction and real estate industries reaped with these mass-produced communities. Environmental diversity was lost. Until vegetation could regrow, erosion increased dramatically. Higher profits could be realized with lower land costs, and land was cheaper farther from the city center. It was also land well removed from municipal sewers and treatment facilities that had been built up over the preceding decades. The new suburbs relied on

Levittown, April 13, 1949. From 1947 through 1949 Levitt and Sons built 10,000 houses on former farmland on the outskirts of suburban Long Island. The low-cost housing development catered to ex-GIs and was an example of the mass production methods of the era. (Bettmann/Corbis)

septic tanks that could seldom accommodate wastes generated by these new centers of human concentration. Streams, creeks, ponds, and lakes that had avoided heavy loads of sewage pollution because of their distance from urban centers became increasingly polluted with nitrate waste from overburdened septic systems and lawn fertilizer.

Most of these new suburban communities were located beyond the reach of public transportation. Shopping was removed from housing, either stretched out along radial highways or centralized in distant malls surrounded by acres of parking lots. Sold on an ideal of freedom and individuality as represented by the automobile, Americans, except for those in the densest urban areas, began to abandon support for public transit systems. Funds for public transportation declined. With declining revenues, transit systems cut costs and services and raised prices and, in turn, lost even more riders.

Americans increasingly became dependent upon private transportation to get to work or stores. Even those who commuted to work by rail often needed to drive to the station to catch their trains. Automobile ownership and automobile size grew with suburbanization. Suburban homes were built to save construction costs, not energy costs. Energy use grew exponentially. Not only were these homes not particularly energy efficient, but habits of consumption favored high-energy use. Electrical appliances grew in number, size, and function. Soon average homes sported air-conditioning, electric dishwashers, washing machines, dryers, coffeemakers, and televisions. Through the 1960s consumption of energy increased 100 percent (Nye 1999, 198–208). With low energy costs, Americans saw little reason to constrain their consumption.

Coal was the source of cheap electricity for most of the middle section of the nation. Although nuclear energy plants were built throughout the late 1950s, 1960s, and early 1970s, environmental concern about nuclear wastes and the potential for accidents grew in the mid-1970s. Groups such as the New England Coalition on Nuclear Power and the Clamshell Alliance in the Northeast, the Paddlewheel Alliance in the Midwest, and Friends of the Earth, the Union of Concerned Scientists, and Critical Mass nationwide opposed the construction of new nuclear plants. With the 1979 accident at the Three Mile Island nuclear plant on the Susquehanna River, popular opposition crystallized against the construction of new nuclear plants. In 1979 the nuclear reactor at Three Mile Island, Pennsylvania, lost water to its cooler system and experienced a partial meltdown. Although the operators of the plant were able to disable the reactor, the plant came close to a serious explosion that could have endangered the live of hundreds of thousands of people living in surrounding regions. Nevertheless, the plant released significant amounts of nuclear contaminants into the atmosphere. The extent of the potential disaster was later revealed to the world when the Soviet Union's nuclear plant at Chernobyl in the Ukraine blew off the top of its reactor, releasing radioactive material for thousands of miles. The Chernobyl disaster instantly killed dozens of people, forced the evacuation of over a hundred thousand residents, and exposed nearly a million people to the increased risk of cancer.

Increased demand for electricity was met by coal-fed, steam-powered generators. The coal used for these plants, which were scattered across the middle section of the nation, was local, cheap high-sulfur bituminous. Local concern over smoke emissions led electric companies to build high smokestacks that sent sulfur-laden smoke high into the atmosphere. There, the sulfur fumes would mix with atmospheric moisture. Prevailing winds would move this increasingly acidic mix east to rain it down on lakes, ponds, rivers, streams, homes, and buildings. By the 1980s, states in the Northeast began protesting against the acid rain pollution coming at them from the middle section of the nation, even as their own power plants sent acid rain northeast into Canada (although eastern power companies were more likely to burn petroleum than high-sulfur coal).

Midwestern states, under pressure from local high-sulfur coal companies, coal miners, and energy companies that appreciated the cheap coal, resisted switching to more expensive and often out-of-state low-sulfur coal and reluctantly offered to consider the option of installing scrubbers to remove sulfur from their smokestacks. Installation of scrubbers reduced some acid rain, but the energy companies have continued to fight against the installation of new pollution-control equipment or demands that they switch to low-pollution fuel on into the twenty-first century, making acid rain a continuing problem for much of the eastern United States.

Although the nation suffers from continued problems of air quality, American political leaders tout the advantages of cheap energy and focus policy around the assumption that cheap energy is a necessary national goal. And to a great extent they have been successful. Gasoline prices in the United States are a tenth of what Europeans pay.

In the 1970s, when the Organization of the Petroleum Exporting Countries (OPEC) cut production and gasoline prices shot rapidly upward, America's energy gluttony became all too obvious. Americans looked to smaller, more efficient cars. Cars' average miles per gallon increased from thirteen to fifteen, but increased efficiency stalled over the last two decades of the twentieth-century and into the twenty-first as the oil shortage crisis passed and Americans switched to bigger and higher horsepowered vehicles. And more cars on the road meant oil consumption continued to grow, and with it air pollution. Although automobile companies argued that they could not meet stricter emission controls, they were at the same time developing and patenting exhaust-control engines. They agreed among themselves in the 1960s not to market these lower-polluting engines and not to compete with each other on pollution reduction. When this story broke in the press, an outraged Congress quickly passed a strengthened 1970 Clean Air Act. Under pressure, the automobile companies developed the catalytic converter. Rather than redesign their engines, the automobile companies added the converter to their existing engines. Although it reduced some pollutants, its biggest contribution to the environment was that it did not work on engines that burned leaded gasoline. Its addition to cars hastened the shift to unleaded gasoline, radically reducing the amount of lead spilling into the atmosphere. Although California's antipollution legislation continues to pressure automobile manufacturers to innovate in low-polluting cars, limited public transportation and a culture built around the car encourages more miles of driving and more tons of hydrocarbons, carbon monoxide, and nitrogen oxides pumped into the air in the twenty-first century.

The opening of new oil supplies outside OPEC control put pressure on OPEC to loosen supplies and the price of oil dropped again in the mid-1980s. With falling gas prices, pressure for greater gas efficiency declined. New SUVs began to appear on the road in the 1990s, and gas consumption again grew dramatically. By the end of the twentieth century, Americans consumed the lion's share of the world's energy supplies and contributed over 25 percent of the world's carbon greenhouse gases that act to trap the solar heat at the Earth's surface.

More than just automobiles produce greenhouse gases. America of the late twentieth century was a nation of climate-controlled environments. Central heating became common in American homes by the middle of the twentieth century. By the end of the century, central air, or at the very least several-room air conditioners became common even in the Midwest and Northeast. In converting hot, humid air into cool, dry air, condensers worked overtime—cooling homes while heating the outdoors. Americans became accustomed to moving from air-conditioned homes to air-conditioned cars to air-conditioned offices to air-conditioned restaurants. Freon, chlorofluorocarbon (CFC), was the material of choice in these condensers. Chlorofluorocarbons were also used to clean tools and to propel aerosols. When released into the air, CFCs float upward to the stratosphere, where they react with ozone. A catalytic reaction occurs in which each CFC molecule causes the destruction of several ozone molecules, depleting the upper atmosphere's protective ozone layer. The ozone of the stratosphere protects plants and animals from damaging radiation from the sun. After several decades of unrestricted release of CFCs into the atmosphere, this chemical reaction resulted in a huge hole of ozone-depleted atmosphere over Antarctica. When news leaked out about the size and significance of the ozone hole, public outcry, and a threatened boycott of CFC products, eventually led to an international accord, the Montreal Protocol of 1987, to reduce the use of CFCs.

Although agreement was reached on CFCs, global-warming gases continue to be pumped into the atmosphere, disrupting weather patterns and raising temperatures in some regions. Cities, particularly, suffer from rising temperatures. Besides the problem of global warming, miles of cement and asphalt sidewalks, roads, streets, and parking lots absorb and hold heat. With few trees to block the sun, cities become gigantic heat pockets. Air conditioners, machines, and cars pump billions of BTUs (British thermal units) of heat into the city environment. During unusually hot summers, this combination can be deadly.

It proved so in Chicago in 1995. In early July of that year, a high-pressure air mass established itself over the southwestern plains and moved slowly northeast, picking up heat as it went. When this warm air mass moved over an extraordinarily wet Midwest, the air picked up humidity. A temperature inversion over Chicago kept this hot, humid air from dissipating upward. On July 13, 1995, Chicago began a week of humidity in the range of 80 percent or more (over 20 percent higher than normal) with temperatures in the nineties. City streets, sidewalks, and buildings absorbed the sun's heat and released it back into an atmosphere that could not dissipate it. The heat held and the city grew hotter. Hundreds of children suffered heat exhaustion; poor residents turned on hydrants in desperate attempts to cool off their neighborhoods. Stores sold out of air conditioners, which put so much demand on the city's electrical infrastructure that it failed. With rolling blackouts, homes and businesses were left without power. Ambulances were overwhelmed with calls. Twenty overburdened hospitals, mostly in poor areas, refused to accept new admissions, and the bodies of the dead began piling up at city morgues. Bodies had to be stored in meatpacking trailers. By the time cooler weather came, 739 people had died from the heat.

Chicagoans died for a number of reasons. Many of the elderly poor were afraid to go outside to seek relief from the heat, or even to open their windows, because of persistent crime in their neighborhoods. Overburdened public services failed to adequately respond with centers for relief from the heat. Poverty was also a cause. Air conditioners require energy to run. The more expensive the air conditioner, the more energy efficient it is, and the lower the cost of running the unit. Homeowners have the choice, if they have the resources, to purchase more costly units that in the long term save on energy use and electrical bills. The poor do not have that luxury. If they buy, they can afford only the upfront cheaper units. Apartment complex builders and landlords did not buy expensive, more efficient units with sophisticated motors and large coils. They bought the cheaper units and let their tenants pay the ensuing high utility bills. High utility bills forced many of the city's poor to avoid using their air conditioners or to allow them to fall into disrepair. Inefficient air conditioners also contributed to the city's high power demands and the rolling brownouts that plagued the city during the heat wave.

Congress attempted to deal with the problem of the preponderance of inefficient units by mandating increased energy efficiency from air conditioners. President Clinton then raised the standard further. President Bush has since lowered the efficiency increase by a third. Less efficiency means more energy must be used, more electrical generation facilities must be built, and more waste pollution in the form of heat, global-warming gases, and acid rain will be discharged. During heat waves like the one that hit Chicago, it could mean fewer air conditioners in use, more brownouts, and more deaths.

America's growing appetite for energy brings with it not only the possibility of more localized problems during heat waves, it also contributes to the worldwide problem of global warming. Before the Industrial Revolution of the nineteenth century, carbon concentrations in the atmosphere were roughly 275 parts per million. Today, carbon dioxide makes up over 389 parts per million. The figure is rising rapidly. The consequence of this increased concentration of carbon in the atmosphere is the trapping of the sun's heat within the atmosphere and the warming of the globe. Scientists have been debating for some time the timing and the relative responsibility of the various agents in this warming, but there is no more debate among the scientific community about the reality of global warming than there is about evolution. It is happening, and its consequences are drastic. And a major—probably *the* major—cause of global warming is the burning of fossil fuel, particularly coal and oil. The potential consequences of global warming for the northeastern United States are alarming. Melting of the Greenland ice cap could send massive amounts of lighter-than-saltwater, cold freshwater into the North Atlantic Ocean. Such an event could alter the Gulf Stream current and plunge Europe and eastern North America into severe cold. Melting of the polar ice caps could also dramatically raise sea levels, flooding the coastal plains where millions of Americans find their homes. Despite the attitude of the U.S. government, which has refused to sign a global warming treaty because it might hurt the American economy, and the Bush administration, whose EPA has deleted global climate change from its annual reports, evidence of global warming grows daily, even as greenhouse gas production continues unabated. Unless the nations of the world move quickly and dramatically to deal with the continued production of greenhouse gases, the United States may be forced to adjust to a world very unpleasant, but very much of its own creation.

THE WAY THINGS ARE

My daughter now lives in Cambridgeport, Massachusetts, a few houses from where my mother visited me thirty years ago. The streets are cleaner now. Garbage is kept in trash receptacles and is picked up at regular intervals. Recycling bins hold cans, bottles, and plastic containers. Dog owners are more likely to pick up their pets' wastes. Two neighborhood parks that thirty years ago were known as "trash park" and "dog shit park" are now clean and comfortable neighborhood gathering places. But Cambridge is also a different city. Tax revenues have skyrocketed with rising housing values. New middle-class and upper-middle-class residents enjoy the amenities that the working class who once lived in Cambridgeport (this section of Cambridge) would have appreciated but could not afford.

Near my sister's house in New Hampshire are small fishing villages. Similar fishing villages dot the New England coast. For centuries men from these villages have gone to sea to harvest its bounty, particularly cod. These fishermen set sail for the Georges, the Grand, and the other fishing banks off the northern

Atlantic coast. On these banks, cod fed on marine life that flourished where the warm Gulf Stream collides with the colder Labrador currents. In the nineteenth and early twentieth centuries, schooners sailed out to the banks and there dropped dories loaded with line and barrels of herring bait overboard. The fishermen would then row out, bait and set their lines, and later haul in the catch. Filled with cod, the dories returned to the schooner. Loaded with cod, the schooners headed for shore. Once in port, the fish were traditionally either dried and salted or sold fresh to coastal markets. The fishing banks supported families across the northwestern Atlantic coast. Boat builders, fishers, fishmongers, house builders, bar owners, barrel builders, and rope and line makers all lived on the bounty of the banks.

In 1925, Clarence Birdseye discovered an alternative to the traditional means of handling fish. Having experimented with freezing vegetables one winter in Labrador, Birdseye also found that cod frozen in brine water kept without curing. In 1928 filleting machinery was adopted in New England, which hastened the process of moving fish from the ocean to the dinner table. Fishermen soon found a ready market for all the catch they could bring in.

Although sails dominated the banks until the 1930s, increasingly thereafter fishermen looked to steam and then diesel trawlers and bottom draggers, or otto trawls, to make their catch. The new diesel trawlers powered out to the banks quickly, dragged their nets just above the ocean floor, filled their hulls, and returned to port.

Throughout the 1950s, small diesel fishing trawlers still set out from coastal villages and towns across the Northeast, but once on the banks they found themselves in competition with large foreign-owned stern trawlers hauling in huge nets loaded with fish that were then frozen while still at sea. Although cod fisheries were collapsing in the North Sea and eastern North Atlantic due to overfishing, the catch on the banks increased. Seeing the opportunity to expand their fishing industries, in 1977 the United States and Canada declared a limit on foreign fishers within 200 miles of the North American coast. This put most of the prolific fishing grounds within Canadian and U.S. waters.

The 200-mile limit kept the modern European fishing ships off of the banks. Smaller, less-efficient American and Canadian boats could once again depart from fishing villages to fish offshore waters, unconcerned about huge factory ships taking their catch. Inefficiency promoted conservation, but inefficiency did not promote economic development. In the 1980s, hoping to encourage both the boat-building industry and the fishing industry, the U.S. government offered low-interest loans to those willing to invest in modernizing the North American fishing fleet. Bigger, more efficient American ships soon moved out over the banks to bring home ever-larger volumes of fish. They hauled in so much fish

Depleted fishing fleet, Cape Cod, Massachusetts, c. 2000. (Courtesy of Kazia Cumbler)

that too few were left behind to reproduce. By the 1990s, the world's richest fishing grounds were broke. The cod population crashed.

Today, few commercial fishers leave in the early morning to get out to the fishing grounds. When I visit my sister and go to the docks of ports near her house to watch the fishing boats pull out to sea, the fishers on board are more likely to be bankers, stockbrokers, doctors, and lawyers. They fish for sport. They are after bluefish, striped bass, and tuna, rather than cod. Commercial fishers are more likely to be going after lobsters or shellfish. Coastal New England grew comfortable on its fishing, but it managed this once prolific resource poorly.

New England now prospers on high-tech companies and investment banking. That wealth comes to the old fishing villages and pays for summer homes, pleasure boats, and sport-fishing equipment. The fishing villages also seem to prosper. Real estate prices are high; fancy restaurants vie to rent space in warehouses, boathouses, and on old wharves. But beneath today's prosperity lies a modern-day failure: the failure to husband yesterday's resources. And this was not yesterday's failure; it is today's. The fish stocks of the banks were not deci-

Pleasure boats at Old Harbor, Cape Cod, Massachusetts. (Courtesy of Kazia Cumbler)

mated in the era of sail power; they were overfished with the complicity of modern science. This was not a failure of knowledge, but a failure of public policy.

The New England fisheries are not the only ones in troubled waters. For over a century and a half, fishers on the Great Lakes have overseen fisheries' destruction. In the late 1850s, the once prolific salmon of Lake Ontario were in rapid decline. Overfishing, pollution, and habitat destruction of spawning streams spelled the end to this once profitable industry. But the fishing industry moved on to other fish and other locations. White fish, trout, herring, and sturgeon replaced salmon in the Great Lakes fish market. Fishing communities fanned out along the shores not only of Lake Ontario, but also of Erie, Huron, Michigan, and Superior. Diesel fishing boats and increased net size soon pushed these fisheries to the brink of destruction as well. Declining fish meant increased prices. Fishing company officials pressured fishers to increase their catch, even in face of declining size and numbers. Pollution of the lakes and destruction of crucial spawning and nursery grounds in the coastal wetlands further pressured declining fish stocks. Public officials responded with hatchery programs and restocking, but the number of fish in the lakes declined as the waters became ever more polluted. In 1925, Lake Erie's herring population crashed. Public outcry forced Congress to allocate funds to research the cause. The final report issued in 1933 found heavy and moderate pollution too great to allow spawning in several regions of the lake. Even where the fish could spawn, the report found significant areas of light pollution. The report concluded that overfishing was the primary contributor to the destruction of Lake Erie's herring. Despite heightened public concern, officials failed to effectively curb either overfishing or pollution. Compounding these problems in the 1930s and early 1940s, the sea lamprey eel invaded the Upper Great Lakes, further decimating an already crippled fish stock. Sea lampreys attach themselves to the sides of fish and suck out their life. As the lampreys wiped out the top predators, fish that had been their prey expanded in numbers. Alewives flourished for several years in the 1950s, only to run afoul of their own success. In the mid-1960s alewives, not able to survive the rapidly changing water temperatures of the Great Lakes, began to die by the billions. Islands of dead, rotting fish thirty to forty miles long floated around the lakes and washed ashore, absorbing precious water oxygen and fouling waters and beaches.

In the 1950s, the federal government began a lamprey control program, initially by capturing spawning lampreys and then in 1958 by using a chemical larvicide to kill immature lampreys. The program worked, and lamprey numbers declined significantly. In 1966, in an effort to control the alewife population, Coho salmon were introduced. Lake trout were reintroduced to Lake Superior to encourage sportfishing. Lamprey eel control efforts, the stocking programs of the 1960s and 1970s, and the policies to protect fish stock by limiting fishing for trout and salmon to sportfishers did improve fish populations. By the mid-1970s, sportfishermen were taking in over 3 million sport fish per year and laying out hundreds of thousands of dollars in recreational expenses. Restricted fishing laws also led to significant political and court battles between sportfishers and Native Americans, who claimed treaty and traditional rights to fish the Great Lakes.

Attempting to resolve the legal and political battles between native and sportfishers was not nearly as difficult as resolving the deeper problem of water pollution. The large predator fish accumulated such significant amounts of toxic materials—heavy metals (particularly mercury), pesticides, DDT, and PCBs—in their meat that simply eating them put one at risk for cancer. In more recent times, the lakes and feeder rivers and streams have been plagued with an invasion of zebra mussels believed to have entered the lake in the bilge of ships traveling the St. Lawrence Seaway.

Government intervention banning fishing on the banks off the New England coast and the introduction of hatchery fish into the Great Lakes may restore fish to these once prolific fishing grounds, but for both fish and humans to truly thrive, far more fundamental actions must be taken to bring humans into greater harmony with the natural surroundings. Americans have long understood that their activities affect the physical world. In the eighteenth and nineteenth centuries, they celebrated that impact. Through their hands forests were replaced with gardens and farm fields and "noxious and dangerous" animals like bears and wolves were driven from the land. Americans were also aware of less beneficial changes. Hunters and farmers lamented the loss of deer, turkeys, ducks, geese, and partridges, yet the loss of these resources was seen as almost beyond man's control. Though they understood that hunting, fishing, and farming were causing the decline in wild game, nineteenth-century Americans considered those activities as entirely natural. Such human activities were viewed, like eating and breathing, as inevitable and impossible to restrain. As in the case of the decline of the fish, the loss of these resources was an unintended consequence of American progress. And by the middle of the nineteenth century, progress in the Northeast had reached an advanced state. But at the same time, there were also voices calling into question the ideal of progress and the inevitability of resource destruction. And these voices first emerged from the Northeast.

Some 150 years ago, two New Englanders viewed the changes in their region with worry. Henry David Thoreau, perhaps the most famous New Englander concerned with the impact of development on the environment, looked out from his Concord, Massachusetts, home and was greatly distressed by what he saw: trains rushing by spewing smoke and cinders, blurring nature's beauty and subtlety; farmers driven by demands for greater production; dams blocking fish migration; and factories polluting waters. Thoreau questioned the virtues of progress, particularly the progress of industrial manufacturing: "I cannot believe that our factory system is the best mode by which men may get clothing" (Thoreau 1866, 20). Nor did he believe corporations were the best means of harnessing power for production: "The principal object [of corporations] is not that mankind may be well and honestly clad, but, unquestionably that the corporation may be enriched" (Thoreau 1866, 20). Thoreau's response to the corporation and the factory system was to return to a simpler life: "Simplicity, simplicity, simplicity" (Thoreau 1866, 67).

But a simpler life and a retreat to a cabin on a friend's land were not an option for a majority of nineteenth-century Americans living in industrial town tenements and sweating over machines. In that world, clean air, water, and soil—let alone abundant fish and wildlife—were dear commodities. In their lives, the science and technology of medicine and sanitation made a material difference. But science and technology did not solve the larger problems of poverty and a constrained environment.

Nineteenth-century author Henry David Thoreau. (Library of Congress)

George Perkins Marsh provided another New England voice questioning progress. Marsh warned his fellow Vermonters to revisit how they used their resources. Marsh argued that human settlement had "diminished" nature and that "human progress" had "shorn and crippled" the natural world. Marsh nonetheless believed that "we may still do something to recover, at least a share of the abundance" of the past (Cumbler 2001, 75–76). Marsh looked to the government to ameliorate the destruction of the natural world. But he also recognized that the culture and values of modern society worked against successful state intervention.

Twentieth-century Americans still clung to the idea that technology, science, and state action could work together to overcome some of the worst environmental consequences of modern progress. Indeed, the success in procuring safe, potable water for American cities, the reduction in mortality and morbidity rates, the increase in the nutritional content of the typical American diet, the protection of parks, and the general rise in the prosperity of average Americans through the combined intervention of trade unions and government programs all gave hope to the notion that the environment itself could also be protected.

But effective environmental protection required more than just wishful thinking. The works of the twentieth-century midwestern environmental thinker, Aldo Leopold, pointed to the importance of embracing a new "land ethic." Leopold warned Americans that progress could be measured in diminution as well as in addition. He believed that material progress without ethical progress would bring harm to nature. To him, ethical progress involved seeing the land itself as having a right and a value—not as property—but as a part of a community. Rachel Carson understood this idea and made it real to Americans with her powerful indictment of the pesticide industry. Pennsylvania-born Edward Abby saw what humans were doing to the earth and called for a deeper ecological response of open resistance to human-engineered environmental damage. Americans have seen manifest Leopold's "land ethic," Carson's scientific critique of modern petrochemical practices, and Marsh's belief that protecting the natural world ultimately requires state action in the successes of modern environmentalism and in the Environmental Protection Agency's work to protect endangered species and reduce air and water pollution. Modern science worked to provide technology that limits pollution and controls toxic wastes. National parks have expanded to incorporate not only recreation but also the protection of the interconnected natural environment itself.

Earlier in this work I wrote about my sister's vacation home on Cape Cod. She visits the Cape because it is a beautiful, wild place where the land meets the sea on a tenuous strip of glacial till. Henry David Thoreau called it that "bared and bended arm of Massachusetts." When Thoreau visited the Cape in 1849 and again in 1850 and 1855 "there [we]re many holes and rents in this weather-beaten garment not likely to be stitched in time" (Thoreau 1987, 4). The Cape was a place of solitary—and not particularly prosperous—fishing villages, cranberry farms, and struggling orchards. It maintained saltworks, a glass factory, and boat-building and repair shops. Whaling ships still sailed out from its deeper harbors. Following the Civil War, occasional sportfishers came down to try their

Rachel Carson appears before a Senate Government Operations subcommittee to discuss pesticides in 1963. (Library of Congress)

luck, but, for the most part, Cape Codders were left alone on their increasingly treeless spit of land, tenaciously holding on in the face of fierce northeasters and the relentless tugging of the tides.

With the increased prosperity of the second half of the nineteenth century, wealthy Bostonians and New Yorkers began to look farther afield for recreation and relaxation. The shores of Rhode Island, Massachusetts, New Hampshire, and southern Maine soon filled with summer homes belonging to the well-to-do of Boston and New York. For most, Cape Cod was still too far away and too desolate for summers of recreation and socializing. As the railroad moved across the Cape in the late nineteenth century, it brought a new form of recreation to the southern cape. Large resort hotels, where visitors stayed for extended holidays, appeared along the beaches from Sandwich to Chatham. Here visitors enjoyed comfortable accommodations, the opportunity for solitary walks along deserted beaches, and hotel-sponsored evening clambakes among the dunes. Cape Cod was not alone in the construction of large, all-inclusive resort hotels. Recreation

Nineteenth-century seaside resort York Beach, Maine. (Library of Congress)

in the middle of the nineteenth century was either confined to areas close to home—evening strolls to the town center for weekend band concerts, afternoon picnics, or ice-skating parties—or involved all-male hunting or fishing retreats. But in the late nineteenth century, recreation for the prosperous upper middle classes shifted. Families began to recreate together for extended periods. Sometimes this would involve only the wife and children, with the husband joining them for the weekends. The very wealthy maintained their own summer retreats, but most prosperous Americans centered their recreation in the resort hotels. These hotels sprang up in the rural countryside, on scenic mountaintops, and along rustic lakes from Wisconsin to Maine. Investors also built them on the rocky shores north and south of Boston and among the sand dunes of Cape Cod, Long Island, and the New Jersey shore. L. L. Bean began a successful company catering to these amateur visitors to nature.

Concern over protecting these scenic locations led to the creation in 1892 of the Adirondack State Park, set among the mountains between the St. Lawrence

River and the Mohawk Valley in upper New York State. Interest in the Adirondacks among wealthy sportsmen was raised by the publication of Joel Headley's *The Adirondacks: or Life in the Woods* (1849) and William Murray's *Adventure in the Wilderness: or Camp-life in the Adirondacks* (1869). Land abandoned by logging operations that came into public ownership through tax default was originally set aside as a forest preserve in 1885. Today, it has become a major resort area for vacationing middle-class Americans. And the L. L. Bean Company's outlet store is a major shopping stop for thousands of middle-class Americans, while its catalogue selling the rugged outdoor image is as ubiquitous today as Montgomery Ward's was a hundred years ago.

The desolation, natural wild beauty, and (due to its generally depressed economic condition) cheap land made Cape Cod ideal for retreats. Summer revival camps sprang up across the bottom length of the Cape, away from the more expensive shore property.

Rachel Carson noted in her famous *The Edge of the Sea* (1941) that, "the shore has a dual nature, changing with the swing of the tides, belonging now to the land; now to the sea." This enigmatic duality holds a fascination for humans. The force of nature is ever present at the ocean's edge, a place of "high winds and driven sand" as Carson noted. It has a stark beauty. It changes also because it is a fragile place: a place where the winds and sea constantly reshape the physical world. Each day presents a different land and seascape. It can be calming in its vastness, in the flow of its grasses and the hum of the wind in its scruffy trees, and in the rhythm of the waves. Strong northeasters pound out their dramatic violence. Cape Cod, as Thoreau noted, thrusts its "bended arm" ready for "boxing with northeast storms," out into the Atlantic, exposed on both sides to the changing power of water and wind. Removed from the crowds of the city, delicately placed amidst the constant challenge of sea and wind, the Cape offered Americans a place radically different from the built world of towns and cities. The Cape represented a world seemingly built and rebuilt by nature.

Beginning in the 1920s, a new group of visitors came to Cape Cod: car owners. Automobile ownership, new roads, and the explosion of small cottage colonies opened the Cape to upper-middle-class Americans. Some owned, others rented, but for both groups the Cape increasingly became a place to spend the summer. Provincetown took on an exotic air as artists flocked to its beautiful dunes and trendy cafes.

The popularity of the Cape Cod vacation increased dramatically following World War II. Paid vacations became a standard part of most employment. Working-class and lower-middle-class Americans, especially those with union jobs, could afford a car and, by joining with relatives, could rent a cottage on the Cape for a week or two. Traffic was bumper to bumper in the height of the season. The

Protecting the Place   213

Cape Cod, Massachusetts, c. 1950. (Photodisc, Inc.)

old hotels of the late nineteenth century fell into ruin and were cleared to make way for newer, sleeker motels and guest cottages. Route 28 became filled with seafood restaurants, miniature golf courses, and saltwater taffy stands. Developers and homebuilders bought all available property near the water from the foot of the bridge to the elbow of the Cape's arm. Tens of thousands of cars crossed the Cape Cod Canal bridges daily.

Farmers and fisherman settled Cape Cod. They pumped wells by hand and dumped their wastes into privies. Running water and flush toilets were a prerequisite for the new homes and motels of the second half of the twentieth century. The wastes flowed into septic systems, eventually seeping into harbors, the bay, or the ocean. Water was drawn from the Cape's aquifer in ever-greater amounts. The momentum for development was great. As private estates and motels grabbed greater patches of beachfront along the southern cape, and public beaches became more crowded, the pristine beauty of sand and surf, wild roses, blueberry bushes, and beach grass quickly gave way to fried-fish stands, boardwalks, and beach umbrellas. By the end of the 1950s, it was clear that the steady march of development would soon reach the Cape's eastern edge and turn northward to embrace the outer cape as well. The wild rugged beauty of the Cape that so enthralled Henry David Thoreau in the middle of the nineteenth century was threatened with disappearance a hundred years later.

Concern for the loss of Cape Cod's original beauty led conservationists to pressure the Massachusetts senators to introduce legislation that would eventually lead the National Park Service to begin buying up land on the outer cape. In 1961 efforts by Senators John Kennedy and Leverett Saltonstall finally created the Cape Cod National Seashore. Although the National Seashore was bitterly opposed by some outer-cape residents who feared the buying up of land by the Park Service would restrict development and future profits, by the mid-1960s, the Park Service began acquiring land. Supporters hoped that the new Cape Cod National Seashore would accomplish several things at once. Its primary purpose was to "conserve and preserve" the Cape's delicate natural resources from further human degradation. As a part of the Park Service, the National Seashore was also dedicated to opening the Cape to the public. These agendas were difficult to balance. Maximizing human enjoyment would threaten the Cape's delicate ecosystem. The Cape Cod National Seashore attempted to balance these needs by purchasing as much land as possible, restricting use in the park, educating park visitors about the Cape's fragile ecosystem, and channeling activities to areas designed to accommodate specific uses.

A casual drive through the Cape Cod towns outside of the park, an afternoon hiking through the park's walking paths, or observing the diversity of bathers on park beaches will convince most cape visitors of the park's success. Vast areas of the outer cape are protected from extensive development. Change there depends more on the whim of nature than the hand of man. While the park does protect much of the outer cape's natural beauty, aesthetics was not the Cape's only problem. Development continues in the towns now surrounded by the park and those outside the park. With significantly less land to develop, the prices of the remaining land parcels have increased. Higher land prices have encouraged larger and more expensive housing on existing sites. Those who can afford the prices expect conveniences like abundant hot water, flourishing gardens and yards, and large driveways accommodating multiple cars. Water usage continues to climb, as does the volume of wastes flowing back into the ecosystem: human wastes as well as fertilizer, pesticides, herbicides, and other yard and garden runoff. The water table for much of the Cape is threatened with salt infusion as well as contamination. The beauty of the cape area protected by the park has encouraged ever more people to seek out this fragile strip of sand jutting out into the raw Atlantic. They come now by the hundreds of thousands, walking on delicate dunes, swimming, eating, and drinking. They bring along their cars and their goods, and while they take their cars home, they typically leave behind the wastes, including burnt petroleum fumes. Boston has recently stopped dumping its untreated wastes into Boston Harbor. (Indeed, it was that over-a-century-old dumping that George Bush Senior used against his presidential rival, Michael Dukakis, in the 1988 campaign.) Boston built primary treatment facilities for its wastes and then pumps that treated waste deep into Cape Cod Bay. Boston Harbor is becoming cleaner. The impact of the outflow pipe's waste on Cape Cod Bay is yet unclear.

How much of this waste and these visits the Cape and its bay can tolerate before they crash is unknown. Those who care are desperately trying to find a way for Cape Cod to thrive, both as an ecosystem and as a place to visit. Others have more immediate concerns—jobs and income. Without a well-thought-out plan for balancing all these competing needs and concerns the Cape may not be able to bear the weight of its success as a beautiful vacation spot. Henry David Thoreau's comment that, "there are many holes and rents in this weather-beaten garment not likely to be stitched in time" may prove to be the Cape's obituary. If so, we will all lose something special. But the dilemma confronting the Cape is not unique. It is the conflict between development and the environment. This conflict is particularly acute within such fragile ecosystems as Cape Cod, Long Island Sound, the New Jersey shore, and the lakes of the Northeast and Midwest, but the problem also confronts the whole of the nation. Henry David Thoreau called on us to live simply. George Perkins Marsh was concerned that our culture could not find in itself the ability to live more simply. Certainly in the past we have found the wherewithal to change our lives, to legislate and enforce laws, and to develop and utilize technologies that helped reduce our harmful impact on the natural world. At the same time, we have also often ignored the need to change or have chosen the easier or more comfortable alternatives, turning a blind eye to their negative impact on the environment.

Thoreau, Marsh, Leopold, and Carson did more than issue warnings about environmental degradation; they also provided us with lessons on how to approach and solve these problems. These are the lessons we need to remember because, in many ways, the problems facing us today they first confronted some 150, 70, and 40 years ago.

IMPORTANT PEOPLE, EVENTS, AND CONCEPTS

Addams, Jane (1860–1935) Founder of the Chicago settlement house, Hull House, and pioneer American social worker. Jane Addams organized as a project of Hull House a campaign for regular garbage removal.

Agricultural Adjustment Administration A New Deal program to cut agricultural production in order to increase prices. It also worked to reduce plowing on marginal agricultural land and worked for soil conservation.

Algonquians Groups of Native Americans living in southern New England and the upper-mid-Atlantic region and speaking languages with a common root, called Algonquian. The Algonquians were made up of several different Native American groupings (such as the Huron, Ottowas, Petuns, and Fox). In the mid-seventeenth century they came under severe attack by the Iroquois, which combined with the ravages of disease introduced by white settlers, pushed most of them out of New England. Very few survive to this day.

Alienation of land The sale of land. The right to alienate land is one in a bundle of rights associated with land ownership and property rights.

Alleghenies The section of the Appalachian mountain range that runs through western Pennsylvania and central New York.

Alluvial soil Rich fertile soil deposited by flood water and made up of deposits of sand, clay, and washed-down topsoil carried to lowlands and then left by rivers and streams. Alluvial soil is usually found in river valleys and coastal estuaries.

Anadromous fish Fish such as salmon, alewives, or shad that live in different environments over their life span. Salmon and shad mature in the ocean but return to fresh water to spawn. The new hatchlings spend their early development in fresh water and then migrate down to the ocean. The movement of these migratory fish upstream to spawn provided fishers an opportunity to catch them.

Anthracite coal A hard, black, lustrous coal composed almost entirely of pure carbon. Various forms of coal have different amounts of fixed carbon, moisture, and volatile matter. Anthracite has the highest carbon and lowest moisture content. It was preferred as a home-heating fuel over the softer bituminous coal because it burned hotter and cleaner. Bituminous coal was more readily available and burns easily. It was preferred as a fuel for industries, railroad locomotives, and making coke. Coal is produced when organic matter decomposes through the action of bacteria. This process of breakdown and compounding produces peat. When peat is further compressed by overlying rock and sediment, moisture and other volatile substances are squeezed out to form compact carbon. Most coal was formed in the Carboniferous period, roughly 290 to 360 million years ago.

Antibody A protein in the blood produced in response to an antigen that fights or counteracts the antigen as part of the immune system's response to exposure to disease-causing organisms.

Appalachian Mountains A mountain range that runs up the east coast of the United States, beginning in northern Georgia and ending in Maine. It encompasses the Smoky and the Blue Ridge mountains in the south, the Allegheny Mountains of the mid-Atlantic region, and the Adirondack, Green, and White mountains in the north.

Arable Land fit for agricultural cultivation. Land suitable for plowing.

Arcadian A term descriptive of a rural lifestyle of apparent harmony with nature. The term comes from Arcadia, a region in ancient Greece surrounded by mountains and inhabited by people renowned for their simple lifestyle connected to nature.

Auto-touring A practice popular in the 1920s of using a car to tour the countryside. Increasingly, state and national parks began to cater to the tourist arriving by automobile.

Balancing doctrine A legal principle employing a cost-benefit analysis for determining whether or not to demand remedial action from polluters. If the benefit to the plaintiff from enjoining the pollution outweighed the cost of the remedial action, courts tended to find for the plaintiff, but if the costs of remedial action were significantly higher than the benefits, courts tended to find for the defendant.

Balloon-frame construction Building construction invented by Augustine Taylor in Chicago in 1833 that used standardized, light, precut pieces of lumber nailed together as sills, floor joists, studs, and roof rafters. The new construction avoided the older beams fitted together with mortise-and-tenon joints by constructing a framework of many lightweight studs and joists. The strength of the frame inhered in the fact that it was linked together by multiple studs all tied together. Because the buildings, before they were covered with siding, created a whole skeletal structure, it was called a balloon frame. The lack of heavy beams and tenon joints meant that it could be easily constructed with fewer workers having less skill. Because the building material consisted of light 2 ∞ 4-inch studs and joists, the wood required for building could be easily and economically shipped to a building site.

Benzene   A clear, volatile liquid by-product of coal tar. It was used in the manufacture of chemicals and dyes, as a solvent, and as a motor fuel.

Bessemer furnace   A special furnace developed in the 1860s to convert pig iron to steel by forcing hot air into the molten iron. *See* pig iron.

Biochemical Oxygen Demand (BOD)   The amount of water used or taken up by microorganisms when they break down waste. Earle Phelps, of the Lawrence experimental station, found that bacteria break down wastes but in doing so they deplete the water of vital oxygen. Organic matter disposed of in water systems (as sewage, for example) acts to take up oxygen, thereby creating a BOD in the water.

Black Hawk (died 1838)   Sauk leader who led a group of Sauk, Fox, and Kickapoo Indians living in Iowa across the Mississippi River to reclaim their ancestral lands in western Illinois. Claiming that they would retain their "right to the soil," Black Hawk soon confronted an American military force better armed and numbering more than double his warriors. In the battle of Bad Axe in August 1832 Illinois militiamen killed dozens of Indian men, women, and children. Although many native groups did not participate in Black Hawk's war, all faced forced treaties that moved them off their midwestern land.

Blickensderfer tile ditching machine   A horse-drawn ditching machine, manufactured in Decatur, Illinois, in the 1880s, that could dig a four-foot trench for laying drainage tiles. *See* drainage tiles.

Bluestem   A wild, high-grass prairie grass, common to the flat and rolling lands of north-central Illinois and southern Wisconsin west to Nebraska. The big bluestem had a massive root system that penetrated as deep as six feet into the soil.

Blunderbuss   Seventeenth-century short gun with a large bore and a flaring muzzle.

Bog   A tract of wet, spongy ground often created by the filling in of old beaver ponds with eroded soil deposited by receding floodwater and decaying vegetable matter.

Boone and Crockett Club   Founded in 1887 by Theodore Roosevelt and George Bird Grinnel to protect wildlife habitat and espouse good sportsmanship in the field. It enlisted a large number of wealthy, male elites into its membership.

Boston Associates (1815–1820)   Investors, many of them merchants in the overseas trade, under the leadership of Francis Cabot Lowell, who pooled their money to create the first integrated, water-powered textile mills in Waltham, Massachusetts.

Boston Manufacturers Association The limited-liability company established by the Boston Associates that built and maintained the mills at Waltham and then Lowell. These mills combined power-driven cotton-carding machines with power looms and spinners. These machines and economies of scale allowed for the production of inexpensive, uniform cotton cloth.

Bottle bill State legislation mandating deposits on bottles and cans of drinking beverages. These bills were designed to reduce litter.

Bradford, William (1590–1657) Leader of the Pilgrims who settled in Plymouth, Massachusetts.

British thermal unit (BTU) A measure of heat production. The quantity of heat required to raise the temperature of one pound of water one degree Fahrenheit, starting at or near 39.2 degrees Fahrenheit.

Broadleafs Deciduous hardwood trees such as maple, oak, chestnut, or hickory.

Buckeye trencher A steam-driven ditch-digging machine developed in 1892 that was capable of digging over 1,400 feet of 4.5-foot trenches a day. These trenches were used for laying drainage tiles.

Burr oak Commonly known as mossycup oak, it is a tree of the heavy woods or prairie boarders. Common in western New England west into Michigan, its bark was used in tanning.

Cape Cod A long sandy peninsula in the shape of a hook that sticks out into the North Atlantic Ocean, formed by the deposit of glacial wash.

Carson, Rachel (1907–1964) A naturalist writer and a marine biologist for the U.S. Department of Fish and Wildlife. Carson was the author of the successful naturalist book, *The Sea Around Us* (1951). Carson's friend Olga Owens Huckins of Duxbury, Massachusetts, called Carson's attention to the increased spraying of DDT and death of songbirds. Carson responded by writing *Silent Spring,* a reasoned attack against the poisoning of the environment with chlorinated hydrocarbon pesticides. Carson argued that human health was linked to environmental health and that without working with the environment, humans put their own health at risk. Carson was attacked by chemical industry spokespeople and called an alarmist and even a communist who was threatening American agriculture and industry.

Catalytic converter A device that could be added to a car engine to reduce exhaust emissions. The catalytic converter allowed automobile manufacturers to meet minimum emissions standards without a major redesign of their car engines. This meant that emission standards were met, but with a loss of engine efficiency and without significant technological improvement in the engines. The catalytic converter did require lead-free gasoline. This reduced the amount of lead discharged into the environment from automobiles.

Cellophane   A plastic wrap developed by the Dupont Company in 1924 that greatly increased packaging on consumer goods and replaced paper as the standard wrapping material. It also increased the amount of slow-degrading material that entered the American waste stream.

Center for Pollution Studies   A 1920s research center sponsored by the Public Health Service and under the direction of Earle Phelps to study water pollution. Located in Cincinnati, the center organized an extensive study of Ohio River pollution.

Central business district   With the development of streetcar lines, cities expanded, and urban areas began to specialize in activity. The regions around the urban periphery tended to concentrate residential housing, while the center of the city specialized in retail businesses, professional services, office work, and some light manufacturing. The central city specializing in those activities became know as the central business district.

Champlain, Samuel (1567–1635)   French explorer, fur trader, and founder of "New France." He explored the coast of New England, making the first detailed charts of the region from Canada to Martha's Vineyard. He also set up French settlements along the St. Lawrence River and west into the interior.

Chemical cleansing   A method of adding chlorine to water to kill harmful bacteria. Chemical cleansing, along with filtration, became the standard treatments for municipal water.

Chicago Board of Trade   Founded in 1848, the Board of Trade was designed to further Chicago's commercial interests. When grain elevator operators objected to having to keep track of the grain of various farmers, the Board of Trade developed a grading system by which each farmer's wheat was labeled according to its type—such as spring, white winter, or red winter—and its quality. Under the new labeling system, a farmer's wheat could be mixed in with other wheat of the same type and quality. The elevator operator, rather than having to keep track of an individual farmer's wheat, could then give the farmer credit for a specific quantity of wheat held for him within the same type and ranking. This began the process of turning wheat into an abstract commodity. The farmer, rather than having specific wheat to be sold, now had a specific amount of wheat being held for sale for his account.

Chicago World's Fair   Also known as the Columbian Exposition of 1893, this exhibition established the idea of the city as a planned and beautiful place. The fair was organized around well-lit boulevards, fountains, and parks. Its example helped establish the "city beautiful movement," and encouraged the building of urban parkways and greenbelts. The park where the event was held was landscaped by Frederick Law Olmsted, America's leading landscape architect.

Chinch bugs Pests that plagued the nineteenth-century wheat crop.

Chlorination and filtration Systems developed at the turn of the twentieth century to purify drinking water. Water from lakes or rivers would be filtered through sand and have chlorine added to kill harmful bacteria. Urban death rates significantly declined with the development of these systems to protect drinking water.

Chlorofluorocarbons (CFCs) Material used as coolant in air conditioners, as a cleaning agent for tools, and as a propellant in aerosols. CFCs released into the atmosphere are lighter than air and drift upward to the stratosphere where they interact with ozone. Ultraviolet radiation breaks chlorine atoms off from the CFC molecule. The chlorine atoms then react with the ozone molecule, destroying the ozone. This reduces the ozone in the upper atmosphere and subjects the earth to significantly higher doses of ultraviolet radiation, increasing the risk of skin cancer, blindness (especially for animals), and disrupting the growth of plants and aquatic organisms. This process was first suspected in 1974 and later seen when a large hole, or depleted region, appeared in the ozone layer of the upper atmosphere in 1984.

Cholera An acute infection manifested by profuse diarrhea, vomiting, muscle cramps, and dehydration. Death was often caused by acute dehydration and organ failure. Cholera was caused by *Vibrio comma* and spread by ingestion of water or foods contaminated by excrement. Cholera usually breaks out in limited but dramatic epidemic episodes.

Church of England A Protestant church, the official church of England, similar in structure and hierarchy to the Roman Catholic Church but without cardinals or the pope. The crown head of England is the official head of the Church of England.

Civilian Conservation Corps (CCC) Franklin Delano Roosevelt's New Deal program to take young men from eighteen to twenty-five years old and put them to work on conservation projects. The CCC workers built campsites, lodges, trails, bridges, and roads in parks. They planted trees and dug drainage ditches. CCC workers also worked on anti-erosion projects.

Clean Air Act Federal law passed in 1970. The Clean Air Act limited the discharge into the air of carbon monoxide, ozone, and lead based on the risk to human health. It set up the National Ambient Air Quality Standards and monitored air quality. It mandated that states come up with plans to reduce polluting emissions in designated air-quality districts.

Clean Water Acts Federal laws of 1960 and 1967 mandating the reduction of pollution in public water. The acts helped fund the building of municipal sewage treatment facilities. See also the Water Pollution Control Act.

Clovis hunters The first peoples to migrate down from Canada to North America, they were named after one of the earliest sites of human activity discovered south of Canada, near Clovis, New Mexico. These hunters were believed to have crossed the Bering Strait when it was above the water line and to have moved below the glacial ice pack around 11,000 B.C.E.

Cod *Gadus callarias.* A large food fish found in the northern seas. Easily dried and stored, cod was the staple fish for Europe for some 800 years between 1100 and 1900 C.E. It was a principal export from New England, but by the 1980s the population of cod off the New England coast was in dramatic decline.

Coke by-product ovens When coking was first developed, the coal was heated in beehive ovens, named for their conical shape, and many of the by-products of the burning were released into the air. In the 1920s, manufacturers, in an attempt to capture some of these wastes for possible use and in response to urban complaints about air pollution, switched to by-product ovens that separated out tar and other by-products but also produced large quantities of liquid effluents high in phenols and ammonia that were dumped into the nearest running stream.

Coked coal Coal that has been slow cooked to burn off the impurities. The process of coking, developed in England by George Darby and perfected by his son in the eighteenth century, allowed coked coal to substitute for charcoal in iron making.

Coliform A bacteria found it water polluted with human and animal feces. George Fuller and George Whipple discovered in the 1890s that coliform bacteria indicated contamination. They developed a method for testing water for coliform bacteria.

Collapsible metal tube A nondegradable, disposable container that became a common means of holding soft dispensable material such as toothpaste. Its increased use in the twentieth century added to the problems of urban garbage and litter.

Crowd diseases Diseases typical of crowded conditions that spread from person to person. Cities of the nineteenth century experienced high rates of mortality due to these diseases. Tuberculosis, typhoid, dysentery, and cholera were epidemic crowd diseases that spread from person to person through contaminated drinking water and close personal contract.

Cryptosporidium parvum A potentially fatal waterborne bacterium that leads to intense gastrointestinal disorders. The bacteria is usually ingested with polluted water.

Crystalline piedmont uplands The geological basis of the upland regions of eastern Pennsylvania and New Jersey.

Culm Waste from coal mines, particularly coal particles too small to be marketable and other wastes that were dumped alongside the mines.

Dam heights Mill machinery required different levels of power. Textile machinery introduced in the nineteenth century had greater power demands than the older gristmills or sawmills needed to turn their stones or work their saws. To gain this added power, the new mills increased dam heights to capture more power from the falling water. The mills also needed power continuously, so they did not lower their dams seasonally as the older mills used to do. Increasing the dam heights and keeping them there also flooded more land and prevented migrating fish from moving upstream.

DDT (dichlordiphenyltrichloroethane) A chemical insecticide developed in 1942, DDT was originally used on soldiers during World War II to fight lice. It was then applied to kill mosquitoes transmitting malaria. When the war ended, DDT manufacturers encouraged its domestic use against pests such as flees and the Japanese beetle. Farmers were encouraged to spray DDT on their farm animals to keep down flies and other insects. Throughout the 1950s, it was widely used against any number of insects. It was also spreading widely through the environment. It was responsible for the thinning of birds' eggshells that led to a decline of a number of popular birds, particularly the brown pelican. And in the 1960s, it was found in human mothers' milk.

Delaware River Basin Commission A multistate commission established in the 1920s to help protect the waters of the Delaware River basin. It was concerned about water pollution and lumbering along the watershed.

Dilution The assumption that pollution would be neutralized through dilution with a large volume of clear water. Believing that dilution took care of pollution encouraged urban leaders to simply dump their sewage wastes into the nearest river.

Diphtheria A contagious bacterial disease characterized by attacks on the respiratory tract. It can be accompanied with toxemia that attacks the peripheral nerves and muscles of the heart.

Dissolved oxygen The amount of oxygen in water. Live plants increase the amount of oxygen in water. Dead plant matter, nitrates, and phosphates, as well as fish, take oxygen out of the water. Fish need dissolved oxygen in the water to breath. Running water also captures oxygen. The levels of dissolved oxygen in water became a measure of its quality. Water-quality experts began measuring dissolved oxygen in water at the turn of the twentieth century.

Dowry Money or other valuables set aside by a bride's family to give to the newly married couple to help them set up a household, homestead, or business.

Drainage districts Areas defined by state legislatures to allow for the building of drainage ditches to encourage the drainage of wetlands.

Drainage tiles Horseshoe-shaped clay tiles developed in the nineteenth century for burying in fields to allow for drainage. Water would flow along these tiles into drainage ditches and streams at the sides of fields. Drainage tiles allowed for the conversion of large areas of wetlands to farmland.

Dynamo A direct-current generator, a machine used to change mechanical energy into electrical energy. The dynamo was originally developed using the principle of electromagnetic induction discovered by Michael Faraday in 1831. When a conductor passes through a magnetic field, a voltage is created across the conductor. The spinning generator or dynamo rapidly moves the conductor through the magnetic field and captures the electricity in an external circuit. The generator that produced the power that lit the 1893 Chicago World's Fair and powered its exhibits was a gigantic dynamo.

Dysentery An infection caused by *Entameba histolytica* that produces a colitis and painful bloody diarrhea discharge. It is caused by ingesting drinking water or food that has been contaminated by amebic cysts.

Earth Day Begun on April 22, 1970, Earth Day is a yearly demonstration in support of ecological issues. A product of Senator Gaylord Nelson's call for a national teach-in on the environment, on that day around the country there were demonstrations, parades, and rallies attended by an estimated 20 million people. The first Earth Day is often seen as the beginning of the mass-based environmental movement.

Elutriate To separate out by washing.

English hay Grasses commonly used in England, particularly white clover and bluegrass fescue, for pasture and haying. These grasses were brought to the New World by early settlers to increase hay production for domesticated animals.

Environmental justice movement A movement to address concerns about the different access to a clean and healthy environment enjoyed by whites and people of color.

Environmental Protection Agency The U.S. government agency created by the National Environmental Policy Act of 1970 that is responsible for monitoring the environment, setting environmental standards, and enforcing environmental regulations.

Environmental racism The practice of putting environmentally hazardous materials in minority communities, or more generally, the phenomenon of minority communities being more subject to environmental risk and degradation than white communities.

Erie Canal A man-made canal that runs along the Mohawk River valley, the only major break in the northern Appalachian mountain range, from the Hudson River at Albany to Lake Erie at Buffalo. It began operation in 1825.

Eutrophication The process by which an excess of nutrients, primarily phosphorus and nitrogen, generate plant and algae which absorb oxygen as they are broken down from the water system and overload its ability to break down organic matter. Excess nutrient levels spur algae, growth that absorbs dissolved oxygen in the water and blocks out sunlight, killing aquatic species. Increased algae and other nutrients in the water create an excessive biochemical oxygen demand and dramatically cut the levels of dissolved oxygen in the water system. When the levels of dissolved oxygen become critically low, the lake eventually dies.

Fauna Animals of any particular region or period.

Fescue A European grass introduced to the Americas as first a meadow grass then as a common lawn grass.

Fish and game commissions Established by states in the late nineteenth and early twentieth centuries to protect game animals and fish from overharvesting, fish and game commissions established limits on game take, issued licenses for hunting and fishing, restricted what could be hunted, maintained fish hatcheries, and acted to protect game habitat.

Fishing banks Any one of a number of relatively shallow regions off the coast of New England where the colliding currents stir up nutrients for marine life to feed upon. These banks, particularly George's, Stelwagen's, Brown's, Stable Island, Banquereau, St. Pierre, and the Grand, were prolific fishing grounds, particularly for cod.

Flax A plant in the genus *Linum*, flax is a cultivated slender annual with narrow pointed leaves and blue flowers. Flax was grown for its seeds from which flaxseed or linseed oil was extracted and for the fiber of its stem. This tough, elastic fiber was used to manufacture linen yarn or thread and woven linen cloth. Flax was a popular crop for early settlers because of its multiple uses. Farm families used the fiber to produce homespun yarn and woven cloth and used the seed for oil. Flaxseeds could be more easily moved to market than other, more bulky surplus production from the farm and they were a commodity that found a ready market.

Flora Native plants of any particular region or period.

Forest and Stream A magazine published in New York beginning in 1873 that propounded the ideal of the "manly arts." It touted the virtues of hunting and fishing, as well as provided tips and helpful advice about these sports.

Forest or long fallow agriculture Horticultural practice of Native Americans and some early settlers to clear forest land for planting, plant the land for three or four years, and then abandon the land to allow it to return to forest, while clearing another tract for planting.

Fourdrinier, Henry (died 1858) and Sealy (died 1847) Inventors of the Fourdrinier machine for making paper. In making paper, rags or wood pulp were pulverized and mixed with water to form a thick liquid "stuff." This was then poured into a Fourdrinier machine, which rolled and matted the material to extract the liquid and create paper. The Fourdrinier machines required large amounts of power to run. The wastes from these machines were dumped back into the rivers that powered the machines.

Fox Like the Sauk and the Potawatomis, the Fox originally occupied the land of northwestern Illinois, southern Wisconsin, and northeastern Iowa. They also were pushed off their land after the Black Hawk War.

Freeholder A landowner who owns land for life, or so that it may be sold or passed on to his or her heirs, in contrast to a tenant who merely has the right to occupy or use land for a specific term upon payment of rent. The majority of heads of families in early New England were freeholders.

Fulling mill A mill that softens, cleans, shrinks, and thickens cloth by moistening, heating, and beating it with a series of mallets.

Futures Market Once wheat was abstracted into types and grades (see Chicago Board of Trade) it could be bought and sold abstractly as well. In the 1850s, traders began to buy wheat for a specific price even before it was produced. The wheat was to be delivered and paid for at some future date. The market for futures spread to other agricultural commodities sold through the agency of the Chicago Board of Trade.

Gangrene The death of body tissue caused by the interruption of the blood supply. It was often a complication of smallpox.

Germ theory The explanation of disease causation developed by Lister, Koch, Pasteur, and Hemle that saw diseases as caused by invisible germs passed between people or between animal or insect vectors and people. This theory of disease causation was developed in the second half of the nineteenth century and replaced the anticontagious theory that held that diseases were caused by bad air or general filth. The germ theory led to specific actions, such as filtering water to protect it against invisible germs, rather than relying on its visual appearance of clarity.

Gill nets Nets with spaces within the weave that allow fish to swim partially in and then be captured by their gills so that they cannot swim out of the net. Gill nets could be stretched across a wide area and catch the fish swimming by. The nets could then be pulled up at the convenience of the gill netter.

Girdling A method of killing trees whereby a circle is cut around the base of the tree penetrating through the bark so that there is no contact between the bark above and below the girdle. This method of killing trees to clear a field was used by Native Americans.

Glacial till   Rocks, stones, boulders, sand, clay, dirt, and gravel gathered up by an advancing glacier and then left behind as the glacier retreats.

Grama grass   A high prairie grass typical of the land west of the ninety-eighth meridian.

Grangers   Farmers who organized in the 1870s against high and discriminatory railroad rates. They pushed for the passage of the granger laws regulating railroad rates. See Long-haul versus short-haul railroad rates.

Gravity-fed system   When municipal water systems were constructed in the nineteenth century, they pumped water to a reservoir located on high ground. Once at a high level, the water would then flow to customers through the power of gravity. The fall of water from its high location to its destination provided the pressure necessary to get water into people's homes and provided the pressure for fire hoses.

Greenhouse gases   Waste gases, mostly from the burning of carbon fuel, that act to trap solar energy at the earth's surface causing the earth's overall temperature to rise. Scientists believe that the increase in greenhouse gases will cause a melting of polar ice caps, an increase in ocean levels, radical climate shifts, and storms. The Kyoto Protocol of 1997, which was not signed by the United States, was designed to have the participating countries reduce their greenhouse gases to their 1990 levels by 2000. Most countries failed to meet their targets.

Grinnell, George Bird (1849–1938)   Editor of *Field and Stream* magazine and organizer of the Audubon clubs to protect songbirds.

Grist stone or grindstone   A circular stone disk, usually made of burstone or other siliceous rocks, with grooves cut into it radiating out from the center. Grist stones were used for grinding grain into flour. Grain was poured between two stones and the rotating of one stone across the other would grind the grain.

Gulf Stream   A warm ocean current flowing out of the Gulf of Mexico into the Arctic Sea in northern Europe. The Gulf Stream is formed when the warm waters of the Florida Current flowing out of the Gulf of Mexico join the warm North Equatorial Current and flow north along the Atlantic coast of North America before turning east toward northern Europe.

Headboard   A board or series of boards that were put across a dam to increase the level of water behind the dam. By raising or lowering the headboards, millers could increase or decrease the flow of water over or under their waterwheels. When they no longer needed the water pressure to work their mills, they could lower the headboards and allow water to drain down from the millpond behind the dam. In early settlements this was done to allow farmers to get to meadow fields that had been flooded by the raised millpond or to allow anadromous fish to swim upstream.

Important People, Events, and Concepts

Hemlock The eastern hemlock is a flat-needled conifer found in both well-drained and moist woods in northern Michigan. Stands of hemlocks were also found scattered through eastern Minnesota, Wisconsin, and Indiana, but very few survive today.

Hessian fly A small, blackish fly or midge that lays its eggs on the leaves of wheat in the spring. The larvae hatch a week later and the pupae migrate down the wheat stem to just above the ground, where they burrow into the stem, weakening it, and often leaving the wheat bent over or dwarfed and unable to be harvested. The fly was named for the Hessian troops that plagued American troops during the Revolutionary War.

Horticulture The practice of cultivating domesticated crops. In reference to Native Americans it involved planting maize, beans, and squash without the use of domesticated animals.

Humus Organic material that has decayed to a stable amorphous state. Humus is an important element in fertile soil. Humus improves soil structure and holds moisture and chemical elements in the soil. It is formed when bacteria and fungi break down animal and vegetable material. During the process of breakdown, nutrients are released that can be used by growing plants. The soil under the thick forests of the North and the prairie grasses of the plains were rich in humus.

Hunting and gathering Maintaining a community through gathering nuts and berries, digging roots, and killing and eating wild animals.

Indentured servants Individuals who gained passage to the New World in exchange for several years of bound labor. It became the common means for bringing labor to New England between 1635 and 1700.

Indian corn Maize. A cereal grain developed in Mesoamerica. It was the staple crop for Native Americans. Europeans planted maize, especially in the early years of settlement, and used the grain for flour.

Indian grass Common grass of the prairie.

Interceptor lines The growing volume of sewage being dumped by nineteenth-century cities began to overwhelm the receiving rivers and streams and pollute water intake sources. In response, cities built interceptor lines that intersected the traditional outflow pipes and redirected sewage wastes either downstream or farther out into harbors or lakes away from urban shorelines and water intake pipes.

Iroquois Confederacy Sometimes known as the Iroquois League, the confederacy was made up of five Native American tribal groups: the Mohawks, Oneidas, Onondagas, Cayugas, and Senecas. They shared a common language stock, Hokan-Siouan, and inhabited the Hudson River valley and the regions west into central New York. The Iroquois grew corn, squash, and beans and hunted deer and small game. Iroquois families lived in distinctive bark-covered structures known as longhouses.

Izaak Walton League An organization established by sportfishers in the 1920s to agitate for habitat and water-quality protection and better management of public resources, particularly in parks and state and national forests.

Kettle holes Small New England ponds formed by a retreating glacier leaving behind large chunks of ice covered with sand and gravel. When these ice chunks melted they formed indented bowls that filled with water.

Kickapoo Native Americans who originally occupied the western Midwest. They were also driven west after the defeat of Black Hawk.

Lawrence Experimental Station A laboratory established in 1887 and jointly run by the Massachusetts Institute of Technology and the Massachusetts Board of Health that looked at water quality and disease causation. Work at the Lawrence Experimental Station established that water filtration systems could be developed that could clean water of harmful germs, particularly those that caused typhoid fever.

Levittowns Suburban communities that became the model for post–World War II mass-produced tract housing. Beginning on Long Island, the Levitt family bought thousands of acres of farmland, brought in bulldozers, leveled everything, and threw up tens of thousands of small identical houses. Levittowns provided cheap, affordable suburban housing for middle-class and lower-middle-class Americans, but they also contributed significantly to urban sprawl.

Loam A mixture of sand, clay, and organic material from decaying vegetable and animal matter. The organic material in loam provides its fertility. Loam with too much clay will be poorly drained, while loam with too much sand cannot hold moisture. Rich, fertile soil with a high content of organic matter is usually found in valleys and where rivers flood and makes the best farmland.

Long-haul versus short-haul railroad rates Because of competition between them, railroads charged significantly higher rates for the short hauls where there was no competition than for the longer hauls over competitive routes. Because the farmers were shipping their produce to market over the short hauls, they saw these differential rates as unfair to them. They organized to pass granger laws to regulate railroad rates. The granger laws had little impact because of the intransigence of the railroads and a series of court decisions favoring the railroads.

Love Canal A suburban community in northern New York State built on a covered-over toxic waste dump. The Hooker Chemical Company dumped over 2,000 tons of toxic wastes, including the cancer-linked dioxin and polychlorinated biphenyls (PCPs), in the dump located outside of Niagara Falls, New York. In 1952, Hooker abandoned the dump and covered it over. Niagara Falls bought the land and built a school on the dump. By the 1960s and 1970s, wastes from the dump seeped into the school, the school yard, and yards and basements of the surrounding homes, serving as the cause, according to many residents, of serious health problems such as epilepsy, liver disease, miscarriages, and birth defects. Local community activists organized protests directed at state and local officials demanding something be done. In response, the government eventually bought out the Love Canal inhabitants.

Lye An alkali solution leached from ashes and used in making soap. Lye is a base that can also be used to neutralize acids. Spread on soil, it counteracts acid in the soil.

Maize A cultivated cereal, commonly called corn, developed in Mesoamerica. It was the staple crop for Native American horticulturalists.

Malaria An acute, sometimes fatal, febrile disease characterized by paroxysms of chills, high fever, and sweating. Malaria is caused by the malarial parasite, plasmodium, that is passed from infected anopheles mosquitoes to humans. Malaria was brought from Africa, where it was common, to the United States through the slave trade.

Manly arts Refers to the sports of hunting and fishing and reflects the belief, especially of the upper classes in the late nineteenth and early twentieth centuries, that these activities would contribute to making men strong and virile and would counter the perceived "softening" of men in modern society.

Marsh A swampy tract of land, often along river estuaries.

Massachusetts State Board of Health The nation's first state board of health organized in 1869 to deal with the problems of disease and the sources of disease. The board pursued an aggressive campaign against polluters.

Meadow A field of low ground and few trees. Often the product of the filling in and drying out of an old beaver pond or an abandoned field.

Measles A contagious, acute disease, *Rubeola morbilli*, manifesting in high fever, cough, and eruptions on the face and buccal mucous membrane. It is caused by a filterable virus that spreads from person to person through droplets of spray from the nose, throat, or mouth. It is typical in urban areas and erupts every three or four years. Measles victims are also susceptible to other infections including streptococci and pneumococci.

Mennonites A sect of Protestant Anabaptists originating in Switzerland but spreading into Holland and Germany in the late sixteenth century. Facing religious persecution in Europe, in the late seventeenth century the Mennonites established settlements in Pennsylvania where religious tolerance was enshrined in the colonial law. In Pennsylvania the Mennonites prospered practicing intensive agriculture.

Merino sheep A large sheep with long thin legs and fine, closely set silky wool. Merino sheep were introduced to this country for their wool by William Jarvis in 1811. Merinos did not produce good meat, but their wool was favored for weaving. The wool became an important marketable commodity for early-nineteenth-century New England.

Mesoamericans Native Americans from the region north of the Isthmus of Panama and south of the northern Mexican desert, the home of the Mayan, and later Aztec, civilizations.

Miasma A vaporous atmosphere or cloud thought to be the cause of disease prior to the acceptance of the germ theory of disease causation in the late nineteenth century. It was believed that noxious vapors from filth, particularly rotting animal and vegetable matter, seared the lungs and caused disease. *See* germ theory.

Microbes Microorganisms, especially bacteria, that spread disease. Microbes became a concern after the emergence of the germ theory of disease causation in the latter nineteenth century.

Mill acts First passed by colonial legislatures in the eighteenth century to encourage the building of mills, the acts allowed for farmers whose fields were flooded by milldams to collect compensation each year as awarded by juries for their damaged fields and thus to discourage the farmers from taking direct action against the dams owners under common law. Mill acts strengthened the hands of millers against aggrieved property owners whose lands were flooded, reflecting the belief that the mills were so essential to the common good that private losses caused by them were to be tolerated, albeit compensated. Under the acts landowners subject to flooded land from milldams effectively became involuntary landlords to the miller.

Millpond A body of water backed up behind a milldam to turn a waterwheel to power the machines of a mill.

Millstone *See* grist stone.

Missouri v. Illinois In 1906, Missouri's court case claiming that Chicago's redirection of sewage from Lake Michigan down the Chicago sanitary canal into the Mississippi River by way of the Illinois River was causing illness in St. Louis came before the U.S. Supreme Court. St. Louis took its drinking water from the Mississippi River and claimed that there had been an increase in typhoid fever since Chicago began dumping its sewage through the canal. Missouri claimed that the natural receptacle for Chicago's sewage was Lake Michigan. Chicago redirected its sewage because it took its drinking water from Lake Michigan and was trying to avoid polluting its source of drinking water. Missouri lost the case because the Court accepted Illinois' argument that it was diluting its sewage with added Lake Michigan water, and hence the water flowing into the Illinois River was cleaner than before, and that Missouri could not prove that it was Chicago's sewage that was causing the additional illness in St. Louis.

Mobility of community Refers to a feature of Native American life in the northern half of the Northern Hemisphere characterized by frequent moves from location to location as a community's fields wore out, or as the Native Americans went in search of richer game. In coastal areas, the communities often moved between camps according to the seasons.

Monoculture A system of agriculture whereby large tracts of land are set aside for a single cash crop, or more generally, any agricultural practice whereby agricultural diversity is limited to a few crops grown on a large scale focused toward larger markets. Such agricultural regions require the importation of other agricultural goods to feed their populations. The cotton region of the American Deep South and the sugar plantations of the Caribbean were examples of monocultures.

Montreal Protocol An international agreement signed in 1987 pledging signing members to reduce CFC production and use. *See* chlorofluorocarbons.

Motor tourists Middle-class Americans who travel by car out of the city to visit rural parks and vacation sites. Motor tourism became popular during the 1920s and continues to this day as the major form of American rural recreation.

Mouldboard plow The classic plow of early America. It initially consisted of large, curved wooden plates, reinforced with a connecting pole, and having an iron tip. In the early nineteenth century, the cast-iron plow was developed. But, like the mouldboard plow, this plow was not able to break the thickly rooted prairie soil.

Mound planting Native American horticulture whereby planters would burn or girdle the trees in a forest, clear it, and gather soil into mounds that was planted with maize, beans, and squash.

Municipal water system Replacing private wells and cisterns in the nineteenth century, these systems usually tapped a large water supply, pumped the water into a reservoir located on high ground, and then fed the water through pipes into urban residences and public hydrants. These water systems provided urban areas with water under pressure to fight fires. Because the water for these systems was often drawn upstream from pollution sources, the introduction of municipal water systems reduced waterborne diseases. Municipal water systems also encouraged increase water use that in turn increased waste discharge and contributed to water pollution. Philadelphia developed the first municipal water system at the turn of the nineteenth century. Its system was then imitated by other municipalities.

Naphtha A volatile, flammable hydrocarbon liquid solvent used in industrial cleaning derived from naphthalene, a by-product of coal tar.

Napoleonic Wars Wars between France and England between 1798 and 1815. During the later phase of these wars, with Napoleon in control of most of Europe, England attempted to cut off the European continent from overseas trade, while Napoleon attempted to cut England off from continental supplies. The actions of both belligerents encouraged both England and the nations of Europe to look to America for basic materials, particularly food, lumber, and naval supplies.

National Park Service Established in 1916 to organize the running and maintenance of the national parks, the service represented a recognition that the nation's parks needed coordinated oversight and protection.

National transportation network A system of trains, rivers, and canals that allowed for the flow of goods across the continent and the creation of an integrated nationwide market system. This network was expanded to include highways in the 1920s.

New York City Women's Municipal League A turn-of-the-century women's club that fought for a cleaner environment. The league led the campaign to clean the city of garbage.

Ohio River Basin Board of Public Health A commission established in 1928 of the nine states within the Ohio River basin to address water-quality issues of the Ohio River.

Oil Pollution Act of 1924 An act of Congress designed to protect coastal water from pollution that killed marine life. It was strongly supported by the members of the Oyster Fishermen's Association who were afraid that oil pollution would destroy their oyster beds. It did not address the problem of oil pollution in rivers and streams and was seldom enforced.

Oil Pollution Control Act A federal law passed in 1924 to reduce the flow of petroleum into the nation's waterways. Although it was weak legislation and not effectively enforced, this act was one of the first national governmental acts to address pollution.

Outflow pipes With the construction of storm sewers and sewage lines, excess rain and sewage was directed into the nearest water system through outflow pipes. These pipes were usually graded and located just above water level so that wastewater would flow out and dump into the receiving harbor, river, pond, or lake.

Overflow waterwheel The traditional gristmill or sawmill used an underflow waterwheel whereby the water flowing under the wheel turned it. The underflow wheel did not capture as much power as the overflow wheel. With the coming of the new textile machinery in the early 1800s, owners of these machines needed more power to operate them. The overflow wheel captured more power, but also required significantly higher dams.

Ozone A molecule made up of three atoms of oxygen. In the upper atmosphere it shields the earth from ultraviolet light. At ground level it is a pollutant caused by the interaction of the sun's energy and nitrogen oxides produced by automobile and industrial emissions. The protective ozone in the stratosphere is depleted by CFC (chlorofluorocarbon) gases that were used by industries and as a propellant.

Paleo-Indians Early hunters and gatherers who crossed over the Bering Strait some 12,000 years ago and migrated into northeastern America some 8,000 years ago.

Paleolithic Refers to hunting and gathering peoples who developed stone weapons and digging tools.

Pathogen A disease-producing organism.

PCB (polychlorinated biphenyls) Cancer-causing chemical waste from industrial processing.

Penn, William (1644–1718) The son of Sir William Penn and founder of the Pennsylvania colony, Penn became involved with the Quakers in 1666. In 1675, Penn became the proprietor of two Quaker communities in West Jersey and in 1681 purchased East Jersey and received a charter from the king for Pennsylvania. Penn then received a grant from the duke of York for Delaware. In 1682, Penn arrived in Pennsylvania, where he drew up a liberal plan of government and established friendly relations with the local Native Americans.

Phenols A carbolic acid that is a common by-product of industrial processing.

Pig iron Iron ore in the United States was frequently hematite, which contained large amounts of impurities. Iron ore was refined in a blast furnace. Into the traditional blast furnaces (huge masonry barrels), coke, limestone, and iron ore were dumped and burned. The molten pig iron then was tapped out at the bottom of the furnace and the waste slag discarded. The product of this process is pig iron that contains about 4 percent carbon and small amounts of silicon, phosphorus, and sulfur. Cast iron is formed when pig iron is remelted in small furnaces and poured into molds. Steel is made by blasting oxygen through the smelted pig iron to burn off more carbon and other impurities.

Pigeon years Periodic episodes of massive passenger pigeon flyovers in the eighteenth century. These huge flocks of pigeons occurred once every seven or eight years and provided a bounty of food and feathers for bedding. The birds, however, became extinct by the 1870s.

Plant succession Process by which the flora of an ecosystem changes. Typically, where a forest was burned out by lightening or horticulturalists, sun-seeking opportunistic plants moved in; first grasses, then fruit-bearing bushes, brambles, and shrubs, then trees—usually conifers first and finally hardwoods.

Pleistocene An epic beginning about 1 million years ago characterized by the spread of glacier ice.

Podzolic soil Gray, ashy soil found in the forest-covered regions south of the tundra in the Northern Hemisphere. Although covered with surface organic material, the soils under the forest regions are highly leached and often acidic. The dark, semipodzolic soils under the prairies are more fertile, owing to the rich vegetative cover of grasses.

Polio An acute viral infection caused by the entrovirus family that was common in urban areas and passed by ingesting contaminated water or food. Until the late nineteenth century, polio was common enough that infection and immunity were usually acquired early in life. As sanitation improved, infection and immunity declined. However, this caused the severity of attacks to increase, particularly when a more virulent strain appeared. In a small number of cases where there had been no previous infection, polio infection caused fever, headache, stiff neck and back, and paralysis of muscle groups that sometimes was permanent.

Populists Adherents of a political movement mostly of western farmers agitating against unfair railroad rates, and banking and monetary practices. The populists had their strongest support from the plains states where farmers needed bank loans to buy not only land, but also lumber and expensive farm equipment to open their farms and where there were few water transportation systems to compete with the railroads.

Potash, pearl ash A caustic alkaline compound, potassium carbonate. This white granular powder is produced from wood ash and used in the manufacture of soaps, glass, and for washing wool. In early settlements, potassium carbonate was produced by leaching ashes mostly from burnt woods. The potash produced by leaching wood ash contained impurities that gave it a red to green color. Further purification yielded white pearl ash.

Potawatomis Native Americans who originally occupied present-day Michigan. In the mid-seventeenth century, attacks by the Iroquois had pushed the Potawatomis into western Michigan. In the early nineteenth century, they were moved west by Euro-Americans into the region of northern Illinois. Following the defeat of Black Hawk, the Potawatomis were forced to cede much of their Illinois land to the U.S. government in the treaties of 1832 and 1833.

Pound sterling Standard weight and purity of English gold and silver coins. English money.

Power looms   Weaving looms that were run by an external power source, usually either waterpower or steam power.

Privy   A small, outdoor shed with one or more holes for defecating or urinating.

Public Health Service   A national public health commission to investigate issues of disease and public health.

Puddling   A process whereby molten iron was separated from its impurities by stirring. A skilled puddler stirred the molten pig iron until the impurities separated. As the iron solidified, the puddler worked it into balls and removed them from the furnace.

Purulent   Characterized by the discharging of pus.

Reasonable use   A legal doctrine through which courts tended to find that even when a riparian user's use of running water reduced its purity and hence negatively affected holders of downstream riparian rights, if this use was nevertheless "reasonable," the injured holder of downstream riparian rights did not have a tort action against the upstream polluter.

Resource Conservation and Recovery Act   Government act of 1976 designed to facilitate recycling and waste recovery.

Ricardo, David (1772–1823)   An early-nineteenth-century English economist who theorized about food costs and land rents. Ricardo was an early advocate of free trade.

Rivers and Harbors Act of 1899   An act of Congress passed to address the problem of waste dumping in the nation's waterways, particularly dumping that tended to fill in harbors and obstruct river navigation.

Roosevelt, Theodore (1858–1919)   President of the United States 1901–1909 who sponsored conservation programs. Roosevelt was also a proponent of hunting and fishing regulations and championed the importance of the sportsman's values in hunting and fishing. He was a founding member of the Boone and Crocket Club.

Sandburg, Carl (1878–1967)   A turn-of-the-twentieth-century midwestern poet who wrote poems about his hometown of Chicago.

Sauk   Native Americans who occupied eastern Iowa before the Black Hawk War. The Sauk had a land concession in southeastern Iowa dating from 1824, but lost most of their Iowa land in a series of treaties between 1832 and 1851.

Separatists   Calvinists who believed the Church of England was beyond reforming and separated themselves formally from it. The Pilgrims who settled in Plymouth were Separatists.

Settlement houses   Urban social reform centers founded at the end of the nineteenth century where social workers worked with and offered assistance to poor communities.

Shay's Rebellion In 1786 falling prices for agricultural goods due to the cutting off of trade to the British West Indies, rising taxes, heavy debt accumulated during the Revolution, and limited currency in circulation led farmers in western Massachusetts into open rebellion. Armed farmers prevented county courts from holding sessions and thus stopped the collection of debts. The rebellion was led by revolutionary war veteran Daniel Shays and was put down by the state militia, manned by eastern residents.

Short-fiber cotton Cotton developed in Mexico that could grow in drier, upcountry southern land. Unlike long-fiber cotton grown on the coast or coastal islands, short-fiber cotton's seeds were deeply embedded within the cotton boll. The development of the cotton gin allowed for more efficient separation of the seeds from the boll. This opened the Deep South to cotton production and created an expanded market for midwestern agricultural produce.

Slater, Samuel (1768–1835) English mechanic who brought the plans for a powered spinning jenny to America. Slater, with the backing of Providence, Rhode Island, merchants set up spinning factories around the southern New England countryside.

Smallpox Common term for Variola. A devastating disease caused by a small penetrating virus known as a filterable virus. The disease causes chills, high fevers, purulent lesions, and prostration. In its third day lesions break out and spread throughout the body. Europeans brought smallpox to the New World, where it struck Native Americans particularly harshly, killing at times 80 or 90 percent of the community infected.

Smith, John (1580–1631) An English military leader who took control of the Virginia colony in its early years. Through enforced military discipline Smith organized the colony and provided a stability that got it through its first years. He visited New England in 1614 and brought back to England a valuable cargo of furs and fish.

Smoke abatement ordinances Acts passed by city governments in the early twentieth century to attempt to reduce smoke pollution. These acts were pushed by municipal women's clubs. They often required boiler inspections and fines for excessive smoke.

de Soto, Hernando (1500–1542) Spanish explorer who traveled across the present-day United States in search of gold and riches.

Spanish influenza An influenza pandemic occurring in 1918, 1919, and 1920. It was a highly contagious, acute disease caused by a filterable virus and characterized by fever, prostration, pains, inflammation of the respiratory mucous membranes, and highly associated with complications of pneumonia. It caused over 20 million deaths worldwide.

Spoil Wastes from coal mines. *See* Culm.

Squanto (life dates unknown) Surviving member of the Pawtuxets, who, having spent time with the English, spoke the language. He had been captured by the English and was the only member of his tribe to survive a smallpox outbreak. Squanto acted as an intermediary between the English settlers and the Native Americans. He also instructed the early Pilgrims on how to raise Indian crops.

Stanley Steamer An early automobile that was powered by steam. The Stanley Steamer was the most common automobile at the turn of the twentieth century. It was cheap and easy to run but heavy and easily bogged down on dirt roads. It was also prohibitively expensive for all but the wealthy.

Storm sewers With the buildup of urban areas, there was less bare soil available to absorb rainwater. Rainwater running off of roofs added to that which fell on streets. This runoff flooded streets, filled in basements, and brought urban commerce to a standstill. In the first half of the nineteenth century urban leaders responded to the problem of rain runoff by raising and grading streets and building storm sewers. Excess rain would then run into the storm sewers that would direct the water into the nearest stream, river, harbor, or pond. When people began building water closets they connected their sewer lines to these storm sewers.

Sumac A tree in the genus *Rhus*. The leaves of sumac trees were used in tanning leather. It was a common shrub in dry, rocky places in the United States and Canada.

Tamarack or American larch A conifer with bunched needles commonly found in wooded swamps and bogs in New England and the wetlands of the upper Midwest. Its lumber was used for posts, beams, and railroad ties. Tannic acid from its bark was used in tanning. Its seeds and needles were eaten by grouse and rabbits.

Tannic acid A brownish astringent acid derived from any number of plants and used for the tanning of leather or for preparing writing ink. Oak bark was also commonly used to produce tannic acid. *See* Sumac.

Tidal in-wash During periods of high tides or river flooding, water would rise up above the openings of sewer outflow pipes and then flow into the pipes, backing up into streets and basements. *See* outflow pipes.

Tow cloth A coarse cloth made from hemp or flax fiber usually with an open weave.

Toxic Substances Control Act Government act of 1976 known as the Superfund law. This act created the Superfund by taxing toxic waste–generating industries. The funds were then used to pay for cleaning up toxic waste sites.

Tramontane West The region west of the Appalachian mountains. Because the mountains provided a barrier to travel and the transporting of goods, trade between eastern cities and the rich farmlands of the west was limited until the development of the Erie Canal and steamboat travel on the Ohio and Mississippi rivers.

Triassic lowlands The geological basis of southeastern Pennsylvania, characterized by rich soil on relatively flat land.

Typhoid fever An infection caused by *Salmonella typhosa* causing fever, malaise, headaches, anorexia, diarrhea, vomiting, and respiratory pneumonia. It is often complicated by intestinal hemorrhage or intestinal perforation. Typhoid is a persistent germ that can survive in extreme temperatures for long periods. Typhoid fever is usually spread through ingesting water contaminated by feces or urine. It can often be fatal, especially with intestinal hemorrhage or perforation. Because the salmonella bacilli are long lasting and persistent, once it is established in a community it easily passes through contaminated drinking water or food to others.

Urban women's clubs Turn-of-the-twentieth-century organizations of middle-class women who fought for cleaner urban environments.

U.S. Forest Service Established in 1905 under the direction of Gifford Pinchot to coordinate forest protection and oversee lumbering on public forestry lands.

Usufruct Using land without the formal ownership of it.

Walking city The city before the development of the streetcar lines. Because of the difficulty of moving people about the city before regular streetcars, most urban areas were concentrated into a space where people could walk from their residences to their places of work. The walking city was a compact, jumbled area of overlapping activities. The development of the streetcar allowed for the segregation of the city by activity and social class.

Walton, Isaac A seventeenth-century British writer who wrote extensively on fishing and published *The Complete Angler* in 1653.

Wampanoags Native American tribal group whose language belongs to the Algonquian linguistic stock. The Wampanoags lived in southeastern Massachusetts. The Noscussets were members of the Wampanoags who inhabited Cape Cod. The Pawtuxets were the Wampanoags who lived in the Plymouth area.

Wampum Beads or disks made from shells of mollusks found on the East Coast used as a medium of exchange or gift. They were worn by communal leaders as symbols of importance or prestige.

Waring, George (1833–1898) A nineteenth-century sanitarian who organized New York City's garbage-collection and street-sweeping programs.

Water filtration systems Systems to filter water through sand in order to remove harmful bacteria. These filtration systems and later the addition of chlorine to water significantly improved the quality and safety of municipal water systems and contributed to the decline in urban mortality rates.

Water Pollution Control Act Governmental act passed in 1972 mandating permits for all businesses discharging pollutants. It required pollution-control devices on discharging industries.

Weirs Traps for fish. Weirs built of woven sticks and poles or stones were put in the way of migrating fish, directing them into a shallow or enclosed area where they could be easily netted or speared.

Whooping cough *Bordetella pertussis.* An acute, highly infectious disease characterized by heaving, uncontrolled coughing. The disease can be fatal to infants from asphyxia. It was one of the diseases Europeans brought to the New World.

Wilderness Society A conservation group founded in 1935 to protect wilderness. Its members, including Aldo Leopold and Robert Marshall, argued against road building in wilderness areas. Of particular interest to the Wilderness Society was the preservation of the wild and rugged areas around the Appalachian Trail and the halting of road construction along the Appalachian ridgeline. Benton MacKaye, an early advocate for the trail, was also instrumental in forming the Wilderness Society and strongly fought development along the ridgeline.

Winthrop, John (1588–1649) A lawyer and member of the English gentry class with strong Puritan leanings, Winthrop was educated at Cambridge. In 1629 he was a member of the Massachusetts Bay Colony and was chosen to lead the group to its new settlement in Boston. Arriving in 1630, Winthrop served as governor of the colony twelve times and gave it direction for almost two decades.

Wisconsin Glacier The last of the great glaciers that covered the northern part of North America and left behind the till that gives the Northeast much of its topographical character.

Yellow fever An acute, infectious viral disease characterized by sudden fever, slow pulse, vomiting of blood, jaundice, and hemorrhage. It is a filterable virus transmitted by the bite of the female *Aedes aegypti* mosquito (only the female mosquitoes bite) that has previously bitten an infected person. The *Aedes aegypti* mosquito was common in urban areas where there were artificial water containers for breeding. Epidemics of yellow fever periodically struck urban areas and took a heavy toll, up to 8 percent of some urban populations in the late eighteenth and early nineteenth centuries. The last yellow fever epidemic in the United States occurred in New Orleans in 1905.

CHRONOLOGY

Pleistocene era to 13,500 years ago   Ice covers most of north-central and northeastern America

14,000 to 18,000 years ago   Arrival of Paleo-Indians in North America

13,500 years ago to present   Holocene Age follows Ice Age

10,000 B.C.E.   Wisconsin Glacier retreats from Midwest and Northeast

8000–6000 B.C.E.   Clovis hunters, hunters of large mammals, arrive in Midwest and Northeast

200 B.C.E.–400 C.E.   Hopewell culture, dependent on native barley, maygrass, and knotweed plus hunting and gathering, develops in lower Midwest

400 C.E.–1100 C.E.   Cahokia, largest city of mound builders of Mississippi and Ohio River valleys, flourishes

900 C.E.   Maize cultivation practiced by Native Americans of Midwest and Northeast

1100 C.E.   Mixed bean, squash, and maize cultivation widely adopted

1492 C.E.   First documented contact between Europe and America

1498 C.E.   John Cabot explores Northeast coast

1580–1780   Extensive fur trade between Native Americans and Europeans causes beaver to be eliminated from most of Northeast

1600s   European diseases spread to Native Americans; Smallpox kills 60–80 percent of victims among Native Americans

1620   First encounter between Pilgrims and Native Americans

1621   Plymouth Colony established

1624   Dutch colony established on Hudson River

1630   Massachusetts Bay Colony established

1630–1800   Europeans introduce animal husbandry to Northeast and Midwest

1649–1660   Iroquois gain access to European weapons and move against Native Americans of Great Lakes and Ohio Valley

1690–1790   Europeans exploit rich fishing grounds off New England coast

1690–1830   American rivers and Great Lakes, rich in salmon, shad, and alewives, provide important source of food

1690–1840   Northern farmers send surpluses to markets in West Indies and import sugar in return

1691 Broad arrow policy initiated to protect best of abundant American lumber for British navy

1700s English settlers move into Pennsylvania and New Jersey

1700–1840 American settlers plant wheat, rye, and orchards and hunt and fish to supplement their diets

1750–1860 New England develops whaling industry and sells oil for lamps in Europe and growing American urban centers

1787 Hessian fly wheat mite attacks American wheat

1785 Northwest Land Ordinance passed opening Midwest to agricultural development

1787 Northwest Ordinance establishes governance for territory

1790–1840 Canals, locks, and dams increase river transport system and inland fish runs decline

1793 Philadelphia yellow fever epidemic

1794 Battle of Fallen Timbers removes Native Americans from Ohio

1796 Massachusetts Mill Act passed giving mill owners enhanced rights to build dams and flood fields

1800 American wheat attacked by wheat blast or "rust" fungus

1812–1815 War with Britain encourages manufacturing

1815 Native Americans driven from Indiana

1817 Steamboat travel on Ohio River and Great Lakes

1820–1900 Growth of large commercial port cities and industrial cities on falls of inland waterways

1821 Boston Associates build large dams and mills on Merrimack River

1825 Erie Canal completed linking ecological system of Midwest to East Coast and European markets

1830–1850 Building of municipal waterworks for large cities

1830–1880 Rural New England suffers economic recession as it can no longer compete with cheaper midwestern staple agricultural goods in West Indies and European markets

1833 Development of balloon-frame housing construction

1839 John Deere develops steel plow, reducing cost of plowing prairie soil

1847 Illinois-Michigan Canal completed

1848 Irish potato blight
Founding of Chicago Board of Trade

1848–1920 Increased immigration from Europe settling midwestern farms and crowding in growing cities

1850 Congress passes Swamp Land Act

1850–1890 Clear-cutting of white pine forest of northern Michigan and Wisconsin to meet lumber demand on prairie

1850–1920 Small truck farms emerge in urban hinterland catering to growing urban populations
1850–1950 Coal substituted for cordwood as heating and cooking fuel
1854 Henry David Thoreau publishes *Walden: A Walk in the Woods*
1856–1900 Frederick Law Olmsted designs Central Park, launching city park movement
1859 Petroleum discovered in northwestern Pennsylvania
1860 Lake salmon fished out of Lake Ontario
1864 George Perkins Marsh publishes *Man and Nature*
1866 Building of Chicago's Union Stock Yards
1870–1890 Building of municipal drain sewer systems
1878 Last large nesting of passenger pigeons in Midwest
1880 Swift develops refrigerated rail cars and ships dressed meat to eastern markets
1880–1890 American acceptance of germ theory
1880–1900 Development of water purification systems
1880–1920 Rural resorts become center for American middle-class recreation
1886 Founding of Boone and Crockett Club
1890–1950 Municipal campaigns for smoke abatement
1892 Adirondack State Park created
1893 Chicago World's Fair
1899 Congress passes Rivers and Harbors Act
1905 Founding of Audubon Society
Forest Service created
1914 Henry Ford produces cheap, reliable automobile on assembly line
1916 Congress passes the Federal Aid Road Act, often referred to as the Federal Highway Act
National Park Service created
1920 U.S. Census reports for first time that over 50 percent of nation's population lives in cities and towns
1920–1940 Creation of state parks and recreational camping
1922 Delaware River Basin created
Izaak Walton League founded
1924 Oil Pollution Act passed
1925 Herring population in Lake Erie crashes because of pollution and overfishing
Clarence Birdseye develops fast freezing technique
1928 Ohio River Basin created
1933 Civilian Conservation Corps created

1933–1940 Civilian Works Administration and Works Projects Administration build roads, city parks, and sewer systems

1935 Soil Conservation Service established as agency of Agriculture Department to prevent soil erosion

Wilderness Society founded

1936 National Wildlife Federation founded

1940–1950 Lake trout wiped out in Great Lakes due to overfishing, pollution, and sea lamprey eel

1940–1970 Widespread spraying of DDT and development and use of chemical herbicides and pesticides

1945–1960 Completion of natural gas pipelines provides cheap natural gas as heating fuel, domestic coal consumption declines, and city air becomes cleaner

1948 Killer smog in Donora, Pennsylvania, kills twenty

1949 Publication of Aldo Leopold's *Sand County Almanac*

1950 Building of first Levittown

1951 Nature Conservancy founded to buy land for conservation

1955 Clean Air Act passed, amended in 1970

1961 Cape Cod National Seashore created

1962 Publication of Rachel Carson's *Silent Spring*

1967 Environmental Defense Fund created as citizen litigation organization

1969 Cuyahoga River catches fire

Friends of the Earth founded

1970 Earth Day celebration involves massive demonstrations for environmental protection

National Environmental Protection Act passed by Congress, creating Environmental Protection Agency

Clean Air Act passed

1971 Vermont passes nation's first bottle bill

1972 Water Pollution Control Act passed

Environmental Pesticide Control Act passed

DDT banned in United States

1973 Endangered Species Act passed

1974 OPEC cuts oil supply, creating shortages of gasoline in United States

1976 Resource Conservation and Recovery Act passed

1977 United States and Canada declare 200-mile coastal limit, restricting fishing of Atlantic coast

1978 Lois Gibbs leads campaign against toxic wastes at Love Canal, New York

1980s Atlantic cod fisheries collapse

1980 Superfund is created to clean up toxic wastes
1985–2000 Environmental justice movement links environmental degradation with racial discrimination
1987 International Montreal Protocol agreement signed restricting production of CFCs that create hole in atmosphere's ozone layer
1988 James Hansen alerts Congress to global warming
1993 Milwaukee's water system contaminated by *Cryptosporidium parvum*
1995 Chicago heat inversion kills over 700 people
1997 Kyoto Protocol adopted by several countries to limit greenhouse gases
2000 186 countries sign the Kyoto Protocol; the United States decides not to sign the protocol in 2001

DOCUMENTS

Collected and annotated by Mark Stoll, series editor

1. AMERICAN INDIANS IN THE NEW ENGLAND LANDSCAPE, 1625

Since the end of the Ice Age 12,000 years ago, humans have occupied the area covered in this book. While they left no written records of their way of living on the land, a number of early European explorers and settlers observed and wrote down their impressions. Thomas Morton was an Englishman who settled in Mount Wollaston, Massachusetts, in 1625, later called Ma-re-Mount or Merrymount, where he lived happily alongside Indians until his scandalized Puritan neighbors forced him out. He published an account of his adventures that included this detailed description of Indian life in New England and of the susceptibility to European disease that would doom Indian possession of the continent.

Thomas Morton, *New English Canaan; or, New Canaan*, 1637

Of Their Houses and Habitations: The Natives of New England are accustomed to build themselves houses much like the wild Irish; they gather poles in the woods and put the great end of them in the ground, placing them in form of a circle or circumference, and, bending the tops of them in form of an arch, they bind them together with the bark of walnut trees, which is wondrous tough, so that they make the same round on the top for the smoke of their fire to ascend and pass through; these they cover with mats, some made of reeds and some of long flags, or sedge, finely sewed together with needles made of the splinter bones of a crane's leg, with threads made of their Indian hemp, which there grows naturally, leaving several places for doors, which are covered with mats, which may be rolled up and let down again at their pleasure, making use of the several doors, according as the wind sits. The fire is always made in the middle of the house, with windfall commonly, yet sometimes they fell a tree that grows near the house, and, by drawing in the end thereof, maintain the fire on both sides, burning the tree by degrees shorter and shorter, until it be all consumed, for it burns night and day.

Their lodging is made in three places of the house about the fire; they lie upon blankets, commonly about a foot or 18 inches above the ground, raised upon rails that are borne upon forks; they lay mats under them, and coats of deer skins, otters, beavers, racoons, and of bears' hides, all which they have dressed and converted into good leather, with the hair on, for their coverings, and in this manner they lie as warm as they desire. In the night they take their rest; in the day time either the kettle is on with fish or flesh, by no allowance, or else the fire is employed in the roasting of fishes, which they delight in. . . .

When they are minded to remove, they carry away the mats with them; other materials the place adjoining will yield. They use not to winter and summer in one place, for that would be a reason to make fuel scarce; but, after the manner of the gentry of civilized natives, remove for their pleasures; sometimes to their hunting places, where they remain keeping good hospitality for that season; and sometimes to their fishing places, where they abide for that season likewise; and at the spring, when fish comes in plentifully, they have meetings from several places, where they exercise themselves in gaming and playing of juggling tricks and all manner of revelries which they are delighted in. . . .

Of the Indians' Apparel: The Indians in these parts do make their apparel of the skins of several sorts of beasts, and commonly of those that do frequent those parts where they do live; yet some of them, for variety, will have the skins of such beasts that frequent the parts of their neighbors, which they purchase of them by commerce and trade. Their skins they convert into very good leather, making the same plume and soft. Some of these skins they dress with the hair on, and some with the hair off; the hairy side in winter time they wear next their bodies, and in warm weather they wear the hair outwards. They make likewise some coats of the feathers of turkeys, which they weave together with twine of their own making, very prettily. These garments they wear like mantels knit over their shoulders, and put under their arms. They have likewise another sort of mantel, made of moose skins, which beast is a great large deer, so big as a horse. These skins they commonly dress bare, and make them wondrous white, and stripe them with furs round about the borders, in form like lace set on by a tailor, and some they stripe with fur in works of fantasies of the workmen, wherein they strive to excel one another. And mantels made of bears' skins is a usual wearing among the natives that live where the bears do haunt.

They make shoes of moose skins, which is the principal leather used to that purpose; and for want of such leather (which is the strongest) they make shoes of deer skins. . . .

Of Their Traffic and Trade With One Another: Although these people have not the use of navigation, whereby they may traffic as other nations, that are civilized, use to do, yet do they barter for such commodities as they have, and have a kind of beads instead of money, to buy withal such things as they want, which they call *Wampampeak* [wampum]; and it is of two sorts, the one is white, the other is of a violet color. These are made of the shells of fish. The white with them is as silver with us; the other as our gold; and for these beads they buy and sell, not only amongst themselves, but even with us. We have used to sell any of our commodities for this Wampampeak, because we know we can have beaver again of them for it: and these beads are current in all the parts of New England, from one end of the coast to the other. . . .

The skins of beasts are sold and bartered, to such people as have none of the same kind in the parts where they live. Likewise they have earthen pots of divers sizes, from a quart to a gallon, two or three to boil their victuals in; very strong, though they be thin like our iron pots. They have dainty wooden bowls of maple, of high price amongst them; and these are dispersed by bartering one with the other, and are but in certain parts of the country made, where the several trades are appropriated to the inhabitants of those parts only. . . .

Of Their Magazines or Store Houses: These people are not without providence, though they be uncivilized, but are careful to preserve food in store against winter; which is the corn that they labor and dress in the summer. And, although they eat freely of it, while it is growing, yet have they a care to keep a convenient portion thereof to relieve them in the dead of winter (like to the ant and the bee), which they put under ground. Their barns are holes made in the earth, that will hold a hogshead of corn a piece in them. In these (when their corn is out of the husk and well dried) they lay their store in great baskets (which they make of bark) with mats under, about the sides, and on the top; and putting it into the place made for it, they cover it with earth; and in this manner it is preserved from destruction or putrefaction. . . .

Of a Great Mortality That Happened Amongst the Natives of New England About the Time That the English Came There to Plant: It fortuned some few years before the English came to inhabit at New Plymouth, in New England, that . . . the hand of God fell heavily upon them, with such a mortal stroke that they died on heaps as they lay in their houses; and the living, that were able to shift for themselves, would run away and let them die, and let their carcasses lie above the ground without burial. For in a place where many inhabited, there had been but one left to live to tell what became of the rest; the living being (as it seems) not able to bury the dead, they were left for crows, kites and vermin to prey upon. And the bones and skulls upon the several places of their habitations made such a spectacle after my coming into those parts, that, as I travelled in that forest near the Massachusetts, it seemed to me a new found Golgatha. . . . By all likelihood the sickness that these Indians died of was the plague. . . . And by this means there is as yet but a small number of savages in New England, to that which had been in former time, and the place is made so much the more fit for the English nation to inhabit it, and erect in it temples to the glory of God. . . .

Of Their Custom in Burning the Country, and the Reason Thereof: The savages are accustomed to set fire of the country in all places where they come, and to burn it twice a year, viz.: at the spring, and the fall of the leaves. The reason that moves them to do so, is because it would otherwise be so overgrown with underweeds that it would be all a coppice wood, and the people would not be able in any wise to pass through the country out of a beaten path. The means that they do it with, is with certain mineral stones, that they carry about them in bags made for that purpose of the skins of little beasts, which they convert into good leather, carrying in the same a piece of touch wood, very excellent for that purpose, of their own making. These mineral stones they have from the Piquenteenes (which is to the southward of all the plantations in New England), by trade and traffic with those people.

The burning of the grass destroys the underwoods, and so scorches the elder trees that it shrinks them, and hinders their growth very much; so that he that will look to find large trees and good timber, must not depend upon the help of a wooden prospect to find them on the upland ground; but must seek for them (as I and others have done), in the lower grounds, where the grounds are wet, when the country is fired, by reason of the snow water that remains there for a time, until the sun by continuance of that has exhaled the vapors of the earth, and dried up those places where the fire (by reason of the moisture) can have no power to do them any harm; and if he would endeavor to find out any goodly cedars, he must not seek for them on the higher grounds, but make his inquest for them in the valleys, for the savages, by this custom of theirs, have spoiled all the rest; for this custom has been continued from the beginning.

And lest their firing of the country in this manner should be an occasion of damnifying us, and endangering our habitations, we ourselves have used carefully about the same time to observe the winds, and fire the grounds about our own habitations; to prevent the damage that might happen by any neglect thereof, if the fire should come near those houses in our absence. For, when the fire is once kindled, it dilates and spreads itself as well against, as with the wind; burning continually night and day, until a shower of rain falls to quench it. And this custom of firing the country is the means to make it passable; and by that means the trees grow here and there as in our parks; and make the country very beautiful and commodious. . . .

That the Savages Live a Contented Life: A gentleman and a traveller, that had been in the parts of New England for a time, when he returned again, in his discourse of the country, wondered (as he said) that the natives of the land lived so poorly in so rich a country, like to our beggars in England. Surely, that gentleman had not time or leisure while he was there truly to inform himself of the state of that country, and the happy life the savages would lead were they once brought to Christianity. I must confess they want the use and benefit of navigation (which is the very finest of a flourishing commonwealth), yet are they supplied with all manner of needful things for the maintenance of life and livelihood. Food and raiment are the chief of all that we make the use of; and of these they find no want, but have, and may have them, in most plentiful manner.

If our beggars of England should, with so much ease as they, furnish themselves with food at all seasons, there would not be so many starved in the streets, neither would so many jails be stuffed, or gallows furnished with poor wretches, as I have seen them. . . .

I have observed that they will not be troubled with superfluous commodities. Such things as they find they are taught by necessity to make use of, they will make choice of, and seek to purchase with industry. So that, in respect that their life is so void of care, and they are so loving also that they make use of those things they enjoy (the wife of one excepted), as common goods, and are therein so compassionate that, rather than one should starve through want, they would starve all. Thus do they pass away the time merrily, not regarding our pomp (which they feel daily before their faces), but are better content with their own, which some men esteem so meanly of.

They may be rather accounted to live richly, wanting nothing that is needful; and to be commended for leading a contented life, the younger being ruled by the elder, and the elder ruled by the Powahs, and the Powahs are ruled by the Devil; and then you may imagine what good rule is like to be amongst them.

Source: Oliver J. Thatcher, ed., *The Library of Original Sources* (Milwaukee: University Research Extension Co., 1907), Vol. V: 9th to 16th Centuries, pp. 360–377.

Scanned by Jerome S. Arkenberg, Cal. State Fullerton. The text has been modernized by Prof. Arkenberg. http://www.fordham.edu/halsall/mod/1637morton.html.

2. PURITANS CONDEMN SINFUL WASTE OF NATURAL RESOURCES, 1741

Jonathan Edwards's "Sinners in the Hands of an Angry God," the most famous American sermon ever written, excoriates parishioners for their sinful abuse of God's creation.

—*Their foot shall slide in due time.* Deut. 32:35

In this verse is threatened the vengeance of God on the wicked unbelieving Israelites, who were God's visible people, and who lived under the means of grace; but who, notwithstanding all God's wonderful works towards them, remained (as ver. 28.) void of counsel, having no understanding in them. Under all the cultivations of heaven, they brought forth bitter and poisonous fruit; as in the two verses next preceding the text. . . .

The use of this awful subject may be for awakening unconverted persons in this congregation. This that you have heard is the case of every one of you that are out of Christ—that world of misery, that lake of burning brimstone, is extended abroad under you. There is the dreadful pit of the glowing flames of the wrath of God; there is hell's wide gaping mouth open; and you have nothing to stand upon, nor any thing to take hold of, there is nothing between you and hell but the air; it is only the power and mere pleasure of God that holds you up.

You probably are not sensible of this; you find you are kept out of hell, but do not see the hand of God in it; but look at other things, as the good state of your bodily constitution, your care of your own life, and the means you use for your own preservation. But indeed these things are nothing; if God should withdraw his band, they would avail no more to keep you from falling, than the thin air to hold up a person that is suspended in it.

Your wickedness makes you as it were heavy as lead, and to tend downwards with great weight and pressure towards hell; and if God should let you go, you would immediately sink and swiftly descend and plunge into the bottomless gulf, and your healthy constitution, and your own care and prudence, and best contrivance, and all your righteousness, would have no more influence to uphold you and keep you out of hell, than a spider's web would have to stop a falling rock. Were it not for the sovereign pleasure of God, the earth would not bear you one moment; for you are a burden to it; the creation groans with you; the creature is made subject to the bondage of your corruption, not willingly; the sun does not willingly shine upon you to give you light to serve sin and Satan; the earth does not willingly yield her increase to satisfy your lusts; nor is it willingly a stage for your wickedness to be acted upon; the air does not willingly serve you for breath to maintain the flame of life in your vitals, while you spend your life in the service of God's enemies. God's creatures are good, and were made for men to serve God with, and do not willingly subserve to any other purpose, and groan when they are abused to purposes so directly contrary to their nature and end. And the world would spew you out, were it not for the sovereign hand of him who hath subjected it in hope. There are black clouds of God's wrath now hanging directly over your heads, full of the dreadful storm, and big with thunder; and were it not for the restraining hand of God, it would immediately burst forth upon you. The sovereign pleasure of God, for the present, stays his rough wind; otherwise it would come with fury, and your destruction would come like a whirlwind, and you would be like the chaff of the summer threshing floor.

Source: Jonathan Edwards, "Sinners in the Hands of an Angry God," frequently reprinted. http://www.jonathanedwards.com/sermons/Warnings/sinners.htm.

3. COMPARISON OF GERMAN FARMING METHODS WITH ENGLISH AND SCOTS-IRISH, 1799

In 1789, physician and American patriot Benjamin Rush of Philadelphia published an account of the farming methods of the numerous German settlers in Pennsylvania. It explicitly praises their conscientious care of their land, and implicitly criticizes the wasteful and careless methods of their English and Scots-Irish neighbors that left the land eroded and exhausted.

The Germans taken, as a body, especially as farmers, are not only industrious and frugal, but skillful cultivators of the earth. I shall enumerate a few particulars, in which they differ from most of the other farmers of Pennsylvania. . . .

First. In settling a tract of land, they always provide large and suitable accomodations for their horses and cattle, before they lay out much money in building a house for themselves. The barn and the stables, are generally under one roof, and contrived in such a manner, as to enable them to feed their horses and cattle, and to remove their dung, with as little trouble as possible. . . .

Second. They always prefer good land, or that land on which there is a large quantity of meadow ground. From an attention to the cultivation of grass, they often double the value of an old farm in a few years, and grow rich on farms, on which their predecessors of whom they purchased them, had nearly starved. They prefer purchasing farms with improvements, to settling on a new tract of land. [*Gypsum,* or sulphate of lime, was used as a fertilizer, by Germans. Jacob Berger, a German, was the first that tried *gypsum,* several years before the Revolutionary war, on a city lot, on the commons of Philadelphia. . . .]

Third. In clearing new land, they do not girdle or belt the trees simply, and leave them to perish in the ground, as is the custom of their English or Irish neighbors; but they generally cut them down and burn them. In destroying under-wood and bushes, they generally grub them out of the ground, by which means, a field is as fit for cultivation the second year after it is cleared, as it is in twenty years afterwards. The advantages of this mode of clearing, consist in the immediate product of the field, and in the greater facility with which it is plowed, harrowed and reaped. The expense of repairing a plow, which is often broken, is greater than the extraordinary expense of grubbing the same field completely, in clearing.

Fourth. They feed their horses and cows well, of which they keep only a small number, in such a manner, that the former perform twice the labor of those horses, and the latter yield twice the quantity of milk of those cows, that are less plentifully fed. . . .

Fifth. The fences . . . of a German farm are generally high, and well built, so that his fields seldom suffer from the inroads of his own, or his neighbors' horses, cattle, hogs or sheep.

Sixth. The German farmers are great economists of their *wood.* Hence, they burn it only in stoves, in which they consume but a 4th or 5th part of what is commonly burnt in ordinary open fire places; besides, their horses are saved by means of this economy, from that immense labour, in hauling wood in the middle of winter, which frequently unfits the horses of their (*Scotch*) neighbors for the toils of the ensuing spring. Their houses are, moreover, rendered so comfortable, at all times, by large close stoves, that twice the business is done by every branch of the family, in knitting, spinning and mending farming utensils, that is done in houses where every member in the family crowds near a common fire-place, or shivers at a distance from it, with hands and fingures [sic] that move, by reason of the cold, with only half their usual quickness.

They discover economy in the preservation and increase of their wood, in several other ways. They sometimes defend it, by high fences, from their cattle; by which means, the young forest trees are suffered to grow, to replace those that are cut down for the necessary use of the farm. But where this cannot be conveniently done, they surround the stump of that tree which is most useful for fences, viz: the chestnut, with a small triangular fence. From this stump, a number of suckers shoot out in a few years; two or three of which in the course of five and twenty years, grow into trees of the same size as the tree from whose roots they derived their origin. . . .

Ninth. The German farmers have large or profitable gardens near their houses. These contain little else but useful vegetables. Pennsylvania is indebted to the Germans, for the principal part of her knowledge in horticulture. There was a time when turnips and cabbage were the principal vegetables that were used in diet, by the citizens of Philadelphia. . . . Since the settlement of a number of German gardeners, in the neighborhood of Philadelphia, the tables of all classes of citizens have been covered with a variety of vegetables, in every season of the year; and to the use of these vegetables, in diet, may be ascribed the general exemption of the citizens of Philadelphia, from diseases of the skin. . . .

Sixteenth. From the history that has been given of the German agriculture, it will hardly be necessary to add that a German farm may be distinguished from the farms of the other citizens of the State, by the superior size of their barns; the plain but compact form of their houses, the height of their inclosures; the extent of their orchards; the fertility of their fields; the luxuriance of their meadows, and a general appearance of plenty and neatness in everything that belonged to them.

Source: Benjamin Rush, An account of the manners of the German inhabitants of Pennsylvania (1789) (Philadelphia: S. P. Town, 1875), pp. 11–19, 23–24, 32.

4. INDIAN LIFE ON THE MISSISSIPPI BEFORE DISPOSSESSION, EARLY 1800s

By the nineteenth century historians can rely not only on European accounts of Indian life but records left by Indians themselves, sometimes, as in the case of Black Hawk, dictated to a white writer. Here Black Hawk recalls Indian life along the Mississippi River on the eve of their loss of land and independence.

Our village was situated on the north side of Rock River, at the foot of the rapids, on the point of land between Rock River and the Mississippi.

In front a prairie extended to the Mississippi, and in the rear a continued bluff gently ascended from the prairie. . . . On the side of this bluff we had our corn fields, extending about two miles up parallel with the larger river, where they adjoined those of the Foxes, whose village was on the same stream, opposite the lower end of Rock Island, and three miles distant from ours. We had eight hundred acres in cultivation including what we had on the islands in Rock River. The land around our village which remained unbroken, was covered with blue-grass which furnished excellent pasture for our horses. Several fine springs poured out of the bluff nearby, from which we were well supplied with good water. The rapids of Rock River furnished us with an abundance of excellent fish, and the land being very fertile, never failed to produce good crops of corn, beans, pumpkins and squashes. We always had plenty; our children never cried from hunger, neither were our people in want. Here our village had stood for more than a hundred years, during all of which time we were the undisputed possessors of the Mississippi valley, from the Wisconsin to the Portage des Sioux, near the mouth of the Missouri, being about seven hundred miles in length. . . .

When we returned to our village in the spring, from our wintering grounds, we would finish bartering with our [white] traders, who always followed us to our village. We purposely kept some of our fine furs for this trade, and, as there was great opposition among them, who should get these furs, we always got our goods cheap. After this trade was met, the traders would give us a few kegs of rum, which were generally promised in the fall, to encourage us to make a good hunt and not go to war. They would then start with their furs and peltries, for their homes, and our old men would take a frolic. At this time our young men never drank. When this was ended, the next thing to be done was to bury our dead; such as had died during the year. This is a great medicine feast. The relations of those who have died, give all the goods they have purchased, as presents to their friends, thereby reducing themselves to poverty, to show the Great Spirit that they are humble, so that he will take pity on them. We would next open the caches, take out the corn and other provisions which had been put up in the fall. We would then commence repairing our lodges. As soon as this was accomplished, we repair the fences around our corn fields and clean them off ready for planting. This work was done by the women. The men during this time are feasting on dried venison, bear's meat, wild fowl and corn prepared in different ways, while recounting to one another what took place during the winter.

Our women plant the corn, and as soon as they are done we make a feast, at which we dance the crane dance in which they join us, dressed in their most gaudy attire, and decorated with feathers. At this feast the young men select the women they wish to have for wives. . . . The crane dance often lasts two or three days. When this is over, we feast again and have our national dance. . . . When our national dance is over, our cornfields hoed, every weed dug up and our corn about knee high, all our young men start in a direction toward sundown, to hunt deer and buffalo and to kill Sioux if any are found on our hunting grounds. A part of our old men and women go to the lead mines to make lead, and the remainder of our people start to fish and get meat stuff. Every one leaves the village and remains away about forty days. They then return, the hunting party bringing in dried buffalo and deer meat, and sometimes Sioux scalps, when they are found trespassing on our hunting grounds. . . . The party from the lead mines brings lead, and the others dried fish, and mats for our lodges. Presents are now made by each party, the first giving to the others dried buffalo and deer, and they in return presenting them lead, dried fish and mats. This is a happy season of the year, having plenty of provisions, such as beans, squashes and other produce; with our dried meat and fish, we continue to make feasts and visit each other until our corn is ripe. . . .

When our corn is getting ripe, our young people watch with anxiety for the signal to pull roasting ears, as none dare touch them until the proper time. When the corn is fit for use another great ceremony takes place, with feasting and returning thanks to the Great Spirit for giving us corn.

I will here relate the manner in which corn first came. According to tradition handed down to our people, a beautiful woman was seen to descend from the clouds, and alight upon the earth, by two of our ancestors who had killed a deer, and were sitting by a fire roasting a part of it to eat. They were astonished at seeing her, and concluded that she was hungry and had smelt the meat. They immediately went to her, taking with them a piece of the roasted venison. They presented it to her, she ate it, telling them to return to the spot where she was sitting at the end of one year, and they would find a reward for their kindness and generosity. She then ascended to the clouds and disappeared. The men returned to their village, and explained to the tribe what they had seen, done and heard, but were laughed at by their people. When the period had arrived for them to visit this consecrated ground, where they were to find a reward for their attention to the beautiful woman of the clouds, they went with a large party, and found where her right hand had rested on the ground corn growing, where the left hand had rested beans, and immediately where she had been seated, tobacco.

Source: *Autobiography of Ma-Ka-Tai-Me-She-Kia-Kiak, or Black Hawk, 1833; Also: Life, Death and Burial of the Old Chief; Together with a History of the Black Hawk War*, by J. B. Patterson (1882), pp. 57–63.

Available at
http://www.1st-hand-history.org/BHawk/album1.html; also
http://www.gutenberg.net/etext04/bhawk10.txt

5. TRANSCENDENTALISM, 1836

By the early nineteenth century, educated New Englanders had left the harsh Calvinism of their Puritan ancestors far behind, and with the generation of Ralph Waldo Emerson abandoned Christian orthodoxy to look for a faith based on the presence of God in nature. A minister himself whose father and grandfather were ministers, Emerson left his pulpit to pen Nature, *published 1836, the founding document of Transcendentalism and inspiration for generations of New Englanders. Like him, many of his fellow New Englanders were also taking to the woods in search of spiritual and moral influences: painters like those of the Hudson River School, poets such as William Cullen Bryant, educators, ministers, naturalists, scientists, tourists in the White Mountains of New Hampshire and later the Adirondacks of New York, and many of the general public.*

Our age is retrospective. It builds the sepulchres of the fathers. It writes biographies, histories, and criticism. The foregoing generations beheld God and nature face to face; we, through their eyes. Why should not we also enjoy an original relation to the universe? Why should not we have a poetry and philosophy of insight and not of tradition, and a religion by revelation to us, and not the history of theirs? Embosomed for a season in nature, whose floods of life stream around and through us, and invite us by the powers they supply, to action proportioned to nature, why should we grope among the dry bones of the past, or put the living generation into masquerade out of its faded wardrobe? The sun shines to-day also. There is more wool and flax in the fields. There are new lands, new men, new thoughts. Let us demand our own works and laws and worship. . . .

To go into solitude, a man needs to retire as much from his chamber as from society. I am not solitary whilst I read and write, though nobody is with me. But if a man would be alone, let him look at the stars. The rays that come from those heavenly worlds, will separate between him and what he touches. One might think the atmosphere was made transparent with this design, to give man, in the heavenly bodies, the perpetual presence of the sublime. Seen in the streets of cities, how great they are! If the stars should appear one night in a thousand years, how would men believe and adore; and preserve for many generations the remembrance of the city of God which had been shown! But every night come out these envoys of beauty, and light the universe with their admonishing smile.

The stars awaken a certain reverence, because though always present, they are inaccessible; but all natural objects make a kindred impression, when the mind is open to their influence. Nature never wears a mean appearance. Neither does the wisest man extort her secret, and lose his curiosity by finding out all her perfection. Nature never became a toy to a wise spirit. The flowers, the animals, the mountains, reflected the wisdom of his best hour, as much as they had delighted the simplicity of his childhood.

When we speak of nature in this manner, we have a distinct but most poetical sense in the mind. We mean the integrity of impression made by manifold natural objects. It is this which distinguishes the stick of timber of the wood-cutter, from the tree of the poet. The charming landscape which I saw this morning, is indubitably made up of some twenty or thirty farms. Miller owns this field, Locke that, and Manning the woodland beyond. But none of them owns the landscape. There is a property in the horizon which no man has but he whose eye can integrate all the parts, that is, the poet. This is the best part of these men's farms, yet to this their warranty-deeds give no title.

To speak truly, few adult persons can see nature. Most persons do not see the sun. At least they have a very superficial seeing. The sun illuminates only the eye of the man, but shines into the eye and the heart of the child. The lover of nature is he whose inward and outward senses are still truly adjusted to each other; who has retained the spirit of infancy even into the era of manhood. His intercourse with heaven and earth, becomes part of his daily food. In the presence of nature, a wild delight runs through the man, in spite of real sorrows. Nature says,—he is my creature, and maugre all his impertinent griefs, he shall be glad with me. Not the sun or the summer alone, but every hour and season yields its tribute of delight; for every hour and change corresponds to and authorizes a different state of the mind, from breathless noon to grimmest midnight. Nature is a setting that fits equally well a comic or a mourning piece. In good health, the air is a cordial of incredible virtue. Crossing a bare common, in snow puddles, at twilight, under a clouded sky, without having in my thoughts any occurrence of special good fortune, I have enjoyed a perfect exhilaration. I am glad to the brink of fear. In the woods too, a man casts off his years, as the snake his slough, and at what period soever of life, is always a child. In the woods, is perpetual youth. Within these plantations of God, a decorum and sanctity reign, a perennial festival is dressed, and the guest sees not how he should tire of them in a thousand years. In the woods, we return to reason and faith. There I feel that nothing can befall me in life,—no disgrace, no calamity, (leaving me my eyes,) which nature cannot repair. Standing on the bare ground,—my head bathed by the blithe air, and uplifted into infinite space,—all mean egotism vanishes. I become a transparent eye-ball; I am nothing; I see all; the currents of the Universal Being circulate through me; I am part or particle of God. The name of the nearest friend sounds then foreign and accidental: to be brothers, to be acquaintances,—master or servant, is then a trifle and a disturbance. I am the lover of uncontained and immortal beauty. In the wilderness, I find something more dear and connate than in streets or villages. In the tranquil landscape, and especially in the distant line of the horizon, man beholds somewhat as beautiful as his own nature.

The greatest delight which the fields and woods minister, is the suggestion of an occult relation between man and the vegetable. I am not alone and unacknowledged. They nod to me, and I to them. The waving of the boughs in the storm, is new to me and old. It takes me by surprise, and yet is not unknown. Its effect is like that of a higher thought or a better emotion coming over me, when I deemed I was thinking justly or doing right.

Yet it is certain that the power to produce this delight, does not reside in nature, but in man, or in a harmony of both. It is necessary to use these pleasures with great temperance. For, nature is not always tricked in holiday attire, but the same scene which yesterday breathed perfume and glittered as for the frolic of the nymphs, is overspread with melancholy today. Nature always wears the colors of the spirit. To a man laboring under calamity, the heat of his own fire hath sadness in it. Then, there is a kind of contempt of the landscape felt by him who has just lost by death a dear friend. The sky is less grand as it shuts down over less worth in the population.

Source: Ralph Waldo Emerson, *Nature,* in *The Prose Works of Ralph Waldo Emerson,* new and rev. ed., v. 1 (Boston: Fields, Osgood & Co., 1870), pp. 5–9.

6. A TRANSCENDENTALIST LOOKS AT THE WILDERNESS, 1863

One of Emerson's New England friends and followers was Henry David Thoreau (1817–1862), whose writings have practically become scripture for American nature lovers. Of his many influential writings, one of the best loved is his essay "Walking," given as a lecture during his lifetime and published in 1863 after his premature death from tuberculosis. Defenders of wilderness have long found inspiration in these words. Thoreau argues here that civilization historically moves westward toward the wild, and draws its vigor from wild nature.

I wish to speak a word for Nature, for absolute freedom and wildness, as contrasted with a freedom and culture merely civil,—to regard man as an inhabitant, or a part and parcel of Nature, rather than a member of society. I wish to make an extreme statement, if so I may make an emphatic one, for there are enough champions of civilization: the minister and the school-committee, and every one of you will take care of that.

I have met with but one or two persons in the course of my life who understood the art of Walking, that is, of taking walks,—who had a genius, so to speak, for *sauntering:* which word is beautifully derived "from idle people who moved about the country, in the Middle Ages, and asked charity, under pretence of going *à la Sainte Terre,*" to the Holy Land, till the children exclaimed, "There goes a Sainte-Terrer," a Saunterer,—a Holy-Lander. They who never go to the Holy Land in their walks, as they pretend, are indeed mere idlers and vagabonds; but they who do go there are saunterers in the good sense, such as I mean. Some, however, would derive the word from *sans terre,* without land or a home, which, therefore, in the good sense, will mean, having no particular home, but equally at home everywhere. For this is the secret of successful sauntering. . . . But I prefer the first, which, indeed, is the most probable derivation. For every walk is a sort of crusade, preached by some Peter the Hermit in us, to go forth and reconquer this Holy Land from the hands of the Infidels. . . .

I think that I cannot preserve my health and spirits, unless I spend four hours a day at least,—and it is commonly more than that,—sauntering through the woods and over the hills and fields, absolutely free from all worldly engagements. You may safely say, A penny for your thoughts, or a thousand pounds. When sometimes I am reminded that the mechanics and shopkeepers stay in their shops not only all the forenoon, but all the afternoon too, sitting with crossed legs, so many of them,—as if the legs were made to sit upon, and not to stand or walk upon,—I think that they deserve some credit for not having all committed suicide long ago. . . .

Nowadays almost all man's improvements, so called, as the building of houses, and the cutting down of the forest and of all large trees, simply deform the landscape, and make it more and more tame and cheap. A people who would begin by burning the fences and let the forest stand! . . .

When I go out of the house for a walk, uncertain as yet whither I will bend my steps, and submit myself to my instinct to decide for me, I find, strange and whimsical as it may seem, that I finally and inevitably settle southwest, toward some particular wood or meadow or deserted pasture or hill in that direction. My needle is slow to settle,—varies a few degrees, and does not always point due southwest, it is true, and it has good authority for this variation, but it always settles between west and south-south-west. The future lies that way to me, and the earth seems more unexhausted and richer on that side. The outline which would bound my walks would be, not a circle, but a parabola, or rather like one of those cometary orbits which have been thought to be non-returning curves, in this case opening westward, in which my house occupies the place of the sun. I turn round and round irresolute sometimes for a quarter of an hour, until I decide, for a thousandth time, that I will walk into the southwest or west. Eastward I go only by force; but westward I go free. Thither no business leads me. It is hard for me to believe that I shall find fair landscapes or sufficient wildness and freedom behind the eastern horizon. I am not excited by the prospect of a walk thither; but I believe that the forest which I see in the western horizon stretches uninterruptedly toward the setting sun, and there are no towns nor cities in it of enough consequence to disturb me. Let me live where I will, on this side is the city, on that the wilderness, and ever I am leaving the city more and more, and withdrawing into the wildness. . . .

The West of which I speak is but another name for the Wild; and what I have been preparing to say is, that in Wildness is the preservation of the World. Every tree sends its fibres forth in search of the Wild. The cities import it at any price. Men plough and sail for it, From the forest and wilderness come the tonics and barks which brace mankind. Our ancestors were savages. The story of Romulus and Remus being suckled by a wolf is not a meaningless fable. The founders of every State which has risen to eminence have drawn their nourishment and vigor from a similar wild source. It was because the children of the Empire were not suckled by the wolf that they were conquered and displaced by the children of the Northern forests who were.

I believe in the forest, and in the meadow, and in the night in which the corn grows. We require an infusion of hemlock-spruce or arborvitæ in our tea. There is a difference between eating and drinking for strength and from mere gluttony. The Hottentots eagerly devour the marrow of the koodoo and other antelopes raw, as a matter of course. Some of our Northern Indians eat raw the marrow of the Arctic reindeer, as well as various other parts, including the summits of the antlers, as long as they are soft. And herein, perchance, they have stolen a march on the cooks of Paris. They get what usually goes to feed the fire. This is probably better than stall-fed beef and slaughter-house pork to make a man of. Give me a wildness whose glance no civilization can endure,—as if we lived on the marrow of koodoos devoured raw. . . .

Ben Jonson exclaims,—
"How near to good is what is fair!"
So I would say,—
How near to good is what is wild!

Life consists with wildness. The most alive is the wildest. Not yet subdued to man, its presence refreshes him. One who pressed forward incessantly and never rested from his labors, who grew fast and made infinite demands on life, would always find himself in a new country or wilderness, and surrounded by the raw material of life. He would be climbing over the prostrate stems of primitive forest-trees.

Hope and the future for me are not in lawns and cultivated fields, not in towns and cities, but in the impervious and quaking swamps. When, formerly, I have analyzed my partiality for some farm which I had contemplated purchasing, I have frequently found that I was attracted solely by a few square rods of impermeable and unfathomable bog,—a natural sink in one corner of it. That was the jewel which dazzled me. I derive more of my subsistence from the swamps which surround my native town than from the cultivated gardens in the village. . . .

When I would recreate myself, I seek the darkest wood, the thickest and most interminable, and, to the citizen, most dismal swamp. I enter a swamp as a sacred place,—a *sanctum sanctorum*. There is the strength, the marrow of Nature. The wild-wood covers the virgin mould,—and the same soil is good for men and for trees. A man's health requires as many acres of meadow to his prospect as his farm does loads of muck. There are the strong meats on which he feeds. A town is saved, not more by the righteous men in it than by the woods and swamps that surround it. A township where one primitive forest waves above, while another primitive forest rots below,—such a town is fitted to raise not only corn and potatoes, but poets and philosophers for the coming ages. In such a soil grew Homer and Confucius and the rest, and out of such a wilderness comes the Reformer eating locusts and wild honey. . . .

In short, all good things are wild and free. There is something in a strain of music, whether produced by an instrument or by the human voice,—take the sound of a bugle in a summer night, for instance,—which by its wildness, to speak without satire, reminds me of the cries emitted by wild beasts in their native forests. It is so much of their wilderness as I can understand. Give me for my friends and neighbors wild men, not tame ones. The wilderness of the savage is but a faint symbol of the awful ferity with which good men and lovers meet. . . .

Here is this vast, savage, howling mother of ours, Nature, lying all around, with such beauty, and such affection for her children, as the leopard; and yet we are so early weaned from her breast to society, to that culture which is exclusively an interaction of man on man,—a sort of breeding in and in, which produces at most a merely English nobility, a civilization destined to have a speedy limit. . . .

I would not have every man nor every part of a man cultivated, any more than I would have every acre of earth cultivated: part will be tillage, but the greater part will be meadow and forest, not only serving an immediate use, but preparing a mould against a distant future, by the annual decay of the vegetation which it supports. . . .

The sun sets on some retired meadow, where no house is visible, with all the glory and splendor that it lavishes on cities, and perchance, as it has never set before,—where there is but a solitary marsh-hawk to have his wings gilded by it, or only a musquash looks out from his cabin, and there is some little black-veined brook in the midst of the marsh, just beginning to meander, winding slowly round a decaying stump. We walked in so pure and bright a light, gilding the withered grass and leaves, so softly and serenely bright, I thought I had never bathed in such a golden flood, without a ripple or a murmur to it. The west side of every wood and rising ground gleamed like the boundary of Elysium, and the sun on our backs seemed like a gentle herdsman driving us home at evening.

So we saunter toward the Holy Land, till one day the sun shall shine more brightly than ever he has done, shall perchance shine into our minds and hearts, and light up our whole lives with a great awakening light, as warm and serene and golden as on a bank-side in autumn.

Source: Henry D. Thoreau, "Walking," *Excursions* (Boston: Ticknor and Fields, 1863).

7. AN IMMIGRANT FARMBOY ENCOUNTERS WILDERNESS IN WISCONSIN, 1849

Born in Dunbar, Scotland, in 1838, John Muir moved with his zealously religious father and brother David to the frontier near Portage, Wisconsin, in 1849, where the rest of his family joined them the next year. Indian tribes had recently been removed from the land in the manner of Black Hawk's people. Muir grew up there, attended the Uni-

versity of Wisconsin in Madison, and in 1868 moved to California. An avid reader of the Transcendentalists, Muir filled his writings about the Sierras and elsewhere with adventures, wild nature, and religious language. Influential in the creation of Yosemite National Park, in 1892 he helped found the Sierra Club and until his death in 1914 was its president. Near the end of his career he recorded his memories of his days as an immigrant boy discovering the American wilderness. As a result of these formative experiences, Muir spent his career publicizing the glorious joys of encountering the wilderness.

. . . With the help of the nearest neighbors the little shanty was built in less than a day after the rough bur-oak logs for the walls and the white-oak boards for the floor and roof were got together.

To this charming hut, in the sunny woods, overlooking a flowery glacier meadow and a lake rimmed with white water-lilies, we were hauled by an ox-team across trackless carex swamps and low rolling hills sparely dotted with round-headed oaks. Just as we arrived at the shanty, before we had time to look at it or the scenery about it, David and I jumped down in a hurry off the load of household goods, for we had discovered a blue jay's nest, and in a minute or so we were up the tree beside it, feasting our eyes on the beautiful green eggs and beautiful birds,—our first memorable discovery. The handsome birds had not seen Scotch boys before and made a desperate screaming as if we were robbers like themselves, though we left the eggs untouched, feeling that we were already beginning to get rich, and wondering how many more nests we should find in the grand sunny woods. Then we ran along the brow of the hill that the shanty stood on, and down to the meadow, searching the trees and grass tufts and bushes, and soon discovered a bluebird's and a woodpecker's nest, and began an acquaintance with the frogs and snakes and turtles in the creeks and springs.

This sudden plash into pure wildness—baptism in Nature's warm heart—how utterly happy it made us! Nature streaming into us, wooingly teaching her wonderful glowing lessons, so unlike the dismal grammar ashes and cinders so long thrashed into us. Here without knowing it we still were at school; every wild lesson a love lesson, not whipped but charmed into us. Oh, that glorious Wisconsin wilderness! Everything new and pure in the very prime of the spring when Nature's pulses were beating highest and mysteriously keeping time with our own! Young hearts, young leaves, flowers, animals, the winds and the streams and the sparkling lake, all wildly, gladly rejoicing together! . . .

We used to wonder how the woodpeckers could bore holes so perfectly round, true mathematical circles. We ourselves could not have done it even with gouges and chisels. We loved to watch them feeding their young, and wondered how they could glean food enough for so many clamorous, hungry, unsatisfiable babies, and how they managed to give each one its share; for after the young grew strong, one would get his head out of the door-hole and try to hold possession of it to meet the food-laden parents. How hard they worked to support their families, especially the red-headed and speckledy woodpeckers and flickers; digging, hammering on scaly bark and decaying trunks and branches from dawn to dark, coming and going at intervals of a few minutes all the livelong day!

We discovered a hen-hawk's nest on the top of a tall oak thirty or forty rods from the shanty and approached it cautiously. One of the pair always kept watch, soaring in wide circles high above the tree, and when we attempted to climb it, the big dangerous-looking bird came swooping down at us and drove us away.

We greatly admired the plucky kingbird. In Scotland our great ambition was to be good fighters, and we admired this quality in the handsome little chattering flycatcher that whips all the other birds. He was particularly angry when plundering jays and hawks came near his home, and took pains to thrash them not only away from the nest-tree but out of the neighborhood. The nest was usually built on a bur oak near a meadow where insects were abundant, and where no undesirable visitor could approach without being discovered. When a hen-hawk hove in sight, the male immediately set off after him, and it was ridiculous to see that great, strong bird hurrying away as fast as his clumsy wings would carry him, as soon as he saw the little, waspish kingbird coming. But the kingbird easily overtook him, flew just a few feet above him, and with a lot of chattering, scolding notes kept diving and striking him on the back of the head until tired; then he alighted to rest on the hawk's broad shoulders, still scolding and chattering as he rode along, like an angry boy pouring out vials of wrath. Then, up and at him again with his sharp bill; and after he had thus driven and ridden his big enemy a mile or so from the nest, he went home to his mate, chuckling and bragging as if trying to tell her what a wonderful fellow he was....

We never tired listening to the wonderful whip-poor-will....

A near relative, the bull-bat, or nighthawk, seemed hardly less wonderful. Towards evening scattered flocks kept the sky lively as they circled around on their long wings a hundred feet or more above the ground, hunting moths and beetles, interrupting their rather slow but strong, regular wing-beats at short intervals with quick quivering strokes while uttering keen, squeaky cries something like pfee, pfee, and every now and then diving nearly to the ground with a loud ripping, bellowing sound, like bull-roaring, suggesting its name; then turning and gliding swiftly up again....

Everything about us was so novel and wonderful that we could hardly believe our senses except when hungry or while father was thrashing us. When we first saw Fountain Lake Meadow, on a sultry evening, sprinkled with millions of lightning-bugs throbbing with light, the effect was so strange and beautiful that it seemed far too marvelous to be real. Looking from our shanty on the hill, I thought that the whole wonderful fairy show must be in my eyes; for only in fighting, when my eyes were struck, had I ever seen anything in the least like it. But when I asked my brother if he saw anything strange in the meadow he said, "Yes, it's all covered with shaky fire-sparks." Then I guessed that it might be something outside of us, and applied to our all-knowing Yankee to explain it. "Oh, it's nothing but lightnin'-bugs," he said, and kindly led us down the hill to the edge of the fiery meadow, caught a few of the wonderful bugs, dropped them into a cup, and carried them to the shanty, where we watched them throbbing and flashing out their mysterious light at regular intervals, as if each little passionate glow were caused by the beating of a heart. . . .

Partridge drumming was another great marvel. When I first heard the low, soft, solemn sound I thought it must be made by some strange disturbance in my head or stomach, but as all seemed serene within, I asked David whether he heard anything queer. "Yes," he said, "I hear something saying boomp, boomp, boomp, and I'm wondering at it." Then I was half satisfied that the source of the mysterious sound must be in something outside of us, coming perhaps from the ground or from some ghost or bogie or woodland fairy. Only after long watching and listening did we at last discover it in the wings of the plump brown bird.

The love-song of the common jack snipe seemed not a whit less mysterious than partridge drumming. It was usually heard on cloudy evenings, a strange, unearthly, winnowing, spiritlike sound, yet easily heard at a distance of a third of a mile. Our sharp eyes soon detected the bird while making it, as it circled high in the air over the meadow with wonderfully strong and rapid wing-beats, suddenly descending and rising, again and again, in deep, wide loops; the tones being very low and smooth at the beginning of the descent, rapidly increasing to a curious little whirling storm-roar at the bottom, and gradually fading lower and lower until the top was reached. . . .

The love-songs of the frogs seemed hardly less wonderful than those of the birds, their musical notes varying from the sweet, tranquil, soothing peeping and purring of the hylas to the awfully deep low-bass blunt bellowing of the bullfrogs. . . .

We reveled in the glory of the sky scenery as well as that of the woods and meadows and rushy, lily-bordered lakes. The great thunder-storms in particular interested us, so unlike any seen in Scotland, exciting awful, wondering admiration. Gazing awe-stricken, we watched the upbuilding of the sublime cloud-mountains,—glowing, sun-beaten pearl and alabaster cumuli, glorious in beauty and majesty and looking so firm and lasting that birds, we thought, might build their nests amid their downy bosses; the black-browed storm-clouds marching in awful grandeur across the landscape, trailing broad gray sheets of hail and rain like vast cataracts, and ever and anon gashing down vivid zigzag lightning followed by terrible crashing thunder. We saw several trees shattered, and one of them, a punky old oak, was set on fire, while we wondered why all the trees and everybody and everything did not share the same fate, for oftentimes the whole sky blazed. After sultry storm days, many of the nights were darkened by smooth black apparently structureless cloud-mantles which at short intervals were illumined with startling suddenness to a fiery glow by quick, quivering lightning-flashes, revealing the landscape in almost noonday brightness, to be instantly quenched in solid blackness.

Source: John Muir, *The Story of My Boyhood and Youth* (Boston: Houghton Mifflin Company, 1913), pp. 61–76.

http://www.sierraclub.org/john_muir_exhibit/frameindex.html?http://www.sierraclub.org/john_muir_exhibit/writings/

8. INDUSTRY, PROSPERITY, AND POLLUTION, 1863

The United States industrialized at a rapid pace in the late nineteenth century and relied heavily on coal for an energy source. Here a traveler gives his impression of Pittsburg, Pennsylvania, whose iron and steel mills worked at full capacity to supply the Union army during the Civil War. Heavy air pollution gave the city for a century a reputation for unhealthy, sooty dinginess.

Pittsburg, where we halted next night, on the Ohio, is certainly, with the exception of Birmingham, the most intensely sooty, busy, squalid, foul-housed, and vile-suburbed city I have ever seen. Under its perpetual canopy of smoke, pierced by a forest of blackened chimneys, the ill-paved streets, swarm with a streaky population whose white faces are smutched with soot streaks—the noise of vans and drays which shake the houses as they pass, the turbulent life in the thoroughfares, the wretched brick tenements,—built in waste places on squalid mounds, surrounded by heaps of slag and broken brick—all these gave the stranger the idea of some vast manufacturing city of the Inferno; and yet a few miles beyond, the country is studded with beautiful villas, and the great river, bearing innumerable barges and steamers on its broad bosom, rolls its turbid waters between banks rich with cultivated crops.

Source: William Howard Russell, *My Diary North and South* (Boston: T.O.H.P. Burnham; 1863), p. 539.

9. AN EARLY CALL FOR CONSERVATION, 1864

Born in Vermont, George Perkins Marsh (1801–1882) was a linguist, congressman, and diplomat, whose groundbreaking book Man and Nature *(1864) won international acclaim and influence for its recognition that humans had the capacity and indeed the tendency to exhaust natural resources and leave deserts in the wake of failed civilizations. Marsh recommended government control of resources for the common good, and in particular the protection of forests for the many benefits they provide.*

Man has too long forgotten that the earth was given to him for usufruct alone, not for consumption, still less for profligate waste. Nature has provided against the absolute destruction of any of her elementary matter, the raw material of her works; the thunderbolt and the tornado, the most convulsive throes of even the volcano and the earthquake, being only phenomena of decomposition and recomposition. But she has left it within the power of man irreparably to derange the combinations of inorganic matter and of organic life, which through the night of æons she had been proportioning and balancing, to prepare the earth for his habitation, when, in the fulness of time, his Creator should call him forth to enter into its possession.

Apart from the hostile influence of man, the organic and the inorganic world are . . . bound together by such mutual relations and adaptations as secure, if not the absolute permanence and equilibrium of both, a long continuance of the established conditions of each at any given time and place, or at least, a very slow and gradual succession of changes in those conditions. But man is everywhere a disturbing agent. Wherever he plants his foot, the harmonies of nature are turned to discords. The proportions and accommodations which insured the stability of existing arrangements are overthrown. Indigenous vegetable and animal species are extirpated, and supplanted by others of foreign origin, spontaneous production is forbidden or restricted, and the face of the earth is either laid bare or covered with a new and reluctant growth of vegetable forms, and with alien tribes of animal life. These intentional changes and substitutions constitute, indeed, great revolutions; but vast as is their magnitude and importance, they are . . . insignificant in comparison with the contingent and unsought results which have flowed from them.

The fact that, of all organic beings, man alone is to be regarded as essentially a destructive power, and that he wields energies to resist which, nature—that nature whom all material life and all inorganic substance obey—is wholly impotent, tends to prove that, though living in physical nature, he is not of her, that he is of more exalted parentage, and belongs to a higher order of existence than those born of her womb and submissive to her dictates.

There are, indeed, brute destroyers, beasts and birds and insects of prey—all animal life feeds upon, and, of course, destroys other life,—but this destruction is balanced by compensations. It is, in fact, the very means by which the existence of one tribe of animals or of vegetables is secured against being smothered by the encroachments of another; and the reproductive powers of species, which serve as the food of others, are always proportioned to the demand they are destined to supply. Man pursues his victims with reckless destructiveness; and, while the sacrifice of life by the lower animals is limited by the cravings of appetite, he unsparingly persecutes, even to extirpation, thousands of organic forms which he cannot consume. . . .

The ravages committed by man subvert the relations and destroy the balance which nature had established between her organized and her inorganic creations; and she avenges herself upon the intruder, by letting loose upon her defaced provinces destructive energies hitherto kept in check by organic forces destined to be his best auxiliaries, but which he has unwisely dispersed and driven from the field of action. When the forest is gone, the great reservoir of moisture stored up in its vegetable mould is evaporated, and returns only in deluges of rain to wash away the parched dust into which that mould has been converted. The well-wooded and humid hills are turned to ridges of dry rock, which encumbers the low grounds and chokes the watercourses with its debris, and—except in countries favored with an equable distribution of rain through the seasons, and a moderate and regular inclination of surface—the whole earth, unless rescued by human art from the physical degradation to which it tends, becomes an assemblage of bald mountains, of barren, turfless hills, and of swampy and malarious plains. There are parts of Asia Minor, of Northern Africa, of Greece, and even of Alpine Europe, where the operation of causes set in action by man has brought the face of the earth to a desolation almost as complete as that of the moon; and though, within that brief space of time which we call "the historical period," they are known to have been covered with luxuriant woods, verdant pastures, and fertile meadows, they are now too far deteriorated to be reclaimable by man, nor can they become again fitted for human use, except through great geological changes, or other mysterious influences or agencies of which we have no present knowledge, and over which have no prospective control. The earth is fast becoming an unfit home for its noblest inhabitant, and another era of equal human crime and human improvidence, and of like duration with that through which traces of that crime and that improvidence extend, would reduce it to such a condition of impoverished productiveness, of shattered surface, of climatic excess, as to threaten the depravation, barbarism, and perhaps even extinction of the species.

Source: George P. Marsh, *Man and Nature: Or, Physical Geography as Modified by Human Action* (New York: C. Scribner, 1864), pp. 35–44 (footnotes omitted).

10. THE PESHTIGO FOREST FIRE, 1871

The growth of Chicago accompanied an accelerated consumption of the resources that were within range of its lumber mills, flour mills, stockyards, and industry. The city doubled its size every decade as it took in the raw materials from the disappearing frontier and sent out manufactured and processed goods from New York to Denver. Settlers and lumbermen pushed into the forests of Wisconsin, but in a prolonged drought in 1871 turned the forest into a tinderbox. Two fires occurred nearly simultaneously: one destroyed Chicago, while the other, deadlier fire overtook a small Wisconsin frontier settlement named Peshtigo. The infamous Peshtigo forest fire likely arose from fires set by locomotive sparks or hunters' guns or campfires. Logs and slash provided further fuel for the fire.

While we were struggling, in our agony, neighboring States and communities were also visited by the raging monster, and suffered a scourge as keenly felt and more destructive of life. The drought, whose pernicious influence had desiccated the air in our own vicinity, and parched everything to a state of preparation for fire, was very general in the western country. Water, for ordinary purposes of family use and for cattle, had become a luxury in many places, and even an expensive one. The streams and springs were dry in large sections, and the people unprotected from such a foe as charged down upon them. Occasional conflagrations were occurring in the woods of Wisconsin and Michigan, caused by the hunter's carelessness, or as a natural consequence of his sport. . . . The smoke from these stray burnings increased until the bosom of the Lake was veiled, and the country inundated by its volume. These things were of common occurrence, and did not seem to be precursors, as they were, of that devastation which has befallen northern Wisconsin and western Michigan. In actual loss of life we suffered less than the people of those districts; while the protracted nature of their visitation and their remoteness from lines of travel made the individual suffering more keenly felt. Here we had every comfort that a sympathizing world could provide speedily brought to our doors; but there aid came more slowly, as the tidings of their calamity lingered longer on the way to the cares of the world. . . .

A friend, who was in a sailboat on Little Sturgeon Bay, describes the fire blowing off the shore as terrific, so much so that trees on an island about half a mile from shore were set on fire, and the island burned over. He says that after the fire he could have picked up a yawlboatful of birds in the bay, that had got burned in their flight, and dropped into the water. . . . One of the Lake boats running across to Green Bay . . . was greatly detained on her upward trip on account of light wind and smoke, and the latter was so dense that the boat had to be steered entirely with the compass. The fire on the east side of the Bay extended in an almost unbroken line from the eastern shore of Lake Winnebago to the northern extremity of the Eastern Peninsula, fully 150 miles, burning up in its course fences, barns, houses, and an endless quantity of cedar telegraph poles and tan bark, the latter of which was piled in immense heaps on the docks. So deep and dismal was the darkness caused by the immense volume of smoke, that the sun was totally obscured for a distance of 200 miles. This midnight darkness continued for a week. The boat, of course, was delayed, but she left Escanaba for Green Bay on the fatal Sunday night at twelve o'clock, but only made her way twelve miles out when forced to return on account of the stormy sea beneath and the sea of fire overhead. The air was red with burning fragments, carried all the way from Peshtigo and other places along the shore, a distance of nearly fifty miles. The boat laid in Escanaba harbor until six o'clock A.M. Monday, when she was again started, the storm having but slightly subsided; but the course was pursued, and Menomonee was reached with great difficulty. As they approached Menomonee they passed vessels loaded with furniture, etc., all being ready to leave if the place took fire. It was here the passengers learned of the destruction of Peshtigo.

A correspondent writes, twelve days after the event:—

This letter, to give it a local habitation and a name, is dated where Peshtigo was. In the glory of this Indian summer afternoon I look out on the ghastliest clearing that ever lay before mortal eyes. The sandy streets glisten with a frightful smoothness, and calcined fragments are all that remain of imposing edifices and hundreds of peaceful homes. This ominous clearing is in the centre of a blackened, withered forest of oak, pine, and tamarack, with a swift river—the Peshtigo—gliding silently through the centre, from northeast to southwest. Situated seven miles from the Green Bay, on the Peshtigo River, the town commanded all the lumber trade of the northern Peninsula, and grew rapidly into importance as a frontier mart of Chicago. . . . The town has had but one purpose, to make money for its founder and keep up the lumber interests. But one industry breeds many, and in time a railroad running seven miles to the bay, connected the little city with the great chain of lakes. Great foundries and machine-shops rose on the banks of the river, and a busy mill stood in ceaseless operation in the centre of the town. The banks of the Peshtigo teem with a rich and various growth of timber, and a trade of years stood always in prospective to her busy people. The great Northern Pacific Railroad was to be tapped by a road even now building to the place where Peshtigo was, and every hamlet and town in Northern Wisconsin envied and admired the wonderful little city.

. . . Fully 2,000 people had established permanent homes. The site was well chosen for beauty as well as business; the river at this point runs through a slight bluff, which breaks into a low flat before the stream escapes from the borders of the town. The excellent water-power as well as the lumber interest had determined the spot. . . .

. . . The surrounding woods were interspersed with innumerable open glades of crisp brown herbage and dried furze, which had for weeks glowed with the autumn fires that infest these regions. Little heed was paid them, for the first rain would inevitably quench the flames. But the rain never came, and finally valiant battle was waged far and near against the slowly increasing fires. . . .

The sharp air of early October had sent the people in from the evening church services more promptly than usual, although, numbers delayed to speculate on a great noise and ado which set in ominously from the west. The housewives looked tremblingly at the fires and lights within, and the men took a last look at the possibilities without; for many it was truly a last glimpse. The noise grew in volume, and came nearer and nearer with terrific crackling and detonations. The forest rocked and tossed tumultuously; a dire alarm fell upon the imprisoned villagers for the swirling blasts came now from every side. In one awful instant, before expectation could give shape to the horror, a great flame shot up in the western heavens, and in countless fiery tongues struck downward into the village, piercing every object that stood in the town like a red-hot bolt. A deafening roar, mingled with blasts of electric flame, filled the air, and paralyzed every soul in the place. There was no beginning to the work of ruin; the flaming, whirlwind swirled in an instant through, the town. There is no diversity in general experience; all heard the first inexplicable roar; some aver that the earth shook, while a credulous few avow that the heavens opened, and the fire rained down from above.

Moved by a common instinct, for all knew that the woods that encircled the town were impenetrable, every habitation was deserted to the flames, and the grasping multitude flocked to the river. On the west the mad horde saw the bridge in flames in a score of places, and turning, sharply to the left, with one accord, plunged into the water. Three hundred people wedged themselves in between the rolling booms, swayed to and fro by the current, where they roasted in the hot breath of flames that hovered above them, and singed the hair on each head momentarily exposed above the water. There despairing, men and women held their children till the cold water came as an ally to the flames, and deprived them of strength.

Meantime the eastern bank was densely crowded by the dying and the dead. Rushing to the river from this direction the swirling blasts met the victims full in the face and mowed a swath through the fleeing throng. Inhalation was annihilation. Scores fell before the first blast. A few were able to crawl to the pebbly flats, but so dreadfully disfigured that death must have been preferable. All could not reach the river; even the groups that fell prone on the grateful damp flats suffered excruciating agony. The fierce blaze, playing, in tremendous counter currents above them on the higher ground, was sufficiently strong to set the clothing aflame, and the flying sand, heated as by a furnace, blistered the flesh wherever it fell. All that could break through the stifling simoon had come to the river. In the red glare they could see the sloping bank covered with the bodies of those that fell by the way. Few living on the back streets succeeded in reaching the river, the hot breath of the fire cutting them down as they ran. But here a new danger befell. The cows, terrified by smoke and flame, rushed in a great lowing drove to the river brink. Women and children were trampled by the frightened brutes and many, losing their hold on the friendly logs, were swept under the waters.

This was the situation above the bridge; below, a no less harrowing thing happened. The burning timbers of the mill, built at the edge of the bridge, blew and floated down upon the multitude assembled near the flats, and inflicted the most lamentable sufferings. The men fought this new death bitterly; those who were fortunate enough to have coats flung them over the heads of wives and children, and dipped water with their hats on the improvised shelter. Scores had every shred of hair burned off in the battle, and many lost their lives in protecting others. The firemen had made an effort to save some of the buildings, and the hose was run from the river to some important edifice. The heat instantly stopped the attempt, but not before the hose, swollen with water, had been burned through in a hundred places. Although the onslaught of fire and wind had been instantaneous, and the destruction almost simultaneous, the fierce, stifling currents of heat careered through the air for hours. These currents were more fatal than the flames of the burning village. Ignorant of the extent of the fire, and the frightful combination of wind and flames, many of the company's workmen, some with wives and children, shut themselves up in the great brick building and perished in the raging heats of the next half hour. Others on the remote streets broke for the clearing beyond the woods, but few ever passed the burning barrier. Within the boundaries of the town and accessible to the multitude the river accommodation was rather limited, and when the animals had crowded in the situation was full of despair. The flats were covered with prone figures with packs ablaze and faces pressed rigidly into the cooling moist earth. The flames played about and above all with an incessant, deafening roar.

The tornado was but momentary, but was succeeded by maelstroms of fire, smoke, cinders, and red-hot sand. Wherever a building seemed to resist the fire, the roof would be sent whirling in the air, breaking into clouds of flame as it fell. The shower of sparks, cinders, and hot sand fell in continuous and prodigious force, and did quite as much in killing the people as the first terrific sirocco that succeeded the fire. The wretched throng, neck deep in the water, and the still more helpless beings stretched on the heated sands, were pierced and blistered by those burning particles. They seemed like lancets of red-hot steel, penetrating the thickest covering.... The next day revealed a picture exceeding in horror any battle field,—mothers with children hugged closely lay in rigid groups, the clothes burned off and the poor flesh seared to a crisp. One mother, solicitous only for her babe, embalms her unutterable love in the terrible picture left on these woeful sands. With her bare fingers she had scraped out a pass as the soldiers did before Petersburg, and pressing the little one into this, she put her own body above it as a shield, and when the daylight came, both were dead,—the little baby, face unscarred, but the mother burnt almost to cinders.

Source: Edgar Johnson Goodspeed, *History of the great fires in Chicago and the West ... a proud career arrested by sudden and awful calamity, towns and counties laid waste by the devastating element: scenes and incidents, losses and sufferings, benevolence of the nations, etc., etc. with a history of the rise and progress of Chicago, the "young giant": to which is appended a record of the great fires in the past* (New York: H. S. Goodspeed & Co., 1871), pp. 550–559.

11. CLEANING AMERICAN CITIES, 1875

In the late nineteenth century America's leading sanitary engineer was George E. Waring Jr., born in 1833 in Poundridge, New York. Beginning as agricultural and drainage engineer for the new Central Park in New York City in 1857, he devised a sewerage system for Memphis, Tennessee, after a terrible yellow fever epidemic in 1879, that became a model for other cities. Appointed New York's commissioner of street cleaning in 1895, he developed organization and publicity methods that for the first time successfully dealt with the city's garbage problem.

It is proposed in these papers to consider a subject which, one might almost say, was born—or reborn—but a quarter of a century ago, and which has contended with much difficulty in bringing itself to the notice of the public. Indeed, it is only within the past ten years that it has made its way in any important degree outside of purely professional literature.

Happily men, and women too, are fast coming to realize the fact that humanity is responsible for much of its own sickness and premature death, and it is no longer necessary to offer an apology for presenting to public consideration a subject in which, more than in any other,—that is, the subject of its own healthfulness and the cleanliness of its own living,—the general public is vitally interested.

The evils arising from sanitary neglect are as old as civilization, perhaps as old as human life, and they exist about every isolated cabin of the newly settled country. As population multiplies, as cabins accumulate into hamlets, as hamlets grow into villages, villages into towns, and towns into cities, the effects of the evil become more intense, and in their appeal to our attention they are reinforced by the fact that while in isolated life fatal or debilitating illness may equally arise, in compact communities each case arising is a menace to others, so that a single centre of contagion may spread devastation on every side.

It is not enough that we build our houses on healthful sites, and where we have pure air and pure water; we must also make provision for preventing these sites from becoming foul, as every unprotected house-site inevitably must by sheer force of the accumulated waste of its occupants.

Houses, even of the best class, which are free from sanitary objections are extremely rare. The best modern appliances of plumbing are made with almost no regard for the tendency of sewer-gas to find its way into living-rooms, and for other insidious but well known defects. So generally is this true, that it is hardly an exaggeration to say that unwholesomeness in our houses is practically universal. Hardly less universal is a curious sensitiveness on the part of the occupants of these houses to any suggestion of their short-comings.

Singularly enough, no one whose premises are subject to malarial influences seems willing to be told the truth with regard to them. No man likes to confess that his own well and his own cess-pool occupy the same permeable stratum in his garden; that the decaying vegetables in his cellar are the source of the ailments in his household; or that an obvious odor from his adjacent pigsty, or from his costly marble-topped wash-stand, has to do with the disease his physician is contending against.

That the imperfections of our own premises are a menace to our neighbors is a still more irritating suggestion, and such criticism seems to invade the domain of our private rights. Yet surely there can be no equitable or legal private right whose maintenance jeopardizes the well-being of others. It is not possible, in a closely-built town or compact neighborhood, for one to retain in his own ground (either on the surface or in a vault or cess-pool) any form of ordure or festering organic matter, without endangering the lives of his neighbors, through either the pollution of the common air or the poisoning of wells fed from strata underlying the whole ground and more or less tainted by household wastes. Even if he might be permitted to maintain a source of injury to his own family, his neighbors may well insist that he shall not endanger them.

It being important for all that each be made to live cleanly, and the requirements of all, so far as the removal of the wastes of life is concerned, being essentially of the same character, the question of drainage is one in which the whole public is interested, and should be decided and carried out by public authority, so that all may have the advantage of the economy of organized work and the security of work well done.

Source: George E. Waring, Jr., "The Sanitary Drainage of Houses and Towns" *Atlantic Monthly* v. 36, n. 215, September 1875, pp. 339–340.

12. HOW THE WILDERNESS FRONTIER FORMED AMERICAN CHARACTER AND INSTITUTIONS, 1893

Frederick Jackson Turner carried with him the experience of the frontier wilderness and a perception that frontier life transformed people and society—a notion perhaps not far removed from those of Thoreau and Muir. Born in 1861 in Portage, Wisconsin, not far from Muir's home, and educated at the University of Wisconsin and Johns Hopkins University, where he received one of the first history doctorates awarded in the United States, Turner delivered "The Significance of the Frontier in American History" at a meeting of historians in Chicago in 1893—perhaps the most significant history essay ever written.

In a recent bulletin of the Superintendent of the Census for 1890 appear these significant words: "Up to and including 1880 the country had a frontier of settlement, but at present the unsettled area has been so broken into by isolated bodies of settlement that there can hardly be said to be a frontier line. In the discussion of its extent, its westward movement, etc., it can not, therefore, any longer have a place in the census reports." This brief official statement marks the closing of a great historic movement. Up to our own day American history has been in a large degree the history of the colonization of the Great West. The existence of an area of free land, its continuous recession, and the advance of American settlement westward, explain American development.

Behind institutions, behind constitutional forms and modifications, lie the vital forces that call these organs into life and shape them to meet changing conditions. The peculiarity of American institutions is, the fact that they have been compelled to adapt themselves to the changes of an expanding people—to the changes involved in crossing a continent, in winning a wilderness, and in developing at each area of this progress out of the primitive economic and political conditions of the frontier into the complexity of city life. . . . Limiting our attention to the Atlantic coast, we have the familiar phenomenon of the evolution of institutions in a limited area, such as the rise of representative government; into complex organs; the progress from primitive industrial society, without division of labor, up to manufacturing civilization. But we have in addition to this a recurrence of the process of evolution in each western area reached in the process of expansion. Thus American development has exhibited not merely advance along a single line, but a return to primitive conditions on a continually advancing frontier line, and a new development for that area. American social development has been continually beginning over again on the frontier. This perennial rebirth, this fluidity of American life, this expansion westward with its new opportunities, its continuous touch with the simplicity of primitive society, furnish the forces dominating American character. The true point of view in the history of this nation is not the Atlantic coast, it is the Great West. . . .

In this advance, the frontier is the outer edge of the wave—the meeting point between savagery and civilization. . . . The most significant thing about the American frontier is, that it lies at the hither edge of free land. . . .

In the settlement of America we have to observe how European life entered the continent, and how America modified and developed that life and reacted on Europe. . . . The frontier is the line of most rapid and effective Americanization. The wilderness masters the colonist. It finds him a European in dress, industries, tools, modes of travel, and thought. It takes him from the railroad car and puts him in the birch canoe. It strips off the garments of civilization and arrays him in the hunting shirt and the moccasin. It puts him in the log cabin of the Cherokee and Iroquois and runs an Indian palisade around him. Before long he has gone to planting Indian corn and plowing with a sharp stick, he shouts the war cry and takes the scalp in orthodox Indian fashion. In short, at the frontier the environment is at first too strong for the man. He must accept the conditions which it furnishes, or perish, and so he fits himself into the Indian clearings and follows the Indian trails. Little by little he transforms the wilderness, but the outcome is not the old Europe. . . . The fact is, that here is a new product that is American. At first, the frontier was the Atlantic coast. It was the frontier of Europe in a very real sense. Moving westward, the frontier became more and more American. As successive terminal moraines result from successive glaciations, so each frontier leaves its traces behind it, and when it becomes a settled area the region still partakes of the frontier characteristics. Thus the advance of the frontier has meant a steady movement away from the influence of Europe, a steady growth of independence on American lines. And to study this advance, the men who grew up under these conditions, and the political, economic, and social results of it, is to study the really American part of our history. . . .

The Atlantic frontier was compounded of fisherman, fur trader, miner, cattle-raiser, and farmer. Excepting the fisherman, each type of industry was on the march toward the West, impelled by an irresistible attraction. Each passed in successive waves across the continent. Stand at Cumberland Gap and watch the procession of civilization, marching single file—the buffalo following the trail to the salt springs, the Indian, the fur trader and hunter, the cattle-raiser, the pioneer farmer—and the frontier has passed by. Stand at South Pass in the Rockies a century later and see the same procession with wider intervals between. The unequal rate of advance compels us to distinguish the frontier into the trader's frontier, the rancher's frontier, or the miner's frontier, and the farmer's frontier. When the mines and the cow pens were still near the fall line the traders' pack trains were tinkling across the Alleghanies, and the French on the Great Lakes were fortifying their posts, alarmed by the British trader's birch canoe. When the trappers scaled the Rockies, the farmer was still near the mouth of the Missouri. . . .

First, we note that the frontier promoted the formation of a composite nationality for the American people. The coast was preponderantly English, but the later tides of continental immigration flowed across to the free lands. . . . In the crucible of the frontier the immigrants were Americanized, liberated, and fused into a mixed race, English in neither nationality nor characteristics. The process has gone on from the early days to our own. . . .

But the most important effect of the frontier has been in the promotion of democracy here and in Europe. As has been indicated, the frontier is productive of individualism. Complex society is precipitated by the wilderness into a kind of primitive organization based on the family. The tendency is anti-social. It produces antipathy to control, and particularly to any direct control. . . . The frontier individualism has from the beginning promoted democracy. The frontier States that came into the Union in the first quarter of a century of its existence came in with democratic suffrage provisions, and had reactive effects of the highest importance upon the older States whose peoples were being attracted there. An extension of the franchise became essential. . . .

From the conditions of frontier life came intellectual traits of profound importance. The works of travelers along each frontier from colonial days onward describe certain common traits, and these traits have, while softening down, still persisted as survivals in the place of their origin, even when a higher social organization succeeded. The result is that to the frontier the American intellect owes its striking characteristics. That coarseness and strength combined with acuteness and inquisitiveness; that practical, inventive turn of mind, quick to find expedients; that masterful grasp of material things, lacking in the artistic but powerful to effect great ends; that restless, nervous energy; that dominant individualism, working for good and for evil, and withal that buoyancy and exuberance which comes with freedom—these are traits of the frontier, or traits called out elsewhere because of the existence of the frontier. Since the days when the fleet of Columbus sailed into the waters of the New World, America has been another name for opportunity, and the people of the United States have taken their tone from the incessant expansion which has not only been open but has even been forced upon them. He would be a rash prophet who should assert that the expansive character of American life has now entirely ceased. Movement has been its dominant fact, and, unless this training has no effect upon a people, the American energy will continually demand a wider field for its exercise. But never again will such gifts of free land offer themselves. For a moment, at the frontier, the bonds of custom are broken and unrestraint is triumphant. There is not *tabula rasa*. The stubborn American environment is there with its imperious summons to accept its conditions; the inherited ways of doing things are also there; and yet, in spite of environment, and in spite of custom, each frontier did indeed furnish a new field of opportunity, a gate of escape from the bondage of the past; and freshness, and confidence, and scorn of older society, impatience of its restraints and its ideas, and indifference to its lessons, have accompanied the frontier. What the Mediterranean Sea was to the Greeks, breaking the bond of custom, offering new experiences, calling out new institutions and activities, that, and more, the ever retreating frontier has been to the United States directly, and to the nations of Europe more remotely. And now, four centuries from the discovery of America, at the end of a hundred years of life under the Constitution, the frontier has gone, and with its going has closed the first period of American history.

Source: Frederick Jackson Turner, "The Significance of the Frontier in American History," in *The Frontier In American History* (New York: Henry Holt and Company, 1921).

13. AMERICAN INDUSTRIAL MIGHT AND AMERICAN INDUSTRIAL POLLUTION, 1899

By the time William Archer visited Chicago near the turn of the twentieth century, the United States yearly produced as much iron and steel as the next two producers, Germany and Britain, combined. From Boston across the continent to Chicago, a string of industrial cities made the country an economic dynamo, but in the process pumped smoke into the air and dumped slag, ash, and other industrial waste into the streams and onto the earth.

On the other hand, I observe no eagerness on the part of New York to contest the supremacy of Chicago in the matter of smoke. In this respect, the eastern metropolis is to the western as Mont Blanc to Vesuvius. The smoke of Chicago has a peculiar and aggressive individuality, due, I imagine, to the natural clearness of the atmosphere. It does not seem, like London smoke, to permeate and blend with the air. It does not overhang the streets in a uniform canopy, but sweeps across and about them in gusts and swirls, now dropping and now lifting again its grimy curtain. You will often see the vista of a gorge-like street so choked with a seeming thundercloud that you feel sure a storm is just about to burst upon the city, until you look up at the zenith and find it smiling and serene. Again and again a sudden swirl of smoke across the street (like that which swept across Fifth-avenue when the Windsor Hotel burst into flames) has led me to prick up my ears for a cry of "Fire!" But Chicago is not so easily alarmed. It is accustomed to having its airs from heaven blurred by these blasts from hell. I know few spectacles more curious than that which awaits you when you have shot up in the express elevator to the top of the Auditorium tower—on the one hand, the blue and laughing lake, on the other, the city belching volumes of smoke from its thousand throats, as though a vaster Sheffield or Wolverhampton had been transported by magic to the shores of the Mediterranean Sea. What a wonderful city Chicago will be when the commandment is honestly enforced which declares, "Thou shalt consume thine own smoke!"

Source: William Archer, *America To-Day: Observations and Reflections* (New York: Charles Scribner's Sons, 1899), pp. 106–108.

14. PUBLICIZING CONSERVATION, 1910

Named after Hudson River School painter Sanford Gifford, Gifford Pinchot (1865–1946) was the son of a wealthy New York merchant of Huguenot ancestry and his New England–born wife. He lived much of his life at the family estate, Grey Towers, in Milford, Pennsylvania. The first native-born professional forester, Pinchot was the foremost advocate of the new idea of "conservation." Among other achievements, he organized the Forest Service in the administration of Theodore Roosevelt, worked to establish a forestry school at Yale University with family money, and later served twice as governor of Pennsylvania. No less than Puritan Jonathan Edwards did he see the wanton waste of earth's resources as a sin. Like George Perkins Marsh, Pinchot believed the government must step in to ensure that natural resources would be used wisely and sustainably to benefit society for the long term. In this passage he explains the basic principles of conservation.

The principles which the word Conservation has come to embody are not many, and they are exceedingly simple. I have had occasion to say a good many times that no other great movement has ever achieved such progress in so short a time, or made itself felt in so many directions with such vigor and effectiveness, as the movement for the conservation of natural resources. . . .

The first great fact about conservation is that it stands for development. There has been a fundamental misconception that conservation means nothing but the husbanding of resources for future generations. There could be no more serious mistake. Conservation does mean provision for the future, but it means also and first of all the recognition of the right of the present generation to the fullest necessary use of all the resources with which this country is so abundantly blessed. Conservation demands the welfare of this generation first, and afterward the welfare of the generations to follow. . . .

In the second place conservation stands for the prevention of waste. There has come gradually in this country an understanding that waste is not a good thing and that the attack on waste is an industrial necessity. I recall very well indeed how, in the early days of forest fires, they were considered simply and solely as acts of God, against which any opposition was hopeless and any attempt to control them not merely hopeless but childish. It was assumed that they came in the natural order of things, as inevitably as the seasons or the rising and setting of the sun. To-day we understand that forest fires are wholly within the control of men. So we are coming in like manner to understand that the prevention of waste in all other directions is a simple matter of good business. The first duty of the human race is to control the earth it lives upon. . . .

In addition to the principles of development and preservation of our resources there is a third principle. It is this: The natural resources must be developed and preserved for the benefit of the many, and not merely for the profit of a few. We are coming to understand in this country that public action for public benefit has a very much wider field to cover and a much larger part to play than was the case when there were resources enough for every one, and before certain constitutional provisions had given so tremendously strong a position to vested rights and property in general. . . .

The conservation idea covers a wider range than the field of natural resources alone. Conservation means the greatest good to the greatest number for the longest time. One of its great contributions is just this, that it has added to the worn and well-known phrase, "the greatest good to the greatest number," the additional words "for the longest time," thus recognizing that this nation of ours must be made to endure as the best possible home for all its people.

Conservation advocates the use of foresight, prudence, thrift, and intelligence in dealing with public matters, for the same reasons and in the same way that we each use foresight, prudence, thrift, and intelligence in dealing with our own private affairs. It proclaims the right and duty of the people to act for the benefit of the people. Conservation demands the application of common-sense to the common problems for the common good. . . .

The outgrowth of conservation, the inevitable result, is national efficiency. In the great commercial struggle between nations which is eventually to determine the welfare of all, national efficiency will be the deciding factor. So from every point of view conservation is a good thing for the American people.

The National Forest Service, one of the chief agencies of the conservation movement, is trying to be useful to the people of this nation. The Service recognizes, and recognizes it more and more strongly all the time, that whatever it has done or is doing has just one object, and that object is the welfare of the plain American citizen. Unless the Forest Service has served the people, and is able to contribute to their welfare it has failed in its work and should be abolished. But just so far as by coöperation, by intelligence, by attention to the work laid upon it, it contributes to the welfare of our citizens, it is a good thing and should be allowed to go on with its work.

The Natural Forests are in the West. Headquarters of the Service have been established throughout the Western country, because its work cannot be done effectively and properly without the closest contact and the most hearty coöperation with the Western people. It is the duty of the Forest Service to see to it that the timber, water-powers, mines, and every other resource of the forests is used for the benefit of the people who live in the neighborhood or who may have a share in the welfare of each locality. It is equally its duty to coöperate with all our people in every section of our land to conserve a fundamental resource, without which this Nation cannot prosper.

Source: Gifford Pinchot, *The Fight for Conservation* (New York: Doubleday, Page & Company, 1910), pp. 40–52.

15. A PLEA FOR VANISHING WILDLIFE, 1913

William Temple Hornaday (1854–1937) was born in Indiana and grew up on a farm in Iowa, where he experienced the natural abundance of a recently settled area. Educated in the natural sciences, his career led him to the creation of the famous Bronx Zoo in 1896. Best known for his campaign to save the American bison from extinction, he wrote many books, mostly on wildlife. Our Vanishing Wild Life *was his call to save America's wildlife, then endangered primarily by market hunting, sport hunting, and the fashion for women's feathered hats.*

In order that the American people may correctly understand and judge the question of the extinction or preservation of our wild life, it is necessary to recall the near past. It is not necessary, however, to go far into the details of history; for a few quick glances at a few high points will be quite sufficient for the purpose in view.

Any man who reads the books which best tell the story of the development of the American colonies of 1712 into the American nation of 1912, and takes due note of the wild-life features of the tale, will say without hesitation that when the American people received this land from the bountiful hand of Nature, it was endowed with a magnificent and all-pervading supply of valuable wild creatures. The pioneers and the early settlers were too busy even to take due note of that fact, or to comment upon it, save in very fragmentary ways.

Nevertheless, the wild-life abundance of early American days survived down to so late a period that it touched the lives of millions of people now living. Any man 55 years of age who when a boy had a taste for "hunting,"—for at that time there were no "sportsmen" in America,—will remember the flocks and herds of wild creatures that he saw and which made upon his mind many indelible impressions.

"Abundance" is the word with which to describe the original animal life that stocked our country, and all North America, only a short half-century ago. Throughout every state, on every shore-line, in all the millions of fresh water lakes, ponds and rivers, on every mountain range, in every forest, *and even on every desert*, the wild flocks and herds held away. It was impossible to go beyond the settled haunts of civilized man and escape them.

It was a full century after the complete settlement of New England and the Virginia colonies that the wonderful big-game fauna of the great plains and Rocky Mountains was really discovered; but the bison millions, the antelope millions, the mule deer, the mountain sheep and mountain goat were there, all the time. In the early days, the millions of pinnated grouse and quail of the central states attracted no serious attention from the American people-at-large; but they lived and flourished just the same, far down in the seventies, when the greedy market gunners systematically slaughtered them, and barreled them up for "the market," while the foolish farmers calmly permitted them to do it. . . .

The game birds of America, as a class and a mass, have not been swept away to ward off starvation or to rescue the perishing. Even back in the sixties and seventies, very, very few men of the North thought of killing prairie chickens, ducks and quail, snipe and woodcock, in order to keep the hunger wolf from the door. The process was too slow and uncertain; and besides, the really-poor man rarely had the gun and ammunition. Instead of attempting to live on birds, he hustled for the staple food products that the soil of his own farm could produce.

First, last and nearly all the time, the game birds of the United States as a whole, have been sacrificed on the altar of Rank Luxury, to tempt appetites that were tired of fried chicken and other farm delicacies. To-day, even the average poor man hunts birds for the joy of the outing, and the pampered epicures of the hotels and restaurants buy game birds, and eat small portions of them, solely to tempt jaded appetites. If there is such a thing as "class" legislation, it is that which permits a few sordid market-shooters to slaughter the birds of the whole people in order to sell them to a few epicures. . . .

And the man who has had a fine day in the painted woods, on the bright waters of a duck-haunted bay, or in the golden stubble of September, can fill his day and his soul with six good birds just as well as with sixty. The idea that in order to enjoy a fine day in the open a man must kill a wheel-barrow load of birds, is a mistaken idea; and if obstinately adhered to, it becomes vicious! The Outing in the Open is the thing,—not the blood-stained feathers, nasty viscera and Death in the game-bag. One quail on a fence is worth more to the world than ten in a bag. . . .

It is the way of Americans to feel that because game is abundant in given place at a given time, it always will be abundant, and may therefore be slaughtered without limit. That was the case last winter in California during the awful slaughter of band-tailed pigeons, as will be noted elsewhere.

It is time for all men to be told in the plainest terms that there never has existed, anywhere in historic times, a volume of wild life so great that civilized man could not quickly exterminate it by his methods of destruction. Lift the veil and look at the stories of the bison, the passenger pigeon, the wild ducks and shore birds of the Atlantic coast, and the fur-seal.

As reasoning beings, it is our duty to heed the lessons of history, and not rush blindly on until we perpetrate a continent destitute of wild life. . . .

Source: William T. Hornaday, *Our Vanishing Wild Life; Its Extermination and Preservation* (New York: C. Scribner's Sons, 1913), pp. 1–2, 3–4, 5–6.

16. EARTH DAY, 1970

Many environmental leaders in Congress represented Northern states, including Senators Edmund Muskie of Maine and Gaylord Nelson of Wisconsin. Born in 1916 in Clear Lake, Wisconsin, Senator Nelson took an interest in outdoor and environmental issues from the beginning of his career. His most famous achievement may well be his instigation of the first and probably most significant and influential Earth Day, in 1970. Environmental concern had become a major issue by the late 1960s, resulting in a flood of environmental legislation between 1965 and 1980. Here Nelson describes the origins of Earth Day.

What was the purpose of Earth Day? How did it start? These are the questions I am most frequently asked.

Actually, the idea for Earth Day evolved over a period of seven years starting in 1962. For several years, it had been troubling me that the state of our environment was simply a non-issue in the politics of the country. Finally, in November 1962, an idea occurred to me that was, I thought, a virtual cinch to put the environment into the political "limelight" once and for all. The idea was to persuade President Kennedy to give visibility to this issue by going on a national conservation tour. I flew to Washington to discuss the proposal with Attorney General Robert Kennedy, who liked the idea. So did the President. The President began his five-day, eleven-state conservation tour in September 1963. For many reasons the tour did not succeed in putting the issue onto the national political agenda. However, it was the germ of the idea that ultimately flowered into Earth Day.

I continued to speak on environmental issues to a variety of audiences in some twenty-five states. All across the country, evidence of environmental degradation was appearing everywhere, and everyone noticed except the political establishment. The environmental issue simply was not to be found on the nation's political agenda. The people were concerned, but the politicians were not.

After President Kennedy's tour, I still hoped for some idea that would thrust the environment into the political mainstream. Six years would pass before the idea that became Earth Day occurred to me while on a conservation speaking tour out West in the summer of 1969. At the time, anti-Vietnam War demonstrations, called "teach-ins," had spread to college campuses all across the nation. Suddenly, the idea occurred to me—why not organize a huge grassroots protest over what was happening to our environment?

I was satisfied that if we could tap into the environmental concerns of the general public and infuse the student anti-war energy into the environmental cause, we could generate a demonstration that would force this issue onto the political agenda. It was a big gamble, but worth a try.

At a conference in Seattle in September 1969, I announced that in the spring of 1970 there would be a nationwide grassroots demonstration on behalf of the environment and invited everyone to participate. The wire services carried the story from coast to coast. The response was electric. It took off like gangbusters. Telegrams, letters, and telephone inquiries poured in from all across the country. The American people finally had a forum to express its concern about what was happening to the land, rivers, lakes, and air—and they did so with spectacular exuberance. For the next four months, two members of my Senate staff, Linda Billings and John Heritage, managed Earth Day affairs out of my Senate office.

Five months before Earth Day, on Sunday, November 30, 1969, *The New York Times* carried a lengthy article by Gladwin Hill reporting on the astonishing proliferation of environmental events:

"Rising concern about the environmental crisis is sweeping the nation's campuses with an intensity that may be on its way to eclipsing student discontent over the war in Vietnam . . . a national day of observance of environmental problems . . . is being planned for next spring . . . when a nationwide environmental 'teach-in' . . . coordinated from the office of Senator Gaylord Nelson is planned. . . ."

It was obvious that we were headed for a spectacular success on Earth Day. It was also obvious that grassroots activities had ballooned beyond the capacity of my U.S. Senate office staff to keep up with the telephone calls, paper work, inquiries, etc. In mid-January, three months before Earth Day, John Gardner, Founder of Common Cause, provided temporary space for a Washington, D.C., headquarters. I staffed the office with college students and selected Denis Hayes as coordinator of activities.

Earth Day worked because of the spontaneous response at the grassroots level. We had neither the time nor resources to organize 20 million demonstrators and the thousands of schools and local communities that participated. That was the remarkable thing about Earth Day. It organized itself.

Source: Senator Gaylord Nelson, "How the First Earth Day Came About," original source unknown, widely reprinted. http://www.nih.gov/od/ors/earthday/history.html.

17. THE TERRIBLE PRICE OF TOXIC WASTE, 1979

Beginning during World War II and continuing through the record economic boom of the 1950s and 1960s, American industry manufactured not only huge numbers of products for the country and the world, but produced pollution and toxic waste on a massive scale. Mixed with this waste were the new products of the postwar chemical, synthetics, and nuclear industries. In the 1970s, as manufacturers shut down their factories or moved production to other countries, the dirty but once-proud cities of the North discovered not only dying communities amidst dying forests and waters, but deadly wastes in their drinking water and even in their homes and schools. The court battle over toxic waste in Woburn, Massachusetts, provided the story for the movie A Civil Action. *The 1979 nuclear accident at Three Mile Island, near Harrisburg, Pennsylvania, both provided a preview of the 1986 Chernobyl tragedy in the Soviet Union and left a cancerous legacy in the near vicinity, according to recent studies. But it was the tragedy of Love Canal, a housing development in Niagara Falls, New York, that caught the sym-*

pathy of the nation, aroused federal action, and resulted in the passage in 1980 of the Comprehensive Environmental Response, Compensation, and Liability Act (CERCLA), commonly known as the EPA Superfund, designed to fund the cleanup of abandoned or uncontrolled toxic waste sites. One resident, Lois Gibbs, organized Love Canal homeowners in 1978 and formed the Citizens' Clearinghouse for Hazardous Waste in 1980, now renamed the Center for Health, Environment and Justice, to help other communities fight for environmental justice. In the 1980s and 1990s the environmental justice movement grew to national and international importance. Here Love Canal resident Anne Hillis describes the anguish, fear, and anger of living in a polluted neighborhood.

. . . I am a Wife, a Mother, I live in Niagara Falls, New York. I also live close to a "Dump." A dump called Love Canal. I don't want to live there anymore. I hate Love Canal, I hate my life at Love Canal. It's a strange life that I lead now, it is filled with disruptions, frustrations, sleepless nights and a grip of fear that only those in similar situations can understand. . . .

We've lived in the home for 13½ years. We lost a child there. My 10-year-old son went to 99th Street school. . . .

Despair, hopelessness, we ask—What are we doing to our children and to our own bodies, staying? The stress alone is enough to break anyone. . . .

Our homes are valueless, we can't sell, who would buy a home like this? . . .

I want to tell you about my son. . . . He's 10, he's a bright boy, he has a 91 average in school, as a baby he never required much sleep, he was on a sedative at about age 7 months to about 18 months, he developed rashes, frequent bouts of diarrhea and respiratory problems—always respiratory problems.

His first year at 99th Street school, Kindergarten, he was admitted to the hospital—very ill. The diagnosis—acute gastroenterites—cause unknown.

After that—more respiratory infections—tonsilitis. At age 6 the tonsils and adenoids were removed, but the respiratory infections did not improve—he developed asthma.

In 1977 we were told to consult an allergist, he was tested, and found to have many allergies. He has been on a desensitizing program now for a year and a half—with no improvement. . . .

He started the school year off with an abscess in his nose—he was on antibiotics—he had repeated respiratory infections, attacks of asthma.

By this time we the people were well aware of Love Canal. . . .

One night last winter I got up to go to the bathroom—I looked in on him, his bed was empty. I looked all over, it was 2:00 a.m. I heard a cry from under the couch, my son was under there with his knees drawn up to his chin, crying, I asked him to come out, and what was wrong. His reply, "I want to die, I don't want to live here anymore—I know you will be sick again and I'll be sick again."

My husband and my son and I cried together that night. . . .

. . . Two weeks ago another bronchial infection. . . . The antibiotics didn't work. It was changed to another—we went to the hospital, more blood work, chest X-rays. The Doctor knows my son is sick, but they don't know how to help him. I do—get him out! Him, and all the sick people out—out of the contaminated Hell we live in. . . .

Many women cannot get pregnant because they've had hysterectomies due to excessive bleeding, tumors and cancer at young ages—ages 20 to 30.

I, myself, am in this group. . . .

I am a sick woman, nursing a sick child. My thoughts are—will he live to have children, if so, will they be sick or deformed?

Source: Testimony of Anne Hillis and Jim Clark, The Joint Senate Subcommittee on Environmental Pollution and Hazardeous [sic] Waste, March 28 and 29, 1979, Co-Chaired by Senator Edmund Muskie and Senator Culver, Love Canal Collection, University Archives, Universities Libraries, State University of New York at Buffalo. http://ublib.buffalo.edu/libraries/projects/lovecanal/index.html.

BIBLIOGRAPHY

Chapter One

For a discussion of the physical geography of the Northeast, see Neil Jorgensen, *A Guild to New England's Landscape* (Chester, CT: Globe-Pequot, 1977); *Southern New England* (San Francisco: Sierra Club, 1978); Betty Thompson, *The Changing Face of New England* (Reprint, Boston: Houghton Mifflin, 1958); Michael Berrill and Deborah Berrill, *The North Atlantic Coast* (San Francisco: Sierra Club, 1981); Oxford *Regional Economic Atlas: The United States and Canada* (Oxford, UK: Oxford University, 1975); William H. MacLeish, *The Gulf Stream: Encounters with the Blue God* (Boston: Houghton Mifflin, 1989); Chet Raymo and Maureen E. Raymo, *Written in Stone: A Geological History of the Northeastern United States* (Chester, CT: Globe-Pequot, 1986). For a discussion of the making of the North American continent, see Ron Redfern, *The Making of a Continent* (New York: Times, 1983). For a discussion of America's early inhabitants, see Richard White, *The Middle Ground: Indians, Empires, and Republics in the Great Lakes Region, 1650–1815* (Cambridge, UK: Cambridge University, 1991); Henry Dobyns, *Their Numbers Become Thinned: Native Population Dynamics in Eastern North America* (Knoxville: University of Tennessee, 1983).

For a history of the relationship between Americans and their geography, see Tim Flannery, *The Eternal Frontier: An Ecological History of North America and Its Peoples* (New York: Atlantic Monthly, 2001).

For a discussion of the importance of fishing and the resources of the Great Lakes, see William Ashworth, *The Late, Great Lakes: An Environmental History* (New York: Alfred A. Knopf, 1986) and Margaret Beattie Bogue, *Fishing the Great Lakes: An Environmental History 1783–1933* (Madison: University of Wisconsin, 2000).

For a discussion of Native American life in New England, see William Cronon, *Changes in the Land: Indians, Colonists, and Ecology of New England* (New York: Hill and Wang, 1983); Carolyn Merchant, *Ecological Revolutions: Nature, Gender, and Science in New*

England (Chapel Hill: University of North Carolina, 1989); and David Bernstein, *Prehistoric Subsistence on the Southern New England Coast: The Record from the Narragansett Bay* (San Diego: Academic, 1992). For a discussion of Native American land use, see J. Donald Hughes, *American Indian Ecology* (El Paso: Texas Western, 1982), and R. Douglas Hunt, *Indian Agriculture in America: Prehistory to the Present* (Lawrence: University Press of Kansas, 1987). For a general description of the native peoples of North America, see Harold E. Driver, *Indians of North America* (Chicago: University of Chicago, 1969), and Christian F. Feest, *Indians of Northeast North America* (Leiden, Netherlands: Brill, 1986); Kathleen J. Bragdon, *Native Peoples of Southern New England, 1500–1650* (Norman: University of Oklahoma, 1996); *The Columbia Guide to American Indians of the North East* (New York: Columbia University, 2001); and Alan Taylor, *American Colonies: The Settling of North America* (New York: Penguin, 2001). See also Diane Muir, *Reflections in Bullough's Pond* (Hanover, NH: University Press of New England, 2000), and Richard Wilkie and Jack Tager, eds. *Historical Atlas of Massachusetts* (Amherst: University of Massachusetts, 1991).

Chapter Two

For a discussion of the spread of disease among Native Americans, see Alfred W. Crosby, *The Columbian Exchange: Biological and Cultural Consequences of 1492* (Westport, CT: Greenwood, 1972); *Ecological Imperialism: The Biological Expansion of Europe 900–1900* (New York: Cambridge University, 1986); and Jonathan B. Tucker, *Scourge: The Once and Future Threat of Smallpox* (New York: Atlantic Monthly, 2001).

For a history of the differing agricultural practices of Native Americans and Europeans, see Gorden G. Whitney, *From Coastal Wilderness to Fruited Plain* (Cambridge, UK: Cambridge University, 1994).

For a discussion of English and New England farm patterns, see Sumner Chilton Powell, *Puritan Village: The Formation of a New England Town* (New York: Anchor, 1965); Kenneth Lockridge, *A New England Town: The First Hundred Years* (New York: Norton, 1970); Howard S. Russell, *A Deep Furrow: Three Centuries of Farming in New England* (Hanover, NH: University Press of New England, 1976); Stephen Innes, *Labor in a New Land: Economy and Society in Seventeenth Century Springfield* (Princeton, NJ: Princeton University, 1983); and Douglas R. McManis, *Colonial New England: A Historical Geography* (New York: Oxford University, 1975).

For a general view of early American agricultural practices, see Willard W. Cochrane, *The Development of American Agriculture: A Historical Analysis* (Minneapolis: University of Minnesota, 1979); and Donald W. Meinig, *The Shaping of America: A Geographical Perspective on 500 Years of History, Vol. 1, Atlantic America, 1492–1800* (New Haven, CT: Yale University, 1986).

Chapter Three

For a description of the geographic and social history of the region south and west of Philadelphia, see James Lemon, *The Best Poor Man's Country: A Geographical Study of Early Southeastern Pennsylvania* (Baltimore, MD: Johns Hopkins Press, 1976). Alan Taylor, *William Cooper's Town: Power and Persuasion on the Frontier of the Early American Republic* (New York: Alfred A. Knopf, 1995) provides an excellent picture of colonial southeastern New York. John Cumbler, *Reasonable Use: The People, the Environment and the State* (Oxford, UK: Oxford University, 2001) provides a view of the changing patterns of land use along the Connecticut River; while Theodore Steinberg, *Nature Incorporated: Industrialization and the Waters of New England* (Cambridge, UK: Cambridge University, 1991) looks at the Merrimack River valley. For a description of everyday life, see Jack Larkin, *The Reshaping of Everyday Life, 1790–1840* (New York: Harper and Row, 1988); Christopher Clark, *Roots of American Capitalism in Western Massachusetts, 1780–1860* (Ithaca, NY: Cornell University, 1990); and David R. Foster, *Thoreau's Country: Journey through a Transformed Landscape* (Cambridge, MA: Harvard University, 1999). See also Carolyn Merchant, *Ecological Revolutions: Nature, Gender, and Science in New England* (Chapel Hill: University of North Carolina, 1989). For a discussion of transportation and the early canal system, see George Rogers Taylor, *The Transportation Revolution 1815–1860* (New York: Holt, Reinhart, and Winston, 1962). For the conflict between transportation needs and nature's reality, see Jack Burbaker, *Down the Susquehanna to the Chesapeake* (University Park: Pennsylvania State University, 2002). For a discussion of the role of timber in early American life, see Charles F. Carroll, *The Timber Economy of Puritan New England* (Providence, RI: Brown University, 1973).

For early American attitudes toward the natural world and the relationship between Protestant religious attitudes toward the natural world and progress, see Mark Stoll, *Protestantism, Capitalism, and Nature in America* (Albuquerque: University of New Mexico, 1997); see also Robert Dorman, *A Word for Nature* (Chapel Hill: University of North Carolina, 1998); David Foster, *Thoreau's Country: A Journey through a Transformed Landscape* (Cambridge, MA: Harvard University, 1999); Donald Worster, *Nature's Economy: A History of Ecological Ideas* (Cambridge, UK: Cambridge University, 1985); and Lawrence Buell, *The Environmental Imagination: Thoreau, Nature Writing and the Formation of American Culture* (Cambridge, MA: Harvard University, 1995).

Chapter Four

See Thomas R. Cox, Robert S. Maxwell, Phillip Drennon Thomas, and Joseph J. Malone, *Well-Wooded Land: Americans and Their Forests from Colonial Times to the Present* (Lincoln: University of Nebraska, 1985) and Michael Williams, *Americans and Their Forests: A Historical Geography* (Cambridge, UK: Cambridge University, 1989) for a history of the relationship between forests and American development. For a discussion of the development of the forest industry in the upper Midwest, see James Willard Hurst, *Law and Economic Growth: The Legal History of the Lumber Industry in Wisconsin, 1836–1915* (Madison: University of Wisconsin, 1984). William Cronon, *Nature's Metropolis: Chicago and the Great West* (New York: Norton, 1991) describes the relationship between lumbering and the development of Chicago.

Cronon also discusses the relationship between the prairies of the West and Chicago's growth. For a more detailed discussion of the prairie and prairie agriculture, see James Malin, *The Grasslands of North America: Prolegomena to Its History* (Gloucester, MA: Peter Smith, 1967); ———, *History and Ecology: Studies of the Grassland*, Robert Swierenga, ed. (Lincoln: University of Nebraska, 1984); and Richard Manning, *Grasslands: The History, Biology, Politics, and Promise of the American Prairie* (New York: Penguin, 1995). Stephan Pyne, *Fire in America: A Cultural History of Wildland and Rural Fire*, rev. ed. (Seattle: University of Washington, 1997) deals with fire and the ecology of the West.

For the history of prairie agriculture, see Allan Bogue, *From Prairie to Cornbelt: Farming on the Illinois and Iowa Prairies in the Nineteenth Century* (Chicago: University of Chicago, 1963). For American agriculture in general, see Paul Wallace Gates, *Agriculture and the Civil War* (New York: Alfred A. Knopf, 1965) and Peter McClland, *Sowing Modernity: America's First Agricultural Revolution* (Ithaca, NY: Cornell University, 1997). Jack Ralph Kloppenburg, *First the Seed: The Political Economy of Plant Biotechnology, 1492–2000* (Cambridge, UK: Cambridge University, 1988) discusses scientific agriculture and particularly the development of species of plants in America.

See Hildegard Binder Johnson, *Order upon the Land: The U.S. Rectangular Land Survey and the Upper Mississippi Country* (Oxford, UK: Oxford University, 1976) for a history of the land survey, and John Opie, *The Law of the Land: Two Hundred Years of American Farmland Policy* (Lincoln: University of Nebraska, 1987) for a history of government policy concerning the allocation of land.

Jimmy Staggs, *Prime Cut: Livestock Raising and Meatpacking in the United States, 1607–1983* (College Station: Texas A & M, 1986) looks at the meat industry and agriculture.

Chapter Five

For a discussion of early industrialization in New England, see Caroline Ware, *The Early New England Cotton Manufacture: A Study of Industrial Beginnings* (Boston: Houghton Mifflin, 1931); Theodore Steinberg, *Nature Incorporated: Industrialization and the Waters of New England* (Cambridge, UK: Cambridge University, 1991); Robert Dalzell, *Enterprising Elite: The Boston Associates and the World They Made* (Cambridge, MA: Harvard University, 1987); Jonathan Prude, *The Coming of Industrial Order: Town and Factory Life in Rural Massachusetts 1810–1860* (Cambridge, MA: Harvard University, 1983); and John Cumbler, *Reasonable Use: The People, the Environment and the State, New England 1790–1930* (Oxford, UK: Oxford University, 2001). For a discussion of industrialization in southeastern Pennsylvania, see Anthony Wallace, *Rockdale: The Growth of an American Village in the Early Industrial Revolution* (Boston: Alfred A. Knopf, 1972).

For a discussion of early urban life, see Carl Bridenbaugh, *Cities in Revolt: Urban Life in America, 1743–1776* (New York: Capricorn, 1964). See Sam Bass Warner, *The Private City: Three Stages of Growth* (Philadelphia: University of Pennsylvania, 1968) for a discussion of Philadelphia's development. For Pittsburgh's early industrial history, see S. J. Kleinberg, *The Shadow of the Mills: Working-Class Families in Pittsburgh 1870–1907* (Pittsburgh: University of Pittsburgh, 1989); Catherine Reiser, *Pittsburgh's Commercial Development, 1800–1850* (Harrisburg, PA: Pennsylvania Historical and Museum Commission, 1951); for Pittsburgh and Cincinnati, see Richard Wade, *The Urban Frontier: The Rise of Western Cities 1790–1830* (Cambridge, MA: Harvard University, 1959). Rochester's history is covered in David Johnson, *Shopkeeper's Millenium* (New York: Hill and Wang, 1978).

For a history of Chicago, see William Cronon, *Nature's Metropolis: Chicago and the Great West* (New York: Norton, 1991) and Harold Platt, *The Electric City: Energy and Growth of the Chicago Area, 1880–1930* (Chicago: University of Chicago, 1991). For a discussion of energy use in America, see David E. Nye, *Consuming Power: A Social History of American Energies* (Cambridge, MA: MIT, 1998). For a discussion of the anthracite coal industry, see Jack Brubaker, *Down the Susquehanna to the Chesapeake* (University Park, PA: Pennsylvania State University, 2002). For a discussion of urban development, see Sam Bass Warner, *Streetcar Suburbs: The Process of Growth in Boston* (Cambridge, MA: Harvard University, 1962). Also see Warner, *The Urban Wilderness: A History of the American City* (New York: Harper and Row, 1972). Kenneth T. Jackson, *Crabgrass Frontier: The Suburbanization of the United States* (Oxford, UK: Oxford University, 1985) is a detailed overview of the history of suburbanization in America.

For a discussion of the development of rural retreats, see Richard Judd, *Common Lands, Common People: The Origins of Conservation in Northern New England* (Cambridge,

MA: Harvard University, 1997). See Paul S. Sutter, *Driven Wild: How the Fight against Automobiles Launched the Modern Wilderness Movement* (Seattle: University of Washington, 2002) for a discussion of the emergence of motor tourism.

For a discussion of immigrants to America, see David Ward, *Cities and Immigrants: A Geography of Change in Nineteenth Century America* (Oxford, UK: Oxford University, 1971).

Ann Vileisis, *Discovering the Unknown Landscape: A History of America's Wetlands* (Washington, DC: Island, 1997) discusses the role of wetlands and farmers' efforts at draining them. For early canal building in America, see George Rogers Taylor, *The Transportation Revolution, 1815–1860* (New York: Harper and Row, 1951) and Peter Way, *Common Labour: Workers and the Digging of North American Canals: 1780–1860* (Cambridge, UK: Cambridge University, 1993).

See Margaret Beattie Bogue, *Fishing the Great Lakes: An Environmental History, 1783–1933* (Madison: University of Wisconsin, 2000) for a history of fishing on the Great Lakes. For a discussion of climate change, see Brian Fagan, *The Little Ice Age: How Climate Made History 1300–1850* (New York: Basic, 2000).

Thomas C. Cochran, *Frontiers of Change: Early Industrialism in America* (Oxford, UK: Oxford University, 1981); Douglas North, *The Economic Growth of the United States, 1790–1860* (New York: W. W. Norton, 1961); and Robert L. Heilbroner, *The Economic Transformation of America* (New York: Harcourt Brace Jovanovich, 1977) all provide an overview of American industrialization.

Chapter Six

For a discussion of urban infrastructure, see Joel Tarr and Gabriel Dupuy, eds., *Technology and the Rise of the Networked City in Europe and America* (Philadelphia: Temple University, 1988).

For a history of waterworks, see Sam Bass Warner, *The Private City: Philadelphia in Three Periods of Its Growth* (Philadephia: University of Pennsylvania, 1968); Fern Nesson, *Great Waters: A History of Boston's Water Supply* (Hanover, NH: University Press of New England, 1983); Sarah S. Elkind, *Bay Cities and Water Politics: The Battles for Resources in Boston and Oakland* (Lawrence: University of Kansas, 1998); Nelson M. Blake, *Water for the Cities: A History of the Urban Water Supply Problem in the United States* (Syracuse, NY: Syracuse University, 1956); Charles H. Weidner, *Water for a City: A History of New*

York City's Problem from the Beginning to the Delaware River System (New Brunswick, NJ: Rutgers University, 1974); and Stanley Schultz, *Constructing Urban Culture: American Cities and City Planning, 1800–1920* (Philadelphia: Temple University, 1989).

For a discussion of pollution and urban wastes, see Martin Melosi, *Effluent America: Cities, Industry, Energy, and the Environment* (Pittsburgh: University of Pittsburgh, 2001); *Garbage in the Cities: Refuse, Reform, and the Environment, 1880–1980* (College Station: Texas A & M University, 1981); *Pollution and Reform in American Cities* (Austin: University of Texas, 1980); Theodore Steinberg, *Nature Incorporated: Industrialization and the Waters of New England* (Cambridge, UK: Cambridge University, 1991); Joel Tarr, *The Ultimate Sink: Urban Pollution in Historical Perspective* (Akron, OH: University of Akron, 1996); and Robert Gottlieb, *Forcing the Spring: The Transformation of the American Environmental Movement* (Washington, DC: Island, 1993).

For discussion of urban industrial centers, see H. Wayne Morgan, *Industrial America: The Environment and Social Problems, 1865–1920* (Chicago: University of Chicago, 1974); John T. Cumbler, *Working Class Community in Industrial America* (Westport, CT: Greenwood, 1979); Theodore Steinberg, *Nature Incorporated: Industrialization and the Waters of New England* (Cambridge, UK: Cambridge University, 1991); and Susan Kleinberg, *The Shadow of the Mills: Working-Class Families in Pittsburgh, 1870–1907* (Pittsburgh: Pittsburgh University, 1989). For industrial development, see Robert F. Dalzell, *Enterprising Elite: The Boston Associates and the World They Made* (Cambridge, MA: Harvard University, 1987).

For a history of sewer works, see Joanne Abel Goodman, *Building New York's Sewers: Developing Methods of Urban Management* (West Lafayette, IN: Purdue University, 1997). See also Joel Tarr, *The Ultimate Sink: Urban Pollution in Historical Perspective* (Akron, OH: University of Akron, 1996).

For disease and the cities, see Charles Rosenberg, *The Cholera Years: The United States in 1832, 1849 and 1866* (Chicago: University of Chicago, 1962). For discussion of the germ theory and its impact on water analysis, see John T. Cumbler, *Reasonable Use: The People, the Environment and the State, 1790–1930* (Oxford, UK: Oxford University, 2001). For a discussion of public health and the urban environment, see Barbara Rosenkrantz, *Public Health and the State: Changing Views in Massachusetts* (Cambridge, MA: Harvard University, 1972); John Duffy, *A History of Public Health in New York City, 1866–1966* (New York: Russell Sage, 1974); *The Sanitarians* (Urbana: Illinois University, 1990); James H. Cassedy, *Charles V. Chapin and the Public Health Movement* (Cambridge, MA: Harvard University, 1962); and George Rosen, *A History of Public Health* (New York: M.D. Publications, 1958).

For a history of energy use, see Martin Melosi, *Coping with Abundance: Energy and Environment in Industrial America* (New York: Alfred A. Knopf, 1985); David Nye, *Consuming Power: A Social History of American Energies* (Cambridge, MA: MIT, 1998); Mark Rose, *Cities of Light and Heat: Domesticating Gas and Electricity in Urban America* (University Park: Pennsylvania State University, 1995); and Harold Platt, *The Electric City: Energy and the Growth of the Chicago Area, 1880–1930* (Chicago: University of Chicago, 1991). For a history of the petroleum industry and its impact on energy use, see J. Stanley Clark, *The Oil Century* (Norman: University of Oklahoma, 1958).

Chapter Seven

For information about sportsmen and their concern over conservation, see John Rieger, *American Sportsmen and the Origins of Conservation* (Winchester, NY: Winchester, 1975). For a more critical look at elite sportsmen in New England, see Richard Judd, *Common Lands, Common People: The Origins of Conservation in Northern New England* (Cambridge, MA: Harvard University, 1997); John Cumbler, *Reasonable Use: The People, the Environment and the State* (Oxford, UK: Oxford University, 2001); and Karl Jacoby, *Crimes against Nature: Squatters, Poachers, Thieves, and the Hidden History of American Conservation* (Berkeley: University of California, 2001).

For a discussion of environmental racism, see Andrew Hurley, *Environmental Inequalities: Race, Class and Industrial Pollution in Gary, Indiana, 1945–1980* (Chapel Hill: University of North Carolina, 1995); see also Charles Lee, *Toxic Wastes and Race in the United States* (New York: Committee for Racial Justice, United Church of Christ, 1987) and Kathlyn Gay, *Pollution and the Powerless: The Environmental Justice Movement* (New York: Watts, 1994).

For a discussion of American environmental activism, see Robert Gottlieb, *Forcing the Spring: The Transformation of the American Environmental Movement* (Washington, DC: Island, 1993); Marc Mowrey and Tim Redmond, *Not in Our Backyard: The People and Events That Shaped America's Modern Environmental Movement* (New York: Prentice Hall, 1993); Hal K. Rothman, *The Greening of a Nation: Environmentalism in the United States since 1945* (New York: Harcourt Brace, 1998); and Sam Hays, *Beauty, Health, and Permanence: Environmental Politics in the United States, 1955–1985* (Cambridge, UK: Cambridge University, 1987).

For information about suburbanization, see Kenneth Jackson's classic study *Crabgrass Frontier: The Suburbanization of the United States* (Oxford, UK: Oxford University, 1985). For a discussion of postwar suburbanization and its impact on the environmental movement, see Adam Rome, *The Bulldozer in the Countryside: Suburban Sprawl and the Rise of American Environmentalism* (Cambridge, UK: Cambridge University, 2001) and Clay McShane, *Down the Asphalt Path: The Automobile and the American City* (New York: Columbia University, 1994). See Paul S. Sutter, *Driven Wild: How the Fight against Automobiles Launched the Modern Wilderness Movement* (Seattle: University of Washington, 2002) for a discussion of the growing concern that automobiles would destroy America's wilderness.

For a description of American energy use, see David Nye, *Consuming Power: A Social History of American Energies* (Cambridge, MA: MIT, 1998) and Mark Rose, *Cities of Light and Heat: Domesticating Gas and Electricity in Urban America* (University Park: Pennsylvania State University, 1995).

For a discussion of modern urban water systems, see Nelson Blake, *Water for the Cities: A History of the Urban Water Supply Problem in the United States* (Syracuse, NY: Syracuse University, 1956). For a discussion of the 1993 crisis in Milwaukee, see Kate Foss-Mollan, *Hard Water: Politics and Water Supply in Milwaukee, 1870–1995* (West Lafayette, IN: Purdue University, 2001). For a discussion of pollution and waterworks, see William Clark and Walter Viessman, *Pollution and Water Supply* (New York: Harper and Row, 1995). See also Eugene Moehring, *Public Works and Urban History: Recent Trends and New Directions* (Chicago: Public Works Historical Society, 1982) and Anthony Penna, *Nature's Boundary: Historical and Modern Environmental Perspectives* (Armonk, NY: M. E. Sharpe, 1999).

For a discussion of toxic wastes and public policy, see Daniel Mazmanian and David Morell, *Beyond Superfailure: America's Toxic Policy for the 1990s* (Boulder: University of Colorado, 1992) and Barry Rabe, *Beyond NIMBY: Hazardous Waste Siting in Canada and the United States* (Washington, DC: Island, 1994). See Adam Markham, *A Brief History of Pollution* (New York: St. Martin's, 1994) for an overview on the issue of pollution.

For a discussion of Rachel Carson, see Sam Bass Warner, *The Providence of Reason* (Cambridge, MA: Harvard University, 1985).

For a discussion of Chicago's heat wave of 1995, see Eric Klinenberg, *Heat Wave: A Social Autopsy of Disaster in Chicago* (Chicago: University of Chicago, 2002). For a discussion of "natural disasters" and the human role in them, see Theodore Steinberg, *Acts of God: The Unnatural History of Natural Disaster in America* (Oxford, UK: Oxford University, 2000).

See William McGucken, *Lake Erie Rehabilitation: Controlling Cultural Eutrophication, 1960s–1990s* (Akron, OH: University of Akron, 2000) for a discussion of Lake Erie and perceptions of pollution.

For a discussion of the history of the cod fisheries and their destruction, see Mark Kurlansky, *Cod: A Biography of the Fish that Changed the World* (New York: Walker, 1997). For a history of the Great Lakes fisheries, see Margaret Beattie Bogue, *Fishing the Great Lakes: An Environmental History, 1783–1933* (Madison: University of Wisconsin, 2000) and Robert Doherty, *Disputed Waters: Native Americans and the Great Lakes Fishery* (Lexington: University of Kentucky, 1990).

For a discussion of Thoreau's ideas about nature and progress, see Lawrence Buell, *The Environmental Imagination: Thoreau, Natural Writing, and the Formation of American Culture* (Cambridge, MA: Harvard University, 1995); Donald Worster, *Nature's Economy: A History of Ecological Ideas* (Cambridge, UK: Cambridge University, 1977); Walter Harding, *The Days of Henry David Thoreau* (New York: Dover, 1965); Leo Stoller, *After Walden: Thoreau's Changing Views on Economic Man* (Stanford, CA: Stanford University, 1957); and James McIntosh, *Thoreau as Romantic Naturalist: His Shifting Stance toward Nature* (Ithaca, NY: Cornell University, 1974). For America's sense of wilderness, see Roderick Nash, *Wilderness and the American Mind*, rev. ed. (New Haven, CT: Yale University, 1973).

See also David Lowenthal, *George Perkins Marsh, Versatile Vermonter* (New York: Columbia University, 1958).

For a discussion of Cape Cod as a recreational center, see James O'Conner, *Becoming Cape Cod: Creating a Seaside Resort* (Somersworth: University of New Hampshire, 2002). For the history of the Cape Cod National Seashore, see Charles Foster, *The Cape Cod National Seashore: A Landmark Alliance* (Hanover, NH: New England Press, 1985). For the ecology of the northeast coast, see Michael Berrill and Deborah Berrill, *A Sierra Club Naturalist's Guide to the North Atlantic Coast, Cape Cod to Newfoundland* (San Francisco: Sierra Club, 1981).

For a general history of the American environment, see Theodore Steinberg, *Down To Earth: Nature's Role in American History* (Oxford: Oxford University, 2002) and John Opie, *Nature's Nation: An Environmental History of the United States* (Fort Worth, TX: Harcourt Brace, 1998).

DOCUMENTS

Chapter One

For a description of Native American life in presettlement times, see Daniel Gookin, *Historical Collections of the Indians of New England* (Boston, 1806; reprint, North Stanford, NH: Ayer, 2000). For a look at early Native American land use, see William Wood, an early colonist, *New England's Prospect* (London: T. Cotes, 1634; reprint, Boston: Boynton, 1898). See also Roger Williams, the founder of Rhode Island, *A Key into the Language of America* (Menston: Scolar Press, 1971).

Chapter Two

See William Bradford, governor of Plymouth Colony, *A History of Plimouth Plantation* (New York: C. Scriber's and Son, 1908) for the Pilgrim's view of early Anglo-American life. See also Francis Higginson, *New England's Plantation* (Salem, 1908; reprint, New York: DaCapo Press, 1970).

Chapter Three

For a description of life on the early New York frontier, see James Fenimore Cooper's, *The Pioneers*, 1823 (New York: Penguin, 1988). For a discussion of early colonial life, see Benjamin Franklin, *The Autobiography of Benjamin Franklin* (New York: Penquin, 1961). For a description of life among the Pennsylvania Dutch, see Theodore E. Schmauk, ed., *An Account of the Manners of the German Inhabitants of Pennsylvania* (1910). See J. Hector St. John de Crevecoeur, *Letters from an American Farmer* (1782) (New York: Dutton, 1925). For a description of early New England, see Henry David Thoreau, *A Week in the Maine Woods* and *A Week on the Concord and Merrimack* (New York: Penguin, 1988); Matthew Patten, *The Diary of Matthew Patten of Bedford, New Hampshire* (1790) (Concord, NH: 1903); Sylvester Judd, *History of Hadley: Including History of Hatfield, South Hadley, Amherst and Granby* (Springfield, MA: 1905); Timothy Dwight, *Travels through New England and New York*, 4 vols. (Cambridge, MA: Belknap Press, 1969); and Theodore Dwight, *Notes of a Northern Traveler* (Hartford, CT: 1831).

Chapter Four

For an early concern about wildlife destruction, see William T. Hornaday, *Our Vanishing Wild Life: Its Extermination and Preservation* (1913). For a description of life in rural Wisconsin in the mid-nineteenth century, see John Muir, *The Story of My Boyhood and Youth* (1913). See Aldo Leopold, *A Sand County Almanac* (1949) (New York: Ballantine, 1970) for a description of that same region a hundred years later.

John Locke, *Two Treatises on Government* (Cambridge: Cambridge University Press, 1990) provided the philosophical underpinnings of American understanding of property.

See Upton Sinclair, *The Jungle* (1906) (New York: Penguin, 1985) for a description of conditions in the turn-of-the-century slaughterhouses of Chicago.

For an understanding of American attitudes toward the frontier, see Frederick Jackson Turner, *The Frontier in America* (1920).

Chapter Five

For a critical look at American modernization and early factories, see Henry David Thoreau, *Walden: A Life in the Woods* (Boston, 1854) (New York: Harper and Row, 1965) and *A Week on the Concord and Merrimack Rivers* (1849) (Princeton: Princeton University Press, 1980). For a non-anti-modernist critical look at early American interaction with the environment, see George Perkins Marsh, *Man and Nature: Or Physical Geography as Modified by Human Action* (1864) (Cambridge, MA: Harvard University Press, 1989).

Chapter Six

For one of the first investigations of conditions of the poor and disease, see Lemuel Shattuck, *Report of the Sanitary Conditions of Massachusetts* (Boston, 1850). For other early sanitary reports, see Henry Bowditch, *Public Hygiene in America* (Boston, 1877); George Warring, *The Sanitary Drainage of Houses and Towns* (New York, 1876); George C. Whipple, *Vital Statistics* (New York, 1919); *State Sanitation*, 2 Vols. (1917); Murray Horwood, *Public Health Surveys* (New York, 1921). See Edwin Chadwick, *Report on the Sanitary Conditions of the Labouring Population of Great Britain* (London, 1842) (New York: Columbia University Press, 1985) for the seminal work on sanitation. See William Sedgwick, "Investigations of Epidemics of Typhoid Fever" in the Massachusetts State Board of Health, *24th Annual Report* (Boston, 1893) for the nation's first systematic study of typhoid transmission. See M. Ravenel, ed. *A Half Century of Public Health*

(1921) for a series of reports by public health experts. For conditions in urban areas, see Jane Addams, *Twenty Years at Hull House,* 1910 (New York: Signet Classics, 1990).

Chapter Seven

See Jane Addams, *Twenty Years at Hull House* (1910) (New York: Singet Classics, 1990) for a discussion of urban filth and the campaigns to clean it up. For a description of changing leisure activity in midwestern America, see Robert Lynd and Helen Lynd, *Middletown,* 1929 (New York: Harcourt, Brace, 1982). See Gifford Pinchot, *Breaking New Ground* (1947) for an autobiography of one of the nation's early conservationists. For a discussion of the land ethic and the wilderness idea, see Aldo Leopold, *A Sand County Almanac* (1949) (New York: Ballantine, 1970). See Rachel Carson, *Silent Spring* (1962) (Boston: Houghton Mifflin, 1987) for the path-breaking exposé of the pesticide industry in this country and the emergence of toxic poisons in our environment.

For discussion of early environmental concerns of Henry David Thoreau, see Henry David Thoreau, *A Week on the Concord and Merrimack* (1849) (New York: Penguin, 1988), and *Walden* (1854) (New York: Harper and Row, 1965). For Thoreau's view of Cape Cod, see *Cape Cod* (1865). For Marsh's view, see George Perkins Marsh, *Man and Nature: Or, Physical Geography as Modified by Human Action* (1864). For Rachel Carson's vision of the seashore, see Rachel Carson, *Under the Sea Wind* (1941) (Boston: Houghton Mifflin, 1979). For Aldo Leopold's concern about the land ethic, see Aldo Leopold, *Sand County Almanac* (1949) (New York: Ballantine, 1970).

INDEX

Note: italic page numbers indicate illustrations/photos

Abby, Edward, 209
An account of the manners of the German inhabitants of Pennsylvania, 255–257
Acorns, 10
Addams, Jane, 175, 217
Adirondack State Park, 211–212, 245
The Adirondacks: or Life in the Woods, 212
Adventure in the Wilderness: or Camp-life in the Adirondacks, 212
African Americans and environmental inequality, 190–192
Agricultural Adjustment Act, 185
Agricultural Adjustment Administration, 217
Agriculture
  and benefits of Seven Years War, 67–68
  Chicago and farm implements, 139
  European settlers' practices, 46–48
  and forest clearing, *34*, *62*
  hill farms, 62, *62*, 70, 86
  long fallow, 42–43, 226
  Mid-Atlantic coast conditions, 53–54
  Native American practices, 42–43, 46–47
  nonfood produce, 70
  surpluses, 58, 59, 60–61
  and trade, 58
  See also Farmers
AIDS, 150
Air conditioning, 200–201

Air pollution
  and air conditioning, 200–201
  and automobiles, 192–193, 199–200
  campaigns against, 172
  coal particulates, 170–171
  legislation, 192–193
  and lung damage, 172
  and 1920s business boom, 175–177
  Pittsburgh, 129–130, 171
  postwar ordinances against, 188
  public attention on, 187–188, 246
  in valley cities, 171
  wood smoke, 170, 238, 245
Algonquians, 12–13, 39, 217
  and French and Indian War, 67
  and settlers in Northwest Territory, 82
Alienation of land, 217
Allegany River, 5–6
  at Pittsburgh, 5–6, *6*, 128
Alleghenies, 217
Alluvial soil, 217
America To-Day: Observations and Reflections, 281–282
American larch, 239
Anadromous fish, 217
Antarctic Circumpolar Current, 1
Antarctica, 1
Anthracite coal, 171, 217–218
Antibody, 218
Appalachian Mountains, 1, 2, 218
  Blue Mountains, 52
  Green Mountains, 2
  and midwestern rivers, 5–6
  and New England rivers, 4–5
  New Jersey, 52

305

Appalachian Mountains *(cont.)*
  Pennsylvania, 52
  Skyline Drive, 186
  South Mountains, 52
  White Mountains, 4, *4*
Appleton, Nathan, 118, 158–159
Arable, defined, 218
Arcadian, defined, 218
Archer, William, 281
Armour & Co., 108
  refrigerated railroad cars, *107*
Astor, Henry, 113
Audubon Society, 181, 184, 245
Autobiography of Ma-Ka-Tai-Me-She-Kia-Kiak, or Black Hawk, 1833, 257–259
Automobiles, 131–132
  and air pollution, 192–193, 199–200
  and Cape Cod, 212–213
  and Ford, 131–133, 245
  and leisure activities, 182–183
  Model T, *132*
  and national parks, 186
  and suburbs, 197–198
Auto-touring, 182, 218
Avian flu, 150

Babbitt, 128, 132, 176–177
Babcock, Stephen, 108
Babcock Milk Tester, 108
Balancing doctrine, 218
Balloon-frame construction, 97–98, 100, 218–219, 244
Baltimore, Maryland
  canal into Pennsylvania, 88
  population growth (early 19th century), 93
Battle of Fallen Timbers, 82, 244
Battle of Tippecanoe, 82
Bean, L. L., 211
Beans, 10–11, 15, 243
  and European settlers, 31, 53
Beavers, 3, 13–14
  European trade in hides, 37–39, 243
  flourishing of in North America, 35
  Native Americans hunting, 36, 38–39
Beekman family, 79
Bell telephones, 141

Benzene, 219
Bering land bridge, 8
Bessemer furnace, 219
Biochemical Oxygen Demand (BOD), 219
Birdseye, Clarence, 203, 245
Black Ball Line, 111
Black Hawk, 93, 219
  excerpt from autobiography, 257–259
Black Political Caucus, 192
Black Year (1871), 96–97
Blickernsderfer tile ditching machine, 134, 219
Bluestem, 219
Blunderbuss, 219
Bog, 219
Boone and Crockett Club, 181, 182, 219, 245
Boston, Massachusetts
  Boston Harbor water pollution, 188, 215
  18th-century growth, 74
  Fenway Green Belt, 179
  garbage tonnage (1920), 174
  and immigrants, 116
  and international trade, 73–74
  and ocean wealth, 75
  population (1790), 151
  population growth (19th century), 111, 115
  segmentation of, 116–117
  storm sewers, 154, 155, 157–158
  street scene (colonial era), *65*
  as trade center, 125
  water system, 153–154, 158, 164
Boston Associates, 159, 219, 244
Boston Manufacturing Company, 118, 220
Bottle bill, 220, 246
Bradford, William, 28–31, 220
  and Squanto, 29
Brandywine River, 159
British thermal unit (BTU), 220
Broadleafs, 220
Brockton, Massachusetts, 167
Broomcorn, 70
Bryant, William Cullen, 259
Bubonic plague, 32

Buckeye trencher, 134, 220
Bucks County, Pennsylvania, 75–76
Buell, Dorothy, 189
Buffalo, New York, 88
  and Cleveland sewage, 164
  wheat shipping and elevators, 100–101
Bur oak, 220
Burroughs, Sara, 75
Burroughs family, 75
Bush, George H. W., 215
Bush, George W., 201

Cabot, John, 243
Cahokia, 11, 243
  decline of, 11–12
Canals, 71–73, 112–113, 244
  boom in, 88
  for coal transport, 114
  Connecticut River, 71–72
  Midwest, 100, 103
  See also Erie Canal; Illinois-Michigan Canal
Cape Cod, 3, 4, 6, 209–215, *213*, 220
  and automobiles, 212–213
  Doane Rock, 7, 8
  fisheries, 40
  and Pilgrims, 31
  and recreation, 210–211
  windmills, 55, *56*
Cape Cod National Seashore, 214, 246
Caribbean, and American produce, 64, 66, 244
Carson, Rachel, 187, 209, *210*, 212, 215, 220, 246
Catalytic converter, 220
Cattle
  and Chicago stockyards, 106, 108
  to market in spring vs. fall, 69–70
  urban cows, 174
Cellophane, 221
Center for Health, Environment, and Justice, 289
Center for Pollution Studies, 175, 221
Central business district, 221
Centralia, Pennsylvania, 194–195, *194*
CERCLA. *See* Comprehensive Emergency Response Compensation and Liability Act
CFCs. *See* Chlorofluorocarbons
Champlain, Samuel, 24, 221
Chapin, Joseph, 147, 158
Charles River, 118–119
Chemical cleansing, 167, 170, 221
Chesapeake Bay, 53
Chicago, Illinois, 92–93
  and balloon-frame construction, 97–98, 218–219
  catalogue retailing, 140
  and coal, 139
  environment, 133, 135, 140
  extreme heat wave (1995), 200–201, 247
  and farm implements, 139
  garbage collection, 175
  as grain market center, 103–104, 105–106
  and Illinois-Michigan Canal, 103, 136
  manufacturing, 139
  meat industry and refrigerated railroad cars, 107–108, *107*, 136
  and native poultry, 106–107
  People for Community Recovery, 192
  as pig and hog market, 106
  population (1890), 108
  population growth (1840–1900), 135–136, 139
  and railroads, 136–138, 139
  raising of grade level, 140
  sewage disposal, 162–164
  as shipping center, 103, 136
  skyline (1909), *104*
  stockyards and slaughterhouses, 106, 108, 136, 174, 245
  suburbs, 142–143, 145
  and telephone and telegraph, 140–141
  as trade center for products from three ecosystems, 135–136
Chicago Board of Trade, 221, 244
Chicago Gas and Light and Coke Company, 141
Chicago World's Fair, 221, 245
Chinch bugs, 222
Chippewa, 12
Chlorination and filtration, 222, 245
Chlorofluorocarbons, 200, 222

Cholera, 162, 164, 222
Church of England, 222
Cincinnati, Ohio, 127
   air pollution, 171, 172
   as pig and hog market, 105, *105*, 127–128
   and Pittsburgh sewage, 164
   soap and related manufacturing, 127–128
Cities
   animal populations, 144–145
   as chief markets for farm goods, 111
   density, 160, *173*
   and disease, 150–152, 164–165
   and domestic animal waste, 172–174
   and effects of industrialization, 179–180
   and environmental racial inequality, 190–192, 225
   and escapes to the countryside, 123–124, 180–181, 245
   and feed for horses, 108–109
   food preparation, 113
   and garbage, 174–175
   and indoor plumbing, 152, 156, 158, 166
   industrial districts and satellites, 143
   and manufacturing, 123, 125–126
   more than 50 percent of U.S. population in (1920), 180, 245
   night lighting, 131, 141–142
   overcrowding, 179
   pollution and disease, 152, 157, 164–165
   ports and international trade, 73–74, 244
   and skyscrapers, 143–144
   slaughterhouses, 174. *See also under* Chicago, Illinois
   storm sewers, 154–158, 239, 245
   and streetcars, 142
   and uneven distribution of wealth, 179–180
   urban ecosystems, 143–145
   waste disposal, 147–149, 151–152, 155
   water systems and pollution from sewage, 152–154, 160–167
   and working poor, 144
   See also Suburbs; *names of specific cities*
Citizens' Clearinghouse for Hazardous Waste, 289
Civilian Conservation Corps (CCC), 185, 222, 245
Civilian Works Administration, 185, 246
Clapp, Asahel and Sarah, 58–59
Clean Air Act, 192, 199, 222, 246
Clean Water Acts, 192, 193, 222
Clermont, 85, 86
Cleveland, Ohio
   Cuyahoga River fires, 188, 246
   oil refining, 131
   sewage disposal, 164
Clinton, Bill, 201
Clinton, Dewitt, 87
Clovis hunters, 8, 222–223, 243
Coal
   anthracite, 171, 217–218
   ash, 168
   bituminous, 171
   and Chicago, 139
   and electrical power, 198–199
   as fuel, 245
   mining, 114–115, 129
   oil, 131
   and pollution, 115, 129–130
   See also Coke
Cod, 40, 223
   fishing, *59*, 60, 74–75, 203, 246
   frozen, 203
Coke
   by-product ovens, 223
   for iron industry, 129
   as water pollutant, 169–170
Coked coal, 223
Coliform, 166–167, 223
Collapsible metal tube, 223
Comprehensive Emergency Response Compensation and Liability Act, 195–196. *See also* Superfund
Concord River, 119
Connecticut River and valley, 5, 58
   damming for textile mills, 120
   dams and canals, 71–72
   and Dutch, 37

Wright-Clapp family, 58–59
Connecticut, 50
  claim on Western Reserve, 81
Constructed environment, xii
Coolidge, Calvin, 185
Cooper, Peter, 113
Cooper, William, 75–76
Cooper family, 75–76, 111
Corn, 10–11, 15, 243
  and European settlers, 31, 53, 54
  as livestock feed, 104–105, 106
  in Midwest, 98–99
Country stores
  and externally supplied necessities, 59–60, 65–66
  and surplus goods from local farms, 58, 59, 64–65, 68–69, 73–74
  Tappen and Fowle, 69
Countryside
  and auto-touring, 182, 218
  degradation of, 183–184
  and hunting and fishing, 180–182, *181*
  and New Deal programs, 185–186
  as place of escape, 123–124, 210–212, 245
Cowpox, 26
Crowd diseases, 151, 223
Cryptosporidium parvum, 193, 223, 247
Crystalline piedmont uplands, 223
Culm, 223
Cultivators, 106
Cuyahoga River, 188, 246

Dams, 71–73, 244
  to bypass river falls, 71–73
  heights of, 224
  for mills, 55, 57, 73, 159
  See also Beavers
Darby, George, 129
Dart, Joseph, 100–101
DDT (dichlordiphenyltrichloroethane), 186–187, 224, 246
Debt, 64, 65–66, 68
Deere, John, 92, 244
Delaware River and valley, 53
  locks and dams, 72
  and mills, 159
  and wastewater, 153
Delaware River Basin, 184, 224, 245
Dermer, Thomas, 24
Detroit, Michigan, 131, 133
  Belair Park, 179
Diarrhea, 162, 164
Dig, Will H., 184
Dilution, 224
Diphtheria, 162, 224
Diseases, 24–26
  and cities, 150–151, 164–165
  effect on native populations, 26–28, 39, 151, 243
  and microbes, 150, 152, 157, 161, 165
Dissolved oxygen, 168–169, 224
Doane Rock, 7, 8
Domesticated animals, 243
  cattle to market in spring vs. fall, 69–70
  effect on landscape, 41, 42
  livestock allowed to wander, 43–44, 62
  as meat, 49, 54
  See also Cattle, Pigs and hogs
Donora, Pennsylvania, smog incident, 187–188, 246
Dowry, 224
Drainage districts, 224
Drainage tiles, 134, 225
Dukakis, Michael, 215
Dupont family and company, 120, 176
Dutch
  settlements, 51, 243
  traders, 37
  and wampum, 38, *38*
Dwight, Timothy, 117
Dynamo, 225
Dysentery, 162, 164, 225

Earth Day, 190, *191*, 192, 225, 246, 286–288
The Edge of the Sea, 212
Edison, Thomas, 141
Edison Electric Company, 171
Edwards, Jonathan, 253
Electricity
  and lighting, 141–142
  and streetcars, 142
Elutriate, 225

Emerson, Ralph Waldo, 259
Endangered Species Act, 246
England, reduction of forests, 32–33
English hay, 225
Environmental Defense Fund, 246
Environmental history, xi
  and interaction of humans and physical world, xii, xiii
Environmental justice movement, 225, 247. *See also* Love Canal, New York
Environmental movement, 190
  influence of civil rights movement, 192
  and nuclear energy, 198
  and recycling, 193–194
  See also Earth Day
Environmental Pesticide Control Act, 246
Environmental Protection Agency, 192, 225
Environmental racism, 190–192, 225
Erie Canal, 2, 87, 225, 244
  effect on shipping and population, 88, 91
  funding and construction of, 87–88, 111
  and Rochester, 126
  and salmon industry, 88–89
  and wheat shipping, 101
European settlers
  agricultural practices, 46–48
  and beaver trade, 37–38
  desire for furs and land, 77
  division of farms, 61
  division of labor, 47
  early trade with Native Americans, 32
  and externally supplied necessities, 58–60
  and fallow fields, 44–45
  and farming, 28, 40–42
  and forests, xi, 33–34
  home building, 63
  importation of European flora, 41–42
  and land tenure, 40–41, 43
  need for additional farm help and growth of communities, 45–46, 54
  ongoing need for more land, 61–62, 68, 69
  and private property, 43, 47
  reasons for thriving of, 60–61
  setting up households, 58–59, 63, 69
  transformation of landscape, 109–110
  and view of wilderness, 42–43
  westward movement, 78–80
Europeans
  as market for American produce, 63–64
  population and disease, 151
  as settled people, 31–32
  sixteenth-century, 23–24
  and sugar, 64
Eutrophication, 226

Fall River, Massachusetts, 120, 125
Farmers
  cities as chief markets for their goods, 111
  Currier and Ives scene, *89*
  decisions based on market demands, 69–70, 79, 98, 244
  and division of farms, 61
  effects of national transportation network, 182
  and Erie Canal, 88
  farm abandonment, *122*
  and land title in Northwest Territory, 81–82
  need for Great Lakes lumber on prairies, 95, 96, 98
  in Ohio, Indiana, and western New York, 90
  ongoing need for more land, 61–62, 68, 69
  Pilgrims as, 28, 40–42
  and railroads, 138
  in Rochester area, 126–127
  shift to truck farming (Northeast), 121–123, 244
  tools, 90–91
  transporting goods to market, 84–85
  and urban seasonal visitors, 124
  and water pollution, 161, 162
  See also Agriculture
Fauna, 226
Federal Aid Road Act of 1916, 182, 245
Fell, Jesse, 114

Fertilizers, 186–187
Fescue, 226
The Fight for Conservation, 282–284
Finger Lakes, 2
First Encounter Beach, 8, 23
Fish and game commissions, 226
Fish and Wildlife Service, 187
Fishing
  Cape Cod, 40
  depletion of Northeastern and Great Lakes waters, 203–206, *204*
  by European settlers (inland), 49–50, 54
  and frozen fish, 203
  by Native Americans, 12, 14–15, 49
  off New England, *59*, 60, 74–75, 202–205, *204*, 243, 246
  off New England by Europeans before settlement, 40, 74
  and toxic materials in fish, 206
  See also Cod; Hunting and fishing (as sport); Salmon; Whaling
Fishing banks, 226
Five Nations, 39
Flax, 70, 226
Flora, 226
Food preparation, urban, 113
Ford, Henry, 131–133, 245
Forest and Stream, 181, 226
Forest or long fallow agriculture, 42–43, 226
Forests
  abundance and consumption of in eastern U.S., 93–95, 97
  animals and forest succession, 13–14
  in Appalachians, 5
  and beavers, 13–14
  Cape Cod, 7
  east of the Mississippi, 8
  effects of clearing, 63
  European cutting of in North America, 33–34, *34*, 53, *62*
  and foodstuffs, 10
  Native American and European views of, xi
  Native American clearing of for horticulture, 16
  north of Ohio River, 10
  of Northeast, 12–13
  northern Great Lakes region, 95–96
  post-lumbering fires in northern Great Lakes region (Black Year, 1871), 96–97
  presettlement mixed, *80*
  reduction in England, 32–33
  and scarcity of cordwood, 113–114
  white pine, *94*, 96–97, *97*, 244
  wood as building material, 48
Fort Orange, 51
Fourdrinier, Henry and Sealy, 226–227
Fox River, Wisconsin, *148*
Fox tribe, 93, 227
Franklin, Benjamin, 66
  wood-burning stove, 114
Freeholder, 227
French
  and Fort Duquesne, 138
  Huguenot immigrants, 115
  traders, 37
French and Indian War, 61
Friends of the Earth, 246
Fuller, George, 166–167
Fulling mill, 227
Fulton, Robert, 85
Fur trade, 37–39, 77, 78, 243
Futures Market, 227

Gangrene, 227
Gary, Indiana, 139
Gasoline. *See* Oil (petroleum)
Genesee River and valley, 2, 79, 126
  and salmon, 89
Germ theory, 154, 157, 161, 165, 227, 245. *See also* Microbes
German immigrants, 115, 116
Gibbs, Lois, 195, 246, 289
Gill nets, 227
Gillette, Genevieve, 190
Girdling, 227
Glacial till, 228
Glaciers
  advent of grasslands after retreat of, 13
  and Midwest landscape, 2, 101
  and New England landscape, 2–4, 6–8
Glass industry, 129
Global warming, 200, 201–202, 247
Glue, 113

Goodspeed, Edgar Johnson, 276
Grains, 41–42, 48, 54
   and blast, 67, 244
   and Chicago, 103–104, 105–106
   and gristmills, 55–57
   in Midwest, 91, 93, 98–100
   shipments to England, 101
   trade, 58
Grama grass, 228
Grange and grangers, 138, 228
Gravity-fed system, 228
Great Depression, 185–186
Great Lakes and region, 2, 20–21
   and beaver, 39
   fish depletion and restoration, 206
   and lumber for prairie buildings, 95
   overfishing, 205–206, 246
   rainfall, 84
   regional protection agency, 184
   and Save the Dunes Council, 189–190
Green Mountains, 2
Greenhouse gases, 228
Grinnell, George Bird, 181, 228
Grist stone or grindstone, 228
Gulf Stream, 1–2, 228

Harmar, Josiah, 82
Headboard, 228
Headley, Joel, 212
Hemlock, 229
Hemp, 70
Henle, Jacob, 165
Herbicides, 186–187, 246
Hessian fly, 67, 229, 244
Hillis, Anne, 289
History
   as historians' creation, xiii–xiv
   and idealization of Native Americans, 21
   and landscape, xi, xiii
   and the present, xii
History of the great fires in Chicago and the West, 272–276
Holmes, Oliver Wendell, 162–164
Holocene, 243
Holyoke, Massachusetts, 120–121, 148–149, *149*
   Canoe Club, *176*

Hoover, Herbert, 185
Hopewell culture, *9*, 10, 243
   end of, 11
   staple foods, 10
Hornaday, William Temple, 284
Horses, urban
   carcasses, 174
   feed for, 108–109
   wastes, 172–174
Horticulture, 229
   and forest clearing, 16
   and hunter-gatherers, 11, 12, 15, 17–18, 19
   Mesoamerica, 10–11
   Midwest, 15–16
   Mississippi River valley, 11
   Native American practices, 16–17
   Northeast, 15–16
"How the First Earth Day Came About," 286–288
Howell, William Dean, 124
Huckins, Olga Owens, 187
Hudson, Henry, 51
Hudson River and valley
   and Dutch, 37, 51, *52*, 243
   and effects of international trade, 66
   and Erie Canal, 2
   goods produced, 111
   large estates, 61
   and slaves, 46
Hull House, 175
Humus, 229
Hunter-gatherers, 14–15, 229
   and horticulture, 11, 12, 15, 17–18, 19
   and land rights, 20
   New England, 10, 12
   northern areas, 19
   transformation of landscape, 13
Hunting and fishing (as sport), 180–182, *181*, 204
   Great Lakes, 206
   and Izaak Walton League, 184–185
Hunting and gathering, defined, 229
Huron, 12, 19
Huron-Petons, 39

Illinois
   and Black Hawk's war, 93

population growth and settlement
   clusters (1810–1840), 86–87
 rich soil, 91
Illinois tribe, 39
Illinois-Michigan Canal, 103, 136, 244
Immigration, 115–116, 244
Indentured servants, 46, 61, 229
Indian corn, 229
Indian grass, 229
Indiana, 244
 population growth and settlement
   clusters (1810–1840), 86–87
Indiana Dunes National Lakeshore,
   189–190
Industrialization
 effects of, 179–180
 See also Air pollution; Glass industry;
   Mill towns; Steel industry; Textile
   industry; Water pollution
Infant mortality, 123
 and polluted water, 164–165
Interceptor lines, 229
Irish
 and Great Hunger, 116
 immigrants, 115–116
Iron, 68
Iroquois
 Confederacy, 229–230
 westward movement, 39, 243
Izaak Walton League, 184–185, 230, 245

Jackson, Patrick Tracy, 118
Jacob's Creek and Mill, 75, 76
Jarvis, William, 70
Jay's Treaty, 69
Jefferson, Thomas
 and influence of Locke, 84
 and Louisiana Purchase, 85
 and Northwest Land Ordinance, 80–81
Jefferson grid, 80–81
Jervis, John, 153, 154
Johnston, John, 134
Judd, Sylvester Jr., 69–70, 84–85
Judd, Sylvester Sr., 58, 69, 84–85

Kennedy, John F., 214, 287
Kennedy, Robert, 287
Kentucky

falls of the Ohio (Louisville), 85
 land title disputes, 81, 82
Kettle holes, 230
Kickapoo, 93, 230
 westward movement, 39
Knox, Henry, 78, 79
Koch, Robert, 165
Kyoto Protocol, 247

L. L. Bean Company, 211, 212
Ladies' Health Protective Association,
   175
Lake Champlain, 79
Lake Erie, 1, 79
 as dying lake, 188–189, 245
 as eastern edge of Northwest
   Territory, 80
 and Erie Canal, 2
 and Ohio settlement, 87
 and petroleum pollution, 131
Lake Huron, 79
Lake Michigan, 2, *20*, 92–93, 96, 133
Lake Ontario, 6, 72
 salmon, 88–90, 205, 245
Lake Winnapesaukee, 4
Land use
 and Jefferson grid, 80–81
 title battles, 81
 and would-be gentry, 77–79
Landscape
 effect of domesticated animals on, 41,
   42
 and history, xi, xiii
 and human modification, xiii, 13, 22
 transformation by European settlers,
   109–110
Lawrence, Massachusetts, 119, 159
 experimental water treatment station,
   165–166, 175, 230
 typhoid epidemic, 165
Leather, 70
Leisure activities
 and automobiles, 182–183
 and conservation actions, 180–182,
   189
 and middle class, 182, 211, 212, 245
 and the wealthy, 180–182, 210–211
 and working class, 182, 189

Leopold, Aldo, 103, 183, 186, 209, 215, 246
Levi, William, 147, 158
Levitt, William, 196, 246
Levittown, New York, *197*, 246
Levittowns, 230
Lewis, Sinclair, 128, 132, 176–177
Lincoln, Todd and Nancy, 81
Lister, Joseph, 165
Loam, 230
Locke, John, 82–84, *83*
Locks, 71–72, 244
Lodge, Henry Cabot, 181
Long-haul versus short-haul railroad rates, 230
Louisville, Kentucky, 171
Love Canal, New York, 195–196, 230–231, 246, 288–290
Lowell, Francis Cabot, 118, 124, 158–159
Lowell, Sylvanus, 57
Lowell, Massachusetts, 119, 125, 159
  storm sewers, 160
  typhoid epidemic, 165
Lumber
  as building material, 48
  as commodity, 33–35
  commercial centers, 35
  industry and abundance of eastern forests, 93–95, 97
  industry in northern Great Lakes region, 96–97, *97*
  for prairie buildings, 95
  and sawmills, 57–58
  sold by farmers, 62
  trade, 58
Lye, 231
Lyman, John, 59
Lyman, Theodore, 117, 121, 158, 179
Lynd, Robert and Helen, 182

Machinists, 127
Maize, 10, 231, 243. *See also* Corn
Malaria, 150, 231
Man and Nature, 245, 270–271
Manchester, New Hampshire, 119
Manly arts, 231
Manufacturing
  Chicago, 139–140
  effect on urban environment, 123
  iron, 128–129
  and location of cities, 125–126
  and machinists, 124–125
  textile industry, 117–121
  town near Bellows Falls, Vermont, *125*
Map of region, *xv*
Marsh, George Perkins, 208–209, 215
  Man and Nature, 245, 270–271
Marsh, 231
Marshall, Robert, 186
Massachusetts Bay Colony, 50
Massachusetts Board of Health, 149, 231
McCormick, Cyrus, 99
  Chicago factory, 139
McCormick self-rake reaper, 99–100, *99*, 106
Meadow, 231
Measles, 231
Meat, 48–49, 54
  changing tastes in, 136
  and Chicago, 106, 107–108, *107*, 136
  and Cincinnati, 105, *105*, 127–128
  heavy consumption of, 174
Mellon family, 188
Melville, Herman, 95
Meningitis, 162
Mennonites, 231
Menominee, 12, 19
Merino sheep, 232
Merrifield v. City of Worchester, 169
Merrimack River and valley, 4–5, 58
  locks and dams, 72
  textile mills, 119, *119*, 159
Mesoamericans, 10–11, 232
  and horticultural revolution, 10–11
Miami tribe, 39
Miasma, 232
Michigan
  forests, 96
  wheat farming, 100
Microbes, 150, 232
  and animals, 150
  coliform bacteria as fecal indicator, 166–167, 223
  and diseases, 150, 152, 157, 161,

164–165, 175
  moving to other hosts, 150
  and rivers, 166
  See also Germ theory
Mid-Atlantic coast
  agricultural conditions, 53–54
  settlement of, 51–54
Middlesex Corporation, 160
Middletown, 182
Midwest
  central, 1
  coal mining, 139
  and corn, beans, and squash, 15–16, 243
  drained wetlands, 133–135
  grain and pork sales to East and South, 93
  grain production, 91, 93, 98–100, 108
  growth (1810–1840), 112
  mound builders, 11
  and Northwest Land Ordinance, 80–81, 244
  rivers, 5–6
  shift in farming focus, 108
Mill acts, 232, 244
Mill towns, 159, 244
  and increased residential density, 160
  sewer systems, 160–161
  and waste disposal, 159–160
  waste disposal into streams, 160–162
  See also Lawrence, Massachusetts; Lowell, Massachusetts; Paterson, New Jersey
Millpond, 232
Mills, Hiram, 165–166
Mills
  gristmills, 55–57
  large mills and pollution, 148–149
  textile, 117–121, 147
  wind- and water-powered, 55
  See also Sawmills
Milwaukee, 105
  Cryptosporidium outbreak, 193, 247
Mississippi River and valley
  Black Hawk's account of Indian life, 257–259
  and Chicago sewage, 162–164, 166
  and corn, beans, and squash, 11, 243

and Illinois settlement, 87
and mound builders, 11
as western border of Northwest Territory, 80
Missouri v. Illinois and the Sanitary District of Chicago, 162–164, 166, 232–233
Mobility of community, 233
Mohawk River and valley, 2
  locks and dams, 72
  and settlers, 79
Molasses, 68
Monoculture, 233
Monongahela River, 5–6
  and mine waste, 189
  at Pittsburgh, 5–6, 6, 128
Montgomery Ward, 140
Montreal Protocol, 200, 233, 247
Moody, Paul, 118, 124, 158–159
Morgan, J. Pierpont, 181
Morton, Thomas, 249
Motor tourists, 233
Mouldboard plow, 90, 233
Mound builders, 11
  decline of in Midwest, 11–12
Mound planting, 233
Muir, Daniel, 101–103, 265
Muir, David, 102, 265
Muir, John, 101, 102, 182
  on settlers' transformation of landscape, 110
  The Story of My Boyhood and Youth, 265–269
Mule, 174
Municipal water systems, 152–154, 233, 244
Murray, William, 212
Muskie, Edmund, 190, 192, 286
My Diary North and South, 269

Naphtha, 234
Napoleonic Wars, 234
Narragansetts, 18–19
Nashoba Thrust Belt, 5
National Ambient Air Quality Standards, 192
National Environmental Protection Act (NEPA), 192, 246

National Homestead Act of 1862, 84
National Park Service, 184, 214, 234, 245
National Patrons of Husbandry, 138
National transportation network, 182, 234
National Wildlife Federation, 246
Native Americans
  agricultural practices, 46–47
  and arrival of Pilgrims, 24, 28, 243
  and beavers, 35, 36, 38–39
  boat-building, 14
  conflict among nations, 39
  division of labor, 46–47
  early encounters with Europeans, 8, 24, 243
  and early landscape, flora, and fauna, 8–10
  and European diseases, 26–28, 39, 151, 243
  and European idea of private property, 43
  fishing, 12, 14–15, 206
  and forests, xi
  and Great Lakes, 21, 206
  horticulture, 10–11, 15–17
  and husbanding of resources, 21–22, 109
  and land rights, 18–20
  landscape knowledge and survival, 14, 109
  and long fallow agriculture, 42–43, 226
  migrations into North America, 8
  and settling of Northwest Territory, 82
  and smallpox, 24–27, 243
  and trade with Europeans, 32, 35, 37–38, 77, 78
  transformation of landscape, 13, 16, 22, 109
  and wampum, 38–39, 38, 240
  westward movement, 39, 77
  See also Algonquians; Cahokia; Chippewa; Clovis hunters; Five Nations; Fox tribe; Hopewell culture; Hunter-gatherers; Huron; Huron-Petons; Illinois tribe; Iroquois; Kickapoo; Menominee; Miami tribe; Mound builders; Narragansetts; Paleo-Indians; Potawatomis; Sac; Sauk; Wampanoags
Natural gas, 133, 246
Nature, 259–261
Nature as idea, xii–xiii
Nature Conservancy, 246
Nelson, Gaylord, 190
  "How the First Earth Day Came About," 286–288
New Amsterdam, 51, 52
New Deal programs, 185–186
New England, 1
  difficult weather of 19th and early 20th centuries, 66, 86
  effects of glaciation, 2–4, 6–8
  effects of international trade, 66
  fishing off coast of, 59, 60, 74–75, 202–205, 204, 243, 246
  goods produced, 111
  and manufacturing, 117–121
  miniature ice age (17th–19th centuries), 66
  precolonial fishing, 40, 74
  rainfall, 4–5
  rivers, 4–5
  small farms, 61
  soil, 3
  stone walls, 30, 41
New English Canaan; or, New Canaan, 249–253
New Jersey, 52–54, 243
  and effects of international trade, 66
New Orleans, Louisiana, 85, 87
New York, 51, 54
  and Erie Canal, 88
New York City
  and Arthur Kill blob, 188
  Central Park, 179, 245
  18th-century growth, 74
  and Erie Canal, 88
  garbage, 174–175
  and immigrants, 116, 175
  and international trade, 73–74
  as national port of entry, 113
  need for external food sources, 113

overcrowding, 179
population (1790), 151
population growth (19th century), 93, 111, 115
port, *112*
segmentation of, 116–117
as shipping center, 111–112, 113
as trade center, 125
water system, 153, 158, 164
New York City Women's Municipal League, 175, 234
Newport, Rhode Island
 18th-century growth, 74
 and international trade, 73–74
Night lighting
 in Chicago, 141–142
 electric, 141–142
 fuels, 131, 141
 in Pittsburgh, 131
North America
 climatic shifts, 1–2
 distinct ecosystem, 23
North Equatorial Current, 1–2
Northeast
 and corn, beans, and squash, 15–16, 243
 demand for wheat, 93
 difficult weather of 19th and early 20th centuries, 66–67
 forests, 12–13
 grains and blast, 67
 port cities and international trade, 73–74, 244
 See also New England
Northwest Land Ordinance, 80–81, 244
Northwest Ordinance, 82, 244
Nuclear energy, 198

Oberholtzer, Ernest, 186
Ohio
 and Erie Canal, 88
 population growth and settlement clusters (1810–1840), 86–87
 use of river for transport, 84
 See also Western Reserve
Ohio River and valley, 1, 127
 and Cincinnati, 127
 forested region north of, 10

formation point, 6, *6*
 and Hopewell culture, 10
 and Indiana settlement, 87
 and Northwest Land Ordinance, 80–81, 244
 and Ohio settlement, 87
 and Pittsburgh, 5–6, *6*, 128
 regional protection agency, 184, 234
 river as transportation artery, 85
 and steamboats, 85–86, 244
 and urban sewage, 164
Ohio River Basin Board of Public Health, 184, 234, 245
Oil (petroleum), 245
 and automobiles, 131–132, 199
 industry, 131
 and pollution, 131
 wells (Titusville, Pennsylvania), *130*, 131
Oil Pollution Act of 1924, 184, 234, 245
Oil Pollution Control Act of 1924, 175, 234
Olcott, James, 167, 179
Olmsted, Frederick Law, 179, 221, 245
Orchards, 47–48, 54
Organization of the Petroleum Exporting Countries (OPEC), 199–200, 246
Oswego River, 89
Otis, Elisha Graves, *144*
Otis Elevator Company, 143, *144*
Our Vanishing Wild Life, 284–286
Outdoor America, 184–185
Outflow pipes, 234
Overflow waterwheel, 234–235
Ozone, 235

Paleo-Indians, 8, 235, 243
Paleolithic, defined, 235
Parks, 179, 182
 national parks and roads, 186
 and New Deal, 185–186
 state, 183, 211–212, 245
Passaic River, 159, 162
Pasteur, Louis, 165
Paterson, New Jersey, 120, 159
Pathogen, 235

Patten, Mathew, 58
Pawtucket Falls, 4, 119, 159
PCB (polychlorinated biphenyls), 235
Penicillin, 188
Penn, William, 53, 235
Pennsylvania, 52–54, 243
   coal mining, 114–115, 129
   and effects of international trade, 66
   goods produced, 111
   settlement and landscape changes (New Hope, Pennsylvania, area), 75–76
   small farms, 61
Peshtigo forest fire, 272–276
Pesticides, 186–187, 246
Phelps, Earle, 175
Phenols, 235
Philadelphia, Pennsylvania
   18th-century growth, 74
   garbage tonnage (1920), 174
   and immigrants, 116
   and international trade, 73–74
   population (1790), 151
   population growth (19th century), 93, 111, 115
   segmentation of, 116–117
   as trade center, 125
   and typhoid, 153
   water system, 152–153, 158, 164
   yellow fever epidemic (1793), 151–152, 153, 244
Pierson v. Post, 84
Pig iron, 235
Pigeon years, 235
Pigeons, 49, 107, 235, 245
Pigs and hogs, 62, 90
   and Chicago, 106
   and Cincinnati, 105, *105*, 127–128
   and corn feed, 104–105, 106
Pilgrims, 24, *25*, 27
   decision to leave for America, 28–31
   early years, 31–32
   encountering Native Americans, 24, 28, 243
   as farmers, 28, 40–42
   and forests, 32–34
   in Holland, 28
   and land tenure, 40–41
   number on Mayflower, 28
   and Squanto, 31
   and trade with England, 32–34
   and trade with Native Americans, 32, 35
   use of native and European methods, 32
Pinchot, Gifford, 282
Pittsburgh, Pennsylvania
   air pollution, 129–130, 171, 172, 188
   and coal, 129–130
   glass industry, 129
   iron industry, 128–129, 130
   and rivers, 5–6, *6*, 128
   sewage disposal, 164
   steel industry, 130–131, *163*
   traveler's description of, 269
Plant succession, 13, 235–236
Pleistocene, 236, 243
Plows
   cast-iron, 90–91, 100, 112
   mouldboard, 90, 233
   and prairie land, 91–92, 93
   pulled by animals, 106
   steel, 92, 93, 100, 244
Plymouth Colony, 243. *See also* Pilgrims
Plymouth Harbor, 31
Podzolic soil, 236
Polio, 188, 236
Pollution. *See* Air pollution; Waste disposal; Water pollution
Populists, 138, 236
Potash, pearl ash, 62–63, 236
Potatos
   adoption by Europe, 115
   development by Andean horticulturalists, 115
   and Great Hunger in Ireland, 115–116, 244
Potawatomis, 19, 92–93, 236
   westward movement, 39
Pound sterling, 236
Power looms, 237
Prairie chickens, 106–107
Prairie Farmer, 98
Prairies
   ecology, 135

farmers' use of clover crops as green fertilizer, 104
and feed for urban horses, 108–109
and fires, 103
forest enclaves, 95, 100
nature of soil, 91
and plows, 91–92
settlers' need for Great Lakes lumber, 95, 96, 98, 100
as superior wheat area, 108
Predators, 44
Presidential Range, *4*
Privies, 147–148, 237
connected to storm sewers, 155
urban, 153, 155
and wastewater disposal, 153
Proctor and Gamble Company, 128
Property
and European settlers, 43, 47
John Locke on, 82–84
Native American reaction to idea of, 43
Providence, Rhode Island, 120
Public domain, 57
Public Health Service, 175, 237
Public Works Administration, 185
Puddling, 237
Puritans
New Jersey, 53
southern New England, 50–51
See also Massachusetts Bay Colony; Pilgrims
Purulent, 237

Quakers, 53
Quincy, Josiah Jr., 154

Rabbits, 95, 103
Railroads, 112–113
and Chicago, 136–138, 139
and farmers, 138
rates, 138, 230
refrigerated cars, 107–108, *107*, 136, 245
Reasonable use, 169, 237
Recycling, 193–194
Reed, Charles, 172
Report of the Sanitary Commission, 165
Resource Conservation and Recovery Act of 1976, 195, 237, 246
Revolutionary War, 68, 74
Reynolds, Malvina, 196
Rhode Island, 50
Ricardo, David, 101, 237
Richards, Helen Swallow, 165
Riparian law, 169
Rivers
and industrial pollution, 167–170, 175
and locks, canals, and dams, 71–73, 244
and microbes, 166
Midwest, 5–6
and mound builders, 11
New England, 4–5
pollution from sewage, 160–167, 175
as transportation arteries, 70–74
Rivers and Harbors Act of 1899, 184, 237, 245
Rochester, New York, 89, 126–127
Rockefeller, John D., 131
Roosevelt, Franklin D., 185
Roosevelt, Theodore, 181, 237
Root, Elihu, 181
Ruether, Walter, 189
Rum, 68
Rush, Benjamin, 255
Russell, William Howard, 269
Rye, 41, 67

Sac, 93
Salmon, 49, 205, 243, 245
and dams, 73
and Erie Canal, 88–90
Salmon River, 89
Saltonstall, Leverett, 214
Sandburg, Carl, 106, 237
"The Sanitary Drainage of Houses and Towns," 276–278
SARS. *See* Severe Acute Respiratory Syndrome
Sauk, 237
Save the Dunes Council, 189–190
Sawmills, 57–58
Schuylkill River and valley, 53
and Philadelphia water system, 152
and wastewater, 153

Schuylkill River and valley *(cont.)*
  locks and dams, 72
Scots immigrants, 115
Sears, Roebuck, and Company, 140
  prefabricated homes, 143
Sedgwick, Judge, 57
Sedgwick, William, 165–166
Separatists, 237. *See also* Pilgrims
Settlement Houses, 237
Seven Years War, 67–68. *See also* French and Indian War
Severe Acute Respiratory Syndrome (SARS), 150
Shad, 49, 73, 243
Shattuck, Lemuel, 165
Shay's Rebellion, 68–69, 238
Sheep, 70
Shepard, Levi, 64–66, 69
Shipbuilding and forests, 94–95, 243
Short-fiber cotton, 238
Sierra Club, 101, 184, 266
"The Significance of the Frontier in American History," 278–281
Silent Spring, 187, 246
"Sinners in the Hands of an Angry God," 253–255
Skyscrapers, 143–144
Slater, Samuel, 117, 238
Slaves
  and sugar, 64
  Hudson River valley, 46
Smallpox, 24–27, 150, 238, 243
  and Squanto, 29
Smith, Abijah, 114
Smith, John, 24, 238
Smoke abatement ordinances, 238, 245
Soil, xi
  New England, 3
Soil Conservation Service, 246
de Soto, Hernando, 238
South Shore, 2–3
Spanish influenza, 150, 238
Spoil, 238
Spring, Seth, 57
Squanto, *29*, 31, 239
Squash, 10–11, 15, 243
  and European settlers, 31, 53
Squatters, 82

St. Clair, Arthur, 82
St. Lawrence River, 6, 37
St. Louis, Missouri
  air pollution, 171
  and Chicago sewage, 162–164, 166
Standard Oil Company, 131
Stanley Steamer, 132, 239
Steamboats, 85–86, *86*, 112, 244
Steel industry
  Gary, 139
  Pittsburgh, 130–131
Storm sewers, 239
  cities, 154–158, 239, 245
  mill towns, 160–161
The Story of My Boyhood and Youth, 265–269
Streetcars, 142
Suburbs, 142–143, 145
  animal populations, 145
  ecosystem, 145
  postwar, and environmental problems, 196–198, *197*
Sugar, 64
Sumac, 239
Superfund, 195–196, 239, 247
Susquehanna River and valley, 53
  first coal mine, 114
  locks and dams, 72
  and mine waste, 189
Swamp Land Act of 1850, 134, 244
Swift, Gustavus, 107, 136, 245

Tamarac, 239
Tannic acid, 239
Telegraph and telephone systems, 140–141
Tenant farmers, 61, 77
Testimony of Anne Hillis and Jim Clark, The Joint Senate Subcommittee on Environmental Pollution and Hazardous [sic] Waste, 288–290
Textile industry, 117–121, *119*
  pollutants, 167–168
Thoreau, Henry David, 13, 207, *208*, 245
  and Cape Cod, 209, 212, 215
  "Walking," 262–265

Tidal in-wash, 239
Titusville, Pennsylvania, *130*
Tow cloth, 239
Toxic Substances Control Act of 1976, 195, 239
Trade, 58
  in beaver hides, 37–39
  between Native Americans and Europeans, 32, 35, 37–38
  between Pilgrims and England, 32–34
  in grains, 58
  international, 63–64, 66, 67–68, 73–75
  and lumber, 58
Tramontane West, 239–240
A Traveler from Altruria, 124
Triassic lowlands, 240
Turkeys, 107
Turner, Frederick Jackson, 278
Two Treatises on Government, 82
Typhoid fever, 153, 162, 164, 166, 240

Upper Mississippi River Wild Life and Fish Refuge, 185
Urban women's clubs, 240
U.S. Forest Service, 184, 240, 245
Usufruct, 240

Valley Forge, Pennsylvania, 68
van Rensselaer, Stephen, 79
Virginia, claim on Ohio land, 81
Virginia Company, 31

Wabash River valley, 1
Walden, 245
"Walking," 262–265
Walking city, 240
Waltham Cotton and Wool Factory Company, 118
Walton, Isaac, 240
Wampanoags, 8, 240
  and arrival of Pilgrims, 24, 28, 243
Wampum, 38–39, *38*, 240
War of 1812, 74, 244
Ward, Aaron Montgomery, 140
Ward, Henry, 185
Waring, George E., Jr., 175, 240
  "The Sanitary Drainage of Houses and Towns," 276–278

Washington, George, 68, 75
Waste disposal
  and Cape Cod, 213, 215
  inadequacy of and disease, 149–150
  mill towns, 159–160
  overwhelming of old systems, 151
  and recycling, 193–194
  underground fire (Centralia, Pennsylvania), 194–195, *194*
  urban, 147–149
Water
  chlorination, 170, 175, 222, 245
  filtration, 165–167, 175, 222, 240, 245
  indoor plumbing and increased consumption, 152, 156, 158, 166
  and reasonable use, 169, 237
  recreational use, *176*, 184, *205*. *See also* Hunting and fishing (as sport)
  regional protection agencies, 184
  storm sewers, 154–158, 239, 245
  urban systems, 152–154, 233
Water filtration systems, defined, 240
Water pollution
  and chemical cleansing additives, 167, 170, 221
  and choice of upstream filtration, 166
  and disease, 154, 161–162, 164–165
  effects of, 168–169
  industrial, 167–170, 175–176, 184
  Izaak Walton League effort against, 185
  legislation, 189, 192–193, 240–241
  and 1920s business boom, 175–177
  postwar era, 188–189
  from sewage, 152–154, 160–167, 175
  studies, 165–166, 175
Water Pollution Control Act, 241, 246
Weirs, 241
West Nile disease, 150
Western Electric Company, 141
Western Reserve, 81
Western Union, 141
Wetlands
  drained for farmland, 133–135
  ecological functions, 135
Whale oil, 131
Whaling, 60, 75, 244
  and wooden ships, 95

Wheat, 41, 42, 48, 54
  abandonment in central Midwest, 108
  and blast (rust), 67, 108, 244
  difficulties of cultivation, 67, 104, 108, 244
  in Midwest, 99–100, 108
  and wearing out of soil, 104
Whipple, George, 166–167
White, Hugh, 79
White Mountains, 4, *4*
Whooping cough, 241
Wilderness Society, 186, 241, 246
Windmills, 55, *56*
Winthrop, John, 50, 241

Wisconsin
  forests, 95–96
  and glaciation, 101
  prairies, 95
Wisconsin Glacier, 1–2, 241, 243
Wisconsin River, 101
Wood, Jethro, 90
Wool, 70
Works Progress Administration, 185, 246
Wright, Preserved, 58–59
Wright, Seth, 69

Yard, Robert Sterling, 186
Yellow fever, 74, 150, 151–152, 153, 241, 244

ABOUT THE AUTHOR

Professor John T. Cumbler received his Ph.D. from the University of Michigan, Ann Arbor, in 1974. He currently teaches in the Department of History, University of Louisville, Louisville, Kentucky, and is an honorary fellow at the University of Warwick, England. He has received numerous teaching awards and has served as a visiting lecturer at universities across the nation, including M.I.T. and San Diego State University. In addition, he has received postdoctoral grants and fellowships from the New Jersey Historical Comission (1983–1984), the American Council of Learned Societies (Senior Fellow, 1985–1986), and the National Endowment for the Humanities (1996–1997). His published works include *Working Class Community in Industrial America: Work, Leisure and Struggle in Two Industrial Cities* (1979), *The Moral Response to Industrialism: The Music Hall Lectures of the Reverend Cook* (1982), *A Social History of Economic Decline: Business, Politics, and Work in Trenton* (1989), and *Reasonable Use: The People, the Environment and the State in New England, 1790–1930* (2001). He recently completed a book on abolitionists and their involvement in reforms after the Civil War.